Salinity and Water Stress

Tasks for Vegetation Sciences 44

SERIES EDITOR

H. Lieth, *University of Osnabrueck, Germany*

For other titles published in this series, go to
www.springer.com/series/6613

M. Ashraf • M. Ozturk • H.R. Athar
Editors

Salinity and Water Stress

Improving Crop Efficiency

Springer

Editors
M. Ashraf
University of Agriculture
Faislabad, Pakistan

M. Ozturk
Ege University,
Bornova Izmir, Turkey

H.R. Athar
Bahauddin Zakariya University
Multan, Pakistan

Cover photographs caption:

Top left: a general view of the saline habitat (Munir Ozturk); top right: Crops grown on marginal lands (M. Ashraf, 2004); bottom left: salt and water stress tolerant plant (Mesembryanthemum spp) (H.R. Athar, 2006); bottom right: screening and selection of radish cultivars for salt tolerance (courtesy of Zahra Noreen).

ISBN 978-1-4020-9064-6 e-ISBN 978-1-4020-9065-3

Library of Congress Control Number: 2008936826

Printed on acid-free paper

springer.com

Preface

New advances in plant sciences particularly related to abiotic stresses are frequently appearing in the literature. It is imperative to keep updated ourselves with advances in plant abiotic stresses such as salinity and water stress to meet the current scientific challenges, particularly to meeting the growing food demand for world population. New technologies are trying to find out ways through which we can better understand how plants respond to environment and how to improve abiotic stress tolerance in crop plants and what effective strategies should be undertaken to overcome/mitigate the adverse effects of different abiotic stresses.

This book is presenting a timely and wide-ranging overview of the salinity and water stresses. In the three sections of this book, advanced knowledge about molecular, biochemical and physiological basis of plant salt and water stress tolerance is presented covering a broad range of topics in this connection:

- Nature of environmental adversaries that affect plant productivity from the viewpoint of three interrelated disciplines; eco-physiology, breeding, and socio-economics
- Potential biochemical and physiological indicators for successful breeding
- Molecular biological approaches to identify key genes responsible for traits involved in salt and water stress tolerance
- Alternative shotgun approaches to induce stress tolerance
- Alternative non-traditional plants that may be grown on stress hit areas and
- Economic utilization of salt affected areas by growing halophytes

In addition, the strategies economically viable for introducing economically important crops in non-agricultural land are discussed, and this will certainly have a great impact on plant productivity. Overall, the aim of this book is to link the rapid advancements in molecular biology with plant physiology and plant ecology. The book will provide a valuable insight into how the area of "plant adaptations to salt and water stresses" has progressed through the application of new technologies. Application of this knowledge through breeding by developing new high yielding varieties under stressful environments will keep the pace with the growing demand for food. In the last, it is no exaggeration to say that this book presents a number of comprehensive tables and figures to facilitate understanding and comprehension of the information presented throughout the text vis-à-vis a large number of new and updated references are provided together with hundreds of index words to promote the accessibility to the desired information throughout the book. The book is thus an indispensable resource for scientists, students and others seeking advancements in this area of research.

M. Ashraf, University of Agriculture, Faislabad, Pakistan
M. Ozturk, Ege University, Bornova Izmir, Turkey
H.R. Athar, Bahauddin Zakariya University, Multan, Pakistan

Acknowledgements

We would like to thank the production editors of Springer-Verlag for their invaluable help and patience during the compilation of this book. Sincere efforts and invaluable contributions of several competent scientists from different countries are highly acknowledged who really made it possible to produce this unique volume for knowledge seekers. Our special thanks go to Pakistan Academy of Sciences (PAS), Higher Education Commission (HEC), Islamabad, Pakistan, National Core Group in Life Sciences (NCGLS), National Commission on Biotechnology (NCB), and Islamic Development Bank (IDB) for the financial assistance that allowed the interactions between the scientists of two countries (Pakistan and Turkey) to initiate the research collaboration and this book project. Finally we thank our spouses Shamsa Parveen, Birsel Ozturk, and Safia Habib for their continuous support and encouragement in our scientific journey.

M. Ashraf, University of Agriculture, Faislabad, Pakistan
M. Ozturk, Ege University, Bornova, Izmir, Turkey
H.R. Athar, Bahauddin Zakariya University, Multan, Pakistan

Contents

About the Editors

M Ashraf is Professor of Botany and Dean Faculty of Sciences at the University of Agriculture, Faisalabad, Pakistan. Dr. Ashraf received his Ph.D. degree in botany from the University of Liverpool, UK and carried out postdoctoral work as a Fulbright Scholar at the University of Arizona. His research is focused on the improvement of stress tolerance in plants using breeding and physiological approaches. He has published over 300 scientific papers and reviews. Furthermore, more than 10 chapters in edited books of international repute and one edited book are to his credit. He is one of the most productive scientists in the Pakistan in all scientific disciplines. He has supervised 20 Ph.D. students. Dr. Ashraf has earned several prestigious awards and honors for his outstanding contributions in the fields of agriculture and biology including two Gold Medals from Pakistan Academy of Sciences, the Salam Prize, the National Book Foundation of Pakistan Awards, and the presidential awards Izaz-e-Fazeelat, Pride of Performance and Sitara-e-Imtiaz. He was elected as a Fellow of Pakistan Academy of Sciences in 2000, and a Fellow of Third World Academy of Sciences (TWAS), Italy in 2003. He earned the title "HEC Distinguished National Professor" in 2005 by the Higher Education Commission, Pakistan. He was appointed as an **Honorary Scientist for Rural Development Administration, Government of the Republic of Korea** for a period of 3 years from 2005 to 2008.

Munir Ozturk is Rtd. Profesor of Botany at Ege University, Izmir, Turkey. Dr. Ozturk has received his Ph.D. & D.Sc. from Ege University and worked at Munich Technical University Germany under Alexander von Humboldt Fellowship, at the Institute of Gene-Ecology Tohoku University-Japan as JSPS Fellow and as NSF Fellow at the Dept. of Biology University of Chapel Hill, NC, USA. His field of specialization is "Plant Eco-Physiology". He has edited 18 books, authored 3 books and published more than 250 papers. Dr. Ozturk has supervised 17 M.S. and 10 Ph.D. theses. He has got some prestigious awards as well. He is Fellow of the World Islamic Academy of Science.

Habib-ur-Rehman Athar is Assistant Professor in Botany at Institute of Pure and Applied Biology, Bahauddin Zakariya University, Multan, Pakistan. He has recently received his Ph.D. degree in Botany from University of Agriculture, Faisalabad, Pakistan. As one of the important members of Plant Stress Biology research group at the University of Agriculture Faisalabad, Pakistan he is involved in developing shotgun approaches (exogenous application of compatible solutes, antioxidants, inorganic salts and plant hormones) to alleviate the adverse effects of abiotic stresses on crop plants and has published 32 scientific papers including one review on these issues. Furthermore, Dr. Athar has three chapters in edited books of international repute and has edited one proceedings of an international symposium. He is one of the productive scientists of Pakistan in Biology. He is also a dynamic, innovative minded person and a productive scientist. Dr. Athar has developed an e-discussion group "**Plantstress**" having more than 1,100 members world-over, and provides a forum where different scientists from world-over exchange their scientific ideas and discuss their problems they are confronting during their research.

Chapter 1
Strategies for Crop Improvement Against Salinity and Drought Stress: An Overview

H.R. Athar and M. Ashraf

Abstract Abiotic stresses such as salinity, drought, nutrient deficiency or toxicity, and flooding limit crop productivity world-wide. However, this situation becomes more problematic in developing countries, where they cause food insecurity for large populations and poverty, particularly in rural areas. For example, drought stress has affected more than 70 million hectares of rice-growing land world-wide. While salt stress and nutrient stress render more than 100 million hectares of agricultural land uncultivable thereby resulting in low outputs, poor human nutrition and reduced educational and employment opportunities. Thus, abiotic stresses are the major factors of poverty for millions of people. In this scenario, it is widely urged that strategies should be adopted which may be used to get maximum crop stand and economic returns from stressful environments. Major strategies include breeding of new crop varieties, screening and selection of the existing germplasm of potential crops, production of genetically modified (GM) crops, exogenous use of osmoprotectants etc. In the last century, conventional selection and breeding program proved to be highly effective in improving crops against abiotic stresses. Therefore, breeding for abiotic stress tolerance in crop plants (for food supply) should be given high research priority. However, extent and rate of progress in improving stress tolerance in crops through conventional breeding program is limited. This is due to complex mechanism of abiotic stress tolerance, which is controlled by the expression of several minor genes. Furthermore, techniques employed for selecting tolerant plants are time consumable and consequently expensive. During the last decade, using advanced molecular biology techniques different researchers showed some promising results in understanding molecular mechanisms of abiotic stress tolerance as well as in inducing stress tolerance in some potential crops. These findings emphasized that future research should focus on molecular, physiological and metabolic aspects of stress tolerance to facilitate the development of crops with an inherent capacity to withstand abiotic stresses. This would help stabilize the crop production, and significantly contribute to food and nutritional security in developing countries and semi-arid tropical regions.

Keywords Abiotic stresses • food • insecurity • molecular breeding • QTLs • salinity • transgenic plants • water stress

1 Introduction

1.1 Current Scenario of World Population and Food Insecurity

In view of different projections, it is expected that human population will increase over 8 billion by the year 2020 that will worsen the current scenario of food insecurity. According to an estimate improved crop productivity over the past 50 years has resulted in increasing world food supplies up to 20% per person and reducing proportion of food-insecure peoples living in developing countries from 57% to 27% of the total population (FAO 2003). Regardless of these fabulous achievements, 800 million people are still under-nourished in the developing

H.R. Athar (✉),
Institute of Pure and Applied Biology, Bahauddin Zakariya University, Multan, Pakistan
e-mail: habibathar@yahoo.com

M. Ashraf
Department of Botany, University of Agriculture, Faisalabad, Pakistan

M. Ashraf et al. (eds.), *Salinity and Water Stress*,

world. Of them, 232 million are in India, 200 million in sub-Saharan Africa, 112 million in China, 152 million elsewhere in Asia and the Pacific, 56 million in Latin America and 40 million in the Near East and North Africa (UN Millennium Project 2003). It is predicted that at least 10 billion people will be hungry and malnourished in the world by the end of this century (FAO 2003). Thus, to reduce the food insecurity, crop production will have to be doubled, and produced in more environmentally sustainable ways (Borlaug and Dowswell 2005). This can be achieved by expanding cultivable land area or by increasing per hectare crop productivity. However, it is well evident from the history of the past century that enhancement in crop production due to expansion in growing area was only observed in the first half of the twentieth century (Slafer and Satorre 1999). Furthermore, during the second half of the past century rise in per hectare crop productivity was due to improved or high yield potential (Araus et al. 2004). Overall, it seems that focus should be on genetic gain to improve crop productivity.

1.2 Crop Production as Affected Abiotic Stresses

In view of current situation of food insecurity, particularly in developing countries, a number of other factors cause a further decrease in crop productivity. Of them, availability of agricultural land, fresh water resources, ever-increasing biotic and abiotic stresses, and low economic activity in agricultural sector are the most important factors. However, it is generally believed that abiotic stresses are considered to be the main source of yield reduction (Boyer 1982; Rehman et al. 2005; Munns and Tester 2008; Reynolds and Tuberosa 2008). The estimated potential yield losses are 17% due to drought, 20% due to salinity, 40% due to high temperature, 15% due to low temperature and 8% by other factors (Rehman et al. 2005; Ashraf et al. 2008).

1.3 Drought Stress

Drought and salinity are two major abiotic stresses that affect various aspects of human lives of one third world population including human health and agricultural productivity. For example, according to an estimate by the United Nations, one third of the world's population lives in areas where water is scarce (FAO 2003). Furthermore, climatic changes also enhanced the frequency and intensity of water shortage in sub-tropical areas of Asia and Africa. According to the UN climatic report (http://www.solcomhouse.com/drought.htm) the Himalayan glaciers that feed to the Asia's largest rivers (Ganges, Indus, Brahmaputra, Yangtze, Mekong, Salween and Yellow) may disappear by 2035 due to

Fig. 1.1 Increasing demand for food production for growing human world population can be met by cultivating crops on all types of available land. In this figure, different vegetables are growing on available roadside places (Photo taken by M. Ashraf during his visit to Korea during 2004)

rise in temperature. In addition, if the present situation prevails over many years, it is expected that by 2025, 1.8 billion people will live in countries or regions with absolute water scarcity. It is already noticed that drought-affected nations of Central Asia used their shared water resources to bargain between the countries. For example, in 1960, it was officially recognized that Indus River is the main source of water for both India and Pakistan. Similarly, in 1999, Kyrgyzstan succeeded in getting much needed coal from Kazakhstan after closing down water reservoirs (http://www.solcomhouse.com/drought.htm). Thus, the availability of fresh water is a major commodity to improve the economy of a country.

1.4 Salinity Stress

Like shortage of water, high concentration of soluble salts is another menace for human lives. The problem of salinity existed long before the human beings and start of agricultural practices. From the historical record of the last 6,000 years of civilization, it is evident that people were unable to continue their colonization due to salinity-induced destruction of resources. For example, it was found that increase in salinity level over 700 years from 2400 BC to 1700 BC caused a decline in agricultural productivity, e.g., 29 bushels per acre of barley to 10 bushels per acre (Gelburd 1985). Although a progressive increase in salinity has caused degradation of arable land over many hundred-years period, cultivated land could be degraded due to salinity during less than 100 years. For example, in California 4.5 out of 8.6 million hectares irrigated agricultural land has become salt affected during the last century (Lewis 1984). At present, its extent throughout the world is increasing regularly (Schwabe et al. 2006) and it has now become a very serious problem for crop production (Munns and Tester 2008), particularly in arid and semi-arid regions. According to an estimate by FAO (2008; http://www.fao.org/ag/agl/agll/spush accessed on April, 2008) over 6% of the world's land is salt affected. In addition, out of 230 million hectares of irrigated land, 45 million hectares (~20%) are salt affected. However, the intensity of salinity stress varies from place to place. Generally, dry land salinity has been categorized into three different types: low salinity (ECe 2–4 dS/m), moderate salinity (ECe 4–8 dS/m) and high salinity (ECe > 8 dS/m) (Rogers et al. 2005). Depending upon the type of source from which soil became salinized, soil salinity can be categorized as primary and secondary salinization. Primary or natural salinization results from weathering of minerals and soil derived from saline parent rocks, and secondary slalinization that is caused by human interference such as irrigation, deforestation, overgrazing, or intensive cropping (Ashraf 1994). According to an estimate, 32 million hectares (2%) out of 1,500 million hectares are affected by secondary salinity to varying degrees depending upon the type of factors causing salinity (FAO 2008). Based on soil and ground water processes causing salinity, Rengasamy (2006) categorized salinity in three groups as (1) ground water associated salinity (GAS), (2) non-ground water associated salinity, and (3) irrigation associated salinity. He suggested that knowledge about the extent of salinity and process of dominant factor of salinization can be updated with the help of most recent geophysical techniques, which will be conducive to evaluate salt tolerant genetic material or to know up to what level of salt tolerance should be induced in crops which is required for economically viable crop production on saline environment.

1.5 Objectives of This Chapter

Both water stress and salt stress occur naturally in habitats where temperature is high. Both water stress and salt stress affected more than 10% of arable land, which results in rapid increase in desertification and salinization world-wide. As a consequence, average yields of major crops reduced by more than 50% (Bray et al. 2000). Due to this reason, there is an increasing demand for new plant cultivars that have a potential for higher yield under such abiotic adversaries. With considerable advancements in the field of plant physiology and molecular biology in the present era, there are high expectations that plant breeders will certainly provide salt tolerant crops with higher yield. Generally, it is believed that stress tolerant plants have the ability to maintain higher rates of growth under saline conditions. However, during the past decade progress made in this area is very slow because there is a great controversy among plant physiologists, plant breeders, and plant molecular biologists about physiological basis of stress tolerance in plants (Yeo 1998; Hasegawa et al. 2000;

Munns 2002; Serraj and Sinclair 2002; Wang et al. 2003; Ashraf 2004; Ashraf and Harris 2004; Flowers 2004; Reynolds et al. 2005; Cuartero et al. 2006; Munns and Tester 2008). Although there is a reasonable consensus on various strategies of improving degree of stress tolerance in crops such as screening for stress tolerant individuals, identification of promising traits conferring stress tolerance in plants, and development of stress tolerant plants through breeding or genetic engineering, there is still no consensus on physiological traits that confer salt tolerance in plants. Comparisons of adaptive responses among various species suggested that some salt-tolerant plants have evolved specialized complex mechanisms. Although genes for cellular based mechanisms of stress tolerance appear to be common in genotypes, development of an adaptive mechanism in plants to tolerate abiotic stresses requires the combination of several morphological, physiological and metabolic processes which depends on a multitude of genes and varies within each target environment. However, among various mechanisms of stress tolerance, mechanisms that regulate ion and water homeostasis are of prime importance (Bartels and Sunkar 2005; Munns and Tester 2008). Thus, nature of various biochemical and physiological characters responsible for determining crop productivity under stress conditions is very complex (Ashraf et al. 2008). It is highly likely that improving crop efficiency under stress environments cannot be achieved without complete understanding the physiological as well as molecular basis of stress tolerance. Thus, "How plants respond to these stresses?", "How and what type of plants can tolerate these stresses?" and "How these principles can be utilized in improving crop production?" are hot issues these days. After general discussion of the current situation of food security and abiotic stresses such as drought and salinity stress, strategies for improving crop efficiency against salt and water stress based on some recent advances in basic plant biology have been reviewed in this chapter that will eventually help plant breeders to develop stress tolerant cultivars of different crops.

2 Strategies for Improving Crops Against Water and Salt Stresses

As mentioned earlier, both water stress reduces plant growth and crop productivity, so it is imperative to reduce yield gaps by increasing crop drought tolerance under these conditions, thereby ensuring food security for the increasing human population as well as for the benefit of poor farmers world-over. In this context, crop stress tolerance is defined in terms of yield stability under abiotic stress conditions. However, yield losses due to abiotic stresses vary depending on timing, intensity and duration of the water stress, coupled with other environmental factors such as high light intensity and temperature. Based on this information, following means are suggested (Parry et al. 2005; Reynolds et al. 2005; Tuberosa et al. 2007a; Neumann 2008):

1. Water management practices that save irrigation water
2. Exploitation of the agronomic practices by which plants can perform well under water stress conditions
3. Selection of crop cultivars that require relatively lower quantity of water for their growth and crop productivity

Strategies involving water saving irrigation technologies or cultural practices to alleviate drought stress, are expensive, inconvenient, and require specific knowledge for its implementation. On the other hand, use of drought resistant crop plants in drought prone environment i.e. biological approach is more feasible and efficient in achieving high crop productivity on drought hit areas. In addition, the biological approach involves, those methodologies which are used to enable plants that can effectively escape, avoid or tolerate drought.

2.1 Use of Naturally Water Stress Tolerant Plants

Plants adapted to arid environments posses inherent drought escape or drought avoidance mechanisms and can be grown in drought hit areas. Drought escape is a phenological phenomenon of plants achieved by early maturity and completion of life cycle, while drought avoidance mechanisms enable the plants to maintain high water potential so as to avoid the damaging effect of water stress (Boyer 1982). Plants using drought avoidance mechanism have deeper and dense root system, greater root penetration ability, higher stomatal conductance, and higher cuticular resistance to prevent water loss, higher pre-dawn leaf water potential, and avoid leaf rolling for longer intervals (Peng and Ismail 2004). In naturally dry habitats, some plant species rapidly mature and produce seeds before the onset of dry season or start reproducing soon after rainfall. For instance, California poppy (*Escholtzia californica* Cham.) completes its life cycle in a few weeks before drought stress starts. In contrast, Coffe (*Coffea arabica* L.) and cacao (*Theobroma cacao* L.) flower and fruit when rains follow a drought period (Alvim 1985). However, some plant species such as agave (*Agave deserti*), and cactus species store water in their buds, stems or leaves during water stress period. These plants utilize this stored water under conditions of severe drought. Other plant species avoid water stress by developing deep root system and/or mechanism involved in low transpirational water loss. Among crops, arid legumes such as cluster bean [*Cyamopsis tetragonoloba* (L.) Taub], dew bean [*Vigna aconitifolia* (Jacq.) Marechal], cowpea [*Vigna unguiculata* (L.) Walp], and (*Cicer arietinum* L.) are characterized by their deep taproot system with slow growth. They all are drought avoiders (Kumar 2005). Similarly, drought tolerance in *Brassica carinata*, *B. napus*, and *B. campestris* is related to their better-developed root system (Liang et al. 1992). Likewise, *Eruca sativa* L. has also deep root system and fleshy leaves to store water particularly when grown in water deficit conditions. Pearl millet is another drought tolerant cereal widely cultivated in arid and semi-arid regions of the world. From all the above reports it can be infer that water stress reduces plant growth and yield of almost all crops by imposing adverse effects on the traits associated with growth and yield, but it depends on the type of species, and intensity and duration of water stress.

Drought tolerance refers to the extent to which plants maintain their metabolic function when leaf water potential is markedly low. Although mechanism of drought tolerance is poorly understood, osmotic adjustment is considered to be associated with dehydration tolerance. Osmotic adjustment is the accumulation of organic or inorganic solutes in response to water stress thereby maintaining tissue turgor potential. However, in view of earlier studies it is believed that plant tolerance to drought is an adaptive feature involving plant responses at cellular and at whole plant level such as synthesis and accumulation of organic compatible solutes, synthesis of stress proteins, up-regulation of antioxidant enzymes, development of deep and dense root system, epicuticular wax, leaf rolling etc. (Chaves et al. 2004; Parry et al. 2005; Reynolds et al. 2005; Neumann 2008). If we analyze all these traits for water stress tolerance, it appears that drought tolerance in crops usually depends on one or more of the following components include avoidance (1) the capacity of plant roots to extract water from soil (2) osmotic adjustment capacity (3) water use efficiency (Chaves et al. 2004; Parry et al. 2005; Reynolds et al. 2005; Neumann 2008). Therefore, crop plants or wild plants having these traits are capable to tolerate water stress and thus they can be grown on drought hit areas.

2.2 Selection and Breeding for Drought Tolerance

The development of drought-resistant cultivars/lines of crops through selection and breeding is of considerable economic value for increasing crop production in areas with low precipitation or without any proper irrigation system (Subbarao et al. 2005). However, availability of genetic variation at inter-specific, intra-specific and intra-varietal levels is of prime importance for selection and breeding for enhanced resistance to any stress (Blum 1985; Ashraf and Sharif 1998; Serraj et al. 2005a). In order to develop drought tolerant cultivars, it is imperative to develop efficient screening method and suitable selection criteria. Various agronomic, physiological and biochemical selection criteria for drought tolerance are being employed to select drought tolerant plants, such as seed yield, harvest index, shoot fresh and dry weight, leaf water potential, osmotic adjustment, accumulation of compatible solutes, water use efficiency, stomatal conductance, chlorophyll fluorescence (Araus et al. 2002; Richards et al. 2002; Flexas et al. 2004; Reynolds et al. 2005; Kauser et al. 2006; Ashraf et al. 2007; Tambussi et al. 2007; Neumann 2008). Development of drought tolerance in adaptation for a plant is the result of overall expression of many traits in a specific environment. Since many adaptative traits are effective only for certain aspects of drought tolerance and over a limited range of drought stress, there is no single trait that breeders can use to improve productivity of a given crop in a water deficit environment. Therefore, alternative potential systematic approach is to pyramid various traits in one plant genotype which can improve its drought tolerance. In this context, Subbarao et al. (2005) suggested that those traits, whether physiological or morphological, that contribute to check water loss through transpiration, and enhance water use efficiency and/yield are traits of interest. While discussing prospects for crop production under drought, Parry et al. (2005) suggested some key traits for breeding for drought tolerance (e.g. phenology, rapid establishment, early vigor, root density and depths, low and high temperature tolerance, ^{13}C discrimination [a measure of the extent to which photosynthesis is maintained while stomatal conductance decreases], root conductance, osmoregulation, low stomatal conductance, leaf posture, habit, reflectance and duration, and sugar accumulation in stems to support later growth of yield components). However, they stressed that priority should be given to those traits that will maintain or increase yield stability in addition to overall yield, because traits for higher yield may in fact decrease yield stability (e.g. longer growth period). Thus, in order to improve crop productivity under water stress conditions, selection of a cultivar with short life span (drought escape), incorporation of traits responsible for well-developed root system, high stomatal resistance, high water use efficiency (drought avoidance), and traits responsible for increasing and stabilizing yield during water stress period (drought tolerance) should be given high priorities.

Although a number of crop cultivars tolerant to drought stress have been developed through this method, this approach has been partly successful because it requires large investments in land, labor and capital to screen a large number of progenies, and variability in stress occurrence in the target environment. In addition, there is an evidence of marginal returns from conventional breeding, suggesting a need to seek more efficient methods for genetic enhancement of drought tolerance.

2.3 Molecular Breeding

Now it is well evident that water stress tolerant traits are mainly quantitative and are controlled by multiple genes. The regions of chromosomes or the loci controlling these traits are called quantitative trait loci (QTLs). In the QTL approach of plant breeding, parents showing extreme phenotypes for a trait are crossed to produce progenies with a capacity of segregation for that trait. This population is then screened for genetic polymorphism using molecular markers technique such as RFLP, RAPD, AFLP and SNPs. Genetic maps were constructed and markers associated with a trait were identified using computer software. Use of molecular markers to identify QTLs for physiological traits responsible for stress tolerance has helped to identify some potential sub-traits for drought tolerance (Chinnusamy et al. 2005; Hussain 2006). Once molecular markers (i.e. for a trait QTLs) are linked to specific sub-traits of drought tolerance, it would be possible to transfer these various traits into other adapted cultivars with various agronomic backgrounds under specific targeted environments through marker-assisted breeding approaches. Thus, identification of areas of a genome that have a major influence on drought tolerance or QTLs for drought tolerance traits could allow to identify the genes for drought tolerance. Thus, use of molecular marker-assisted selection (MAS) seems to be a more promising approach because it enabled us to dissect quantitative traits into their single genetic components thereby helping in selecting and breeding plants that are resistant to water stress (Chinnusamy et al. 2005; Hussain 2006).

The identification of QTLs for economically important traits has been achieved by developing linkage mapping to anonymous markers (segregation mapping) or through association studies (association mapping or candidate gene approach) involving candidate genes (Araus et al. 2003). Although most of data for QTLs for drought tolerance available in the literature is based on segregation mapping studies (Cattivelli et al. 2008), association mapping or candidate gene approach is more vigorous than segregation mapping (Syvänen 2005). Because single genes controlling a trait such as flowering time, plant height, ear development and osmotic adjustment may have more important role in adaptation to drought-prone environment. For example, a single candidate gene (*or* gene) conferring osmotic adjustment in wheat was mapped on the short arm of chromosome 7A (Morgan and Tan 1996) and breeding for *or* gene improved yield in wheat under water deficit conditions (Morgan 2000). While critically analyzed of the reports on the application of QTL analysis Cattivelli et al. (2008) pointed out that more efforts have been dedicated to understand the genetic basis of physiological traits responsible for drought tolerance, and little attention has been given to understand high yield stability in water deficit conditions. For example, more reports are available on genetic variation for osmotic adjustment, genetic basis of phenological traits, the ability of roots to exploit deep soil moisture, water use efficiency, limitation of non-stomatal water loss, and leaf elongation rate under varying degrees of water stress. Detailed information on QTLs for drought tolerance is available as GRAMENE (http://www.gramene.org/) or GRAINGENES (http://wheat.pw.usda.gov.GG2/). However, despite theoretical advantages of utilizing MAS to improve quantitative traits during the past decade, the overall impact of MAS on the direct release of drought-tolerant cultivars remains non-significant (Reynolds and Tuberosa 2008). In view of this information available in the literature, identification of QTLs responsible for improving drought tolerance and yield potential is the main goal for the present and future research (Maccaferri et al. 2008). Thus, it was suggested that deliberate selection for secondary traits related to drought tolerance is likely to achieve better results than direct selection for yield *per se* under stress (Araus et al. 2004; Bohnert et al. 2006; Tuberosa et al. 2007b).

Marker assisted selection becomes more efficient if available markers are tightly linked to loci for stress related traits. For instance, while working with rice, Babu et al. (2003) found that QTLs for plant yield under drought were coincided with QTLs for root traits and osmotic adjustment. Likewise, Lanceras et al. (2004) found that favorable alleles for yield components were located in a region of rice chromosome 1 where QTLs for many drought related traits (root dry weight, relative water content, leaf rolling and leaf drying) were previously identified (Zhang et al. 2001). However, in this strategy, parents of extreme contrasting traits (yield and drought tolerance) are required which may cause a cost on grain yield by decreasing yield components. From all this discussion, it seems that with the advent of this high throughput molecular biology technique, we are probably on the threshold of breakthroughs in our ability to understand and manipulate plant physiological responses to water deficit. Although use of molecular marker-assisted selection (MAS) seems to be more promising and meaningful, the contribution of molecular breeding to the development of drought tolerant cultivars has so far been marginal and a few reports are available in this regard (Slafer et al. 2007; Cattivelli et al. 2008; Reynolds and Tuberosa 2008; Zhao et al. 2008). For example, while introgressing favorable alleles at five QTLs expressing 38% of total phenotypic variation in maize, Ribaut and Ragot (2007) reported that grain yield of best selected maize hybrid with molecular markers was 50% higher than control hybrids under severe water stress conditions. Furthermore, under non-stress condition no yield penalty was observed. Likewise, Serraj et al. (2005b) reported that drought sensitive genotypes of pear millet carrying introgression of a major QTL for grain yield under terminal drought stress at the target QTL showed a consistent grain yield advantage. Recently, Harris et al. (2007) developed near isogenic lines of sorghum each containing one of the four previously identified stay green QTLs. Favorable alleles in each of the four loci contributed to the lower rate of leaf senescence under post-anthesis water deficit. In view of all these reports mentioned above it is amply clear that efficiency of molecular breeding is not so significant.

Another important application of molecular breeding is cloning of genes/DNA sequences associated with QTLs for drought tolerance. A number of strategies are being used to clone candidate genes/DNA sequences (Salvi and Tuberosa 2005), which can be selected from the available literature, by mapping of known stress responsive genes (Tondelli et al. 2006). For example, Masle et al. (2005) cloned *ERECTA* gene in *Arabidopsis thaliana*, a DNA sequence beyond a QTL for transpiration efficiency. However, there is no report available in the literature on cloning of genes underlying QTLs in any crop species. For identification of QTL corresponding gene (QTN –quantitative trait nucleotide), generation of molecular-linkage maps based on candidate genes (molecular function maps) is suggested to avoid time consuming fine mapping by a number of researchers. For example, this strategy has been applied to find genes for drought tolerance in barley and rice (Zheng et al. 2003; Nguyen et al. 2004; Diab et al. 2004; Tondelli et al. 2006).

By summarizing all the reports mentioned earlier, it can be easily perceived that molecular breeding work has not been extended beyond the detection of a given trait under water stress conditions. However, whether QTL identified in a given mapping population will improve the drought tolerance in high yielding elite genotypes upon introduction is still a great challenge for researchers.

2.4 Transgenic Approach

Drought is primarily manifested as osmotic stress, resulting in the disruption of homeostasis and nutrient distribution in the cell. As a consequence, it activates cell signaling pathways and cellular responses, such as the production of stress proteins, up-regulation of antioxidants and accumulation of compatible solutes (Bartels and Sunkar 2005). Thus, the ongoing research on engineering water stress tolerant plants is mainly based on transfer of one or several genes that are either involved in signaling and regulatory pathways, or that encode enzymes present in pathways leading to the synthesis of functional and structural protectants, such as osmolytes and antioxidants, or that encode stress tolerance conferring proteins (Wang et al. 2003; Vinocur and Altman 2005) All these genes are categorized in three major groups by Wang et al. (2003): (i) genes involved in signaling pathways and in transcriptional control, (ii) genes involved in protection of membranes and proteins, such as heat shock proteins (Hsps) and chaperones, late embryogenesis abundant (LEA) proteins, osmoprotectants and free-radical scavengers; (iii) genes involved in water and ion uptake and transport such as aquaporins and ion transporters (Wang et al. 2003). However, Vinocur and Altman (2005) added one more group i.e., genes involved in metabolism. Under this heading, they discussed the role of osmoprotectants in stress tolerance such as amino acids, amines, proline, sugars, sugar alcohols, glycinebtaine.

Transgenic plants have been developed initially in model plants *Arabidopsis* and tobacco. However, relatively little work has been published on crop plants. Most successful examples of transgenic crops for drought tolerance are transgenics of DREBs/CBFs transcription factors in different crops such as in tomato (Hsieh et al. 2002), rice (Dubouzet et al. 2003; Ito et al. 2006) and wheat (Pellegrineschi et al. 2004). However, over-expression of *DREB2* in *Arabidopsis thaliana* plants did not enhance the stress tolerance probably because of lack of post-translational modification (Sakuma et al. 2006). In a comprehensive review, Wang et al. (2003) concluded from a large number of available reports that over expression of transcription factors may also activate additional non-stress related genes that adversely affect normal agronomic characteristics of a crop thereby resulting in reduced yield. Common adverse effects due to constitutive expression of genes are growth retardation, and reduced fruit, seed number and fresh weight of transgenic plants under normal conditions. Although use of stress-inducible promoter minimizes the adverse effects and enhances stress tolerance, threshold stress under which a promoter activates the gene in target environment needs to be determined.

Metabolic engineering of osmolytes is another successful approach in developing transgenic plants tolerant to water stress. However, real advantage of this strategy in terms of yield is always controversial (Serraj and Sinclair 2002; Araus et al. 2004). First transgenic for drought tolerance by over producing proline was reported in tobacco (Kavi-Kishore et al. 1995) and rice (Zhu et al. 1998). Garg et al. (2002) developed drought tolerant transgenic rice by over producing trehalose, which showed higher photosynthetic capacity and low photo-oxidative damage under both non-stress and stress conditions. A considerable enhancement in water stress tolerance in wheat was achieved by Abebe et al. (2003) through ectopic expression of the mannitol-1-phosphate dehydrogenase (mtlD) gene that caused a small increase in mannitol.

Normal stomatal regulation is believed to improve plant water use efficiency under drought environment, over-expression of a maize NADP-malic enzyme, the primary decarboxylating enzyme in C_4 photosynthesis, produced tobacco plants with reduced stomatal conductance and improved water use efficiency (Laporte et al. 2002). Over expression of AVP1 in Arabidopsis and tomato resulted in more pyrophosphate driven cation transport into root vacuolar fraction which enhanced root biomass and water stress tolerance (Gaxiola et al. 2001; Park et al. 2005). In another study, De Block et al. (2005) produced *Brassica napus* plants tolerant to multiple stresses by preventing over-activation of mitochondrial respiration and high energy consumption.

Overall, it is possible to engineer stress tolerance in plants using different "stress" genes. However, it seems that often the amount of gene product is not enough to provide tolerance, and that the gene has another function in stress tolerance that is not fully understood (Bajaj et al. 1999). For example Abebe et al. (2003) engineered wheat (cv. Bobwhite) to over-express mannitol (an osmolyte).

Although mannitol has been shown to improve stress tolerance, the amounts produced in this study were not enough to confer tolerance through osmotic adjustment, and thus the authors concluded that mannitol may have other stress protective functions. A similar case was found in the overexpression of trehalose in tobacco (Serrano et al. 1999). Although the current efforts to improve water stress tolerance in plants by gene transformation have resulted in important achievements, however, the nature of the genetically complex mechanisms of abiotic stress tolerance, and the potential detrimental side effects, make this task extremely difficult (Wang et al. 2003; Bartels and Sunkar 2005; Vinocur and Altman 2005; Bohnert et al. 2006; Cattivelli et al. 2008).

3 Strategies for Improving Crop Efficiency Against Salt Stress

Various strategies can be adopted to cope with salinity stress. However, farmers and plant biologists are quite familiar with the two major strategies to utilize salt affected lands, i.e., technological approach and biotic approach (Epstein et al. 1980; Ashraf 1994). In the technological approach, one can alter the salty soil through reclamative measures and management practices which enable the plants to grow and produce a reasonable yield. However, these methods are expensive and are not always a practical solution to the problem of soil salinity. Long ago, Epstein proposed that we must adopt biotic approach, rather than solely depending technological approach to counteract the salinity problem (Epstein et al. 1980). This was proposed mainly due to two major reasons, (i) uptake and assimilation of mineral nutrients including Na^+ and Cl^- are genetically controlled and can be manipulated (Ashraf 1994, 2004; Apse et al. 1999; Tester and Davenport 2003; Flowers 2004; Munns 2005; Munns et al. 2006), (ii) some plants have ability to grow under high saline conditions (Greenway and Munns 1980; Ashraf 1994, 2004; Flowers 2004). Biotic approach has considerable promise in mitigating the problem of soil salinity world over. However, recently, current status of some potential biological strategies has been reviewed by which salinity tolerance of potential crops can be maximally increased (Ashraf et al. 2008). Although all biological strategies for crop improvement against salt stress are same as for water stress tolerance such as screening and selection, breeding and use of transgenics, the biochemical, physiological traits for salt tolerance are different from plant water stress tolerance. It is largely believed that the adverse effects of salt stress on plant growth are mainly due to its toxic and osmotic effects, therefore major focus is

Fig. 1.2 Salt and water stress tolerant plant (Mesembryanthemum spp) growing on costal sandy bank of Mediterranean sea at Gammarth, Tunisia. This plant has a number of adaptations to conserve water such as osmotic adjustment, higher photosynthetic capacity. This plant is considered as model plant for exploring mechanism of water and salt stress tolerance in plants using DNA microarrays, transcripteomics and proteomic studies (Photograph taken by Habib-ur-Rehman Athar during his visit to Tunisia in 2006)

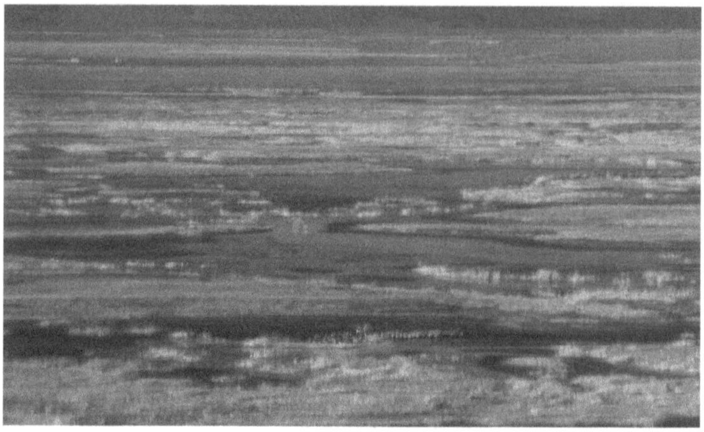

Fig. 1.3 Salinity occurs through natural or human-induced processes that result in the accumulation of dissolved salts in the soil to an extent that inhibits plant growth. Saline/sodic soils are widespread in arid and semi-arid lands of the world. According to FAO estimate, salinity affects over 6% of the world's land (Munir Ozturk)

on selective ion accumulation or exclusion, control of sodium uptake and its distribution within the plant, compartmentation of ions at cellular or at whole plant level (Flowers 2004; Munns 2005; Munns and Tester 2008; Ashraf 1994, 2004).

3.1 Screening and Selection for Salt Tolerance

In recent years there has been much interest in the development of salt tolerant crop varieties. For this purpose, genetic improvement of salinity tolerance in the cultivated genotypes has been proposed as the most effective strategy to solve salinity problems. As is well evident from the literature on the existence of inter- and intra-specific genetic variability for salt tolerance, it could be exploited judiciously for screening and breeding for higher salt tolerance. For example, Moreno et al. (2000) found a great magnitude of genotypic variability in bean cultivars (*Phaseolus vulgaris* L.) for salt tolerance at the seedling stage. They identified some salt tolerant cultivars with higher root growth and mineral nutrient accumulations. In another study, Mano and Takeda (2001) found some salt tolerant wheat cultivars at the seedling stage that maintained their salt tolerance at later growth stages. While screening 100 genotypes of sorghum at the seedling stage, Krishnamurthy et al. (2007) identified 46 genotypes as salt tolerant, that were further confirmed as salt tolerant at the later growth stage using Na^+ exclusion as a potential selection criterion. However, while assessing the value of tissue Na^+ concentration as a criterion for salt tolerance using a diverse collection of bread wheat germplasm, Genc et al. (2007) suggested that Na^+ exclusion and tissue tolerance varies independently, and there was no significant relationship between Na^+ exclusion and salt tolerance in bread wheat. They also suggested that salt tolerance may be achieved through different combinations of Na^+ exclusion and tissue tolerance.

It is now well evident that, improving salt tolerance of genotypes is often inhibited by the lack of effective evaluation growth stage to identify salt tolerant genotypes (Munns 2002, 2005). For instance, in a number of crop species, salt tolerance is a developmental stage specific phenomenon. Thus, salt tolerance should be evaluated at germination, seedling and adult stages (Ashraf 2004). In contrast, while evaluating salt tolerance in tomato at the seedling stage and maturity stage, Dasgan et al. (2002) suggested the screening at the seedling stage is not only less laborious, less time consuming and less expensive, but also has a high reliability. Furthermore, screening process under natural field conditions is not feasible due to the high degree of soil heterogeneity.

While establishing appropriate salinity screening techniques, it is also important to understand which of the physiological or biochemical processes is more sensitive to salt stress that can be used as effective selection criterion (Ashraf 2004; Ashraf and Harris 2004).

While discussing various prospects of increasing salt tolerance, Munns et al. (2002) suggested that screening for a trait associated with a specific mechanism of salt tolerance is a preferable method, as measuring the effect of salt on biomass or yield of a large number of lines is not feasible. Thus, our knowledge of physiological mechanisms of salt tolerance should be used to identify traits that can be employed for rapid and cost-effective selection techniques. Therefore, it is very important to develop an effective evaluation approach for screening salt-tolerant genotypes, which should be reliable, quick, easy, practical and economic.

3.2 Conventional Breeding for Crop Improvement

Development of crop plants tolerant to salt stress is very important to meet the growing food demand. It has been suggested to exploit naturally occurring inter- and intra-specific genetic variability by hybridization of selected salt tolerant genotypes with high yielding genotypes adapted with target environment (Munns et al. 2006). Although considerable progress has been made in achieving this goal through conventional breeding, this progress is not satisfactory in view of current demand to increase crop productivity in saline environment (Flowers 2004). For example, he pointed out that although it is possible to breed and select salt tolerant lines on the basis of some physiological criteria such as Na^+ exclusion in some crop species e.g. (Yeo et al. 1988), and *Trifolium* (Rogers and Noble 1992; Rogers et al. 1997), this strategy is not useful for other crops, e.g. in tomato (Saranga et al. 1992). In a comprehensive review, Ashraf (1994) listed a few salt-tolerant lines/cultivars of different crops that had been developed through conventional breeding. During the last 3 years, many researchers concluded from a large number of published reports that major obstacle in developing salt tolerant plants is due to complex nature of the mechanism of salt tolerance (Flowers 2004; Colmer et al. 2005; Cuartero et al. 2006; Munns et al. 2006; Munns 2007). In view of Munns (Munns 2008; Munns and Tester 2008), genetic diversity for salt tolerance within a species is not fully exploited, because it is very difficult to assess salt tolerance in crops by screening large number of individuals for small, repeatable and quantifiable differences in biomass production. However, Ashraf et al. (2008) summarized reasons for limited success in improving crop salt tolerance through conventional breeding method (1) it is time-consuming and labor intensive, (2) undesirable genes are often transferred along with desirable traits, and (3) reproductive barriers restrict transfer of favorable alleles from inter-specific and inter-generic sources.

3.3 Molecular Biology Approaches to Increase Crop Salt Tolerance

As mentioned earlier that salt tolerance in plants is determined by a number of physiological and biochemical traits (Ashraf 2004; Ashraf and Harris 2004). It is well evident that salt tolerance is a complex trait involving the function of many genes (Hasegawa et al. 2000; Bartels and Sunkar 2005; Munns 2005; Munns and Tester 2008). Furthermore, successful screening and selection of salt tolerant cultivars in conventional breeding program is limited by the significant influence of environmental factors (Ashraf et al. 2008). In view of this argument, it is suggested to identify the molecular markers tightly linked to the genes governing salt tolerance and could be used to select plants in segregating populations because molecular markers are unaffected by the environment. Thus, the use of QTLs has improved the efficiency of selection, in particular, for those traits that are controlled by several genes and are highly influenced by environmental factors (Flowers 2004). As mentioned earlier, salt tolerance in plants varies with the change in growth stage that cause problem in selecting salt tolerant genotypes. However, QTLs associated with salt tolerance at the germination stage in barley (Mano and Takeda 1997), tomato (Foolad et al. 1999) and *Arabidopsis* (Quesada et al. 2002) were different from those associated with salt tolerance at the early stage of growth. Therefore, plants selected by their ability to germinate at high salinity did not display similar salt tolerance during vegetative growth (Yamaguchi and Blumwald 2005). Although QTLs for salinity tolerance have been identified in a number of potential cereal crops such as rice, barley and wheat, robust markers that can be used across a range of germplasm are very few (Munns 2008).

Since 1993, a number of reports are available in the literature showing enhanced salt tolerance in different crop plants by over-expressing genes that are involved in

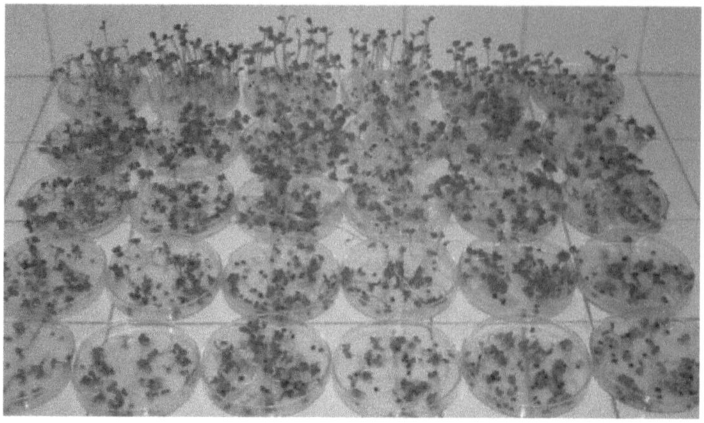

Fig. 1.4 Screening and selection is one of the most effective methods to develop salt tolerant crop cultivars. In this figure, various cultivars of radish are growing on varying levels of salt stress (Courtesy of Zahra Noreen)

controlling traits responsible for salt tolerance (Flowers 2004; Bartels and Sunkar 2005; Munns 2005; Cuartero et al. 2006; Ashraf et al. 2008). Munns (2005) categorized these salt tolerant genes into three categories (1) those that control salt uptake and transport; (2) those that have an osmotic or protective function; and (3) those that could make a plant grow more quickly in saline soil. However, large number of successful reports from transformation experiments have come from manipulating genes responsible for Na^+ exclusion or tissue Na^+ tolerance (Munns and Tester 2008). These claims of improved salt tolerance were highly criticized because of poor experimental designs, inappropriate choices of methods to evaluate salt tolerance (Flowers 2004; Munns 2005; Cuartero et al. 2006; Ashraf et al. 2008).

4 Conclusion and Future Prospects

Although it is widely recognized that salt and drought stresses are major constraints for crop productivity, knowledge about nature and magnitude of both stresses is scanty to develop an economically viable/sustainable agriculture. For example, a great gap exists in knowledge about the level of stress tolerance to be developed in crops intended to be grown on a targeted environment. Such kind of knowledge will certainly be helpful in prioritizing traits/selection criteria and developing screening techniques for improved stress tolerance.

During the last two decades, plant breeders have been able to successfully develop cultivars with at least some tolerance for a number of abiotic stresses by exploiting genetic variation that exists among the cultivated varieties. Inter- and intra-specific genetic variation for stress tolerance in the present germplasm has resulted from long-term farmer selection or from wild relatives of crop plants that have evolved abiotic stress tolerance as a means to allow colonization of marginal and extreme habitats. However, desired diversity for improving stress tolerance is not available though small increase in stress tolerance feasible by exploiting existing genetic variation. In order to increase the extent of existing genetic variation for stress tolerance, use of wide hybridization, molecular breeding or transgenic approaches are suggested. Although wide hybridization can enhance the stress tolerance, it may cause a significant penalty in terms of yield. Development of transgenic plants for transcription factors, antiporters and compatible solutes resulted in enhanced stress tolerance in plants. However, such types of reports on enhanced stress tolerance are highly criticized due to adoption of poor evaluation methodology in carrying out such studies.

At present, we are still unaware about stress-induced changes in metabolism in plants – a major gap in our understanding of stress tolerance. With the advancement in functional genomics, it is possible to identify key genes and their immediate functions at cellular as well as at whole plant level. Thus, detailed analysis of underlying physiological and molecular mechanisms for salt tolerance using functional genomics is an important area of future research, which will eventually assist in developing transgenic plants for stress tolerance. Therefore, the improvement in abiotic stress tolerance in agricultural plants can only be achieved practically by combining traditional and molecular breeding approaches. In the meantime, it would be sensible to use shotgun approaches (exogenous application of compatible solutes, plant growth regulators, antioxidant compounds, inorganic salts) to increase salt tolerance in potential crops.

Acknowledgements The presented paper is part of Ph.D. thesis of Habib-ur-Rehman Athar PIN No. 1999-ILB-0345086, whose Ph.D. study is funded by the Higher Education Commission through Indigenous Ph.D. Scheme.

References

Abebe T, Guenzi AC, Martin B, Chushman JC (2003) Tolerance of mannitol-accumulating transgenic wheat to water stress and salinity. Plant Physiol 131: 1748–1755.

Alvim P deT (1985) Theobroma cacao. In: Halvey AH (ed) Handbook of Flowering. CRC Press, Boca Raton, FL, pp 357–365.

Apse MP, Aharon GS, Snedden WS, Blumwald E (1999) Salt tolerance conferred by overexpression of a vacuolar Na^+/H^+ antiporter in *Arabidopsis*. Science 285: 1256–1258.

Araus JL, Slafer GA, Reynolds MP, Royo C (2002) Plant breeding and drought in C_3 cereals: what to breed for? Ann Bot 89: 925–940.

Araus JL, Bort J, Steduto P, Villegas D, Royo C (2003) Breeding cereals for Mediterranean conditions: ecophysiological clues for biotechnology application. Ann Appl Biol 142: 129–141.

Araus JL, Slafer GA, Reynolds MP, Royo C (2004) Physiology of yield and adaptation in wheat and barley breeding. In: Blum A, Nguyen H (eds) Physiology and Biotechnology Integration for Plant Breeding. Marcel Dekker, New York, pp 1–49.

Ashraf M (1994) Breeding for salinity tolerance in plants. Crit Rev Plant Sci 13: 17–42.

Ashraf M (2004) Some important physiological selection criteria for salt tolerance in plants. Flora 199: 361–376.

Ashraf M, Harris PJC (2004) Potential biochemical indicators of salinity tolerance in plants. Plant Sci 166: 3–16.

Ashraf M, Sharif R (1998) Assessment of inter-cultivar/line variation of drought resistance in a potential oil-seed crop, Ethiopian mustard (*Brassica carinata*). J Plant Nutr 43: 251–265.

Ashraf M, Nawazish S, Athar HR (2007) Are chlorophyll fluorescence and photosynthetic capacity, potential physiological determinants of drought tolerance in maize (*Zea mays* L.)? Pak J Bot 39(4): 1123–1131

Ashraf M, Athar HR, Harris PJC, Kwon TR (2008) Some prospective strategies for improving crop salt tolerance. Adv Agron 97: 45–110.

Babu CR, Nguyen BD, Chamarerk V, Shanmugasundaram P, Chezhian P, Juyaprakash P, Ganesh SK, Palchamy A, Sadasivam S, Sarkarung S, Wade LJ, Nguyen TH (2003) Genetic analysis of drought resistance in rice by molecular markers: association between secondary traits and field performance. Crop Sci 43: 1457–1469.

Bajaj S, Targolli J, Liu LF, Ho THD, Wu R (1999) Transgenic approaches to increase dehydration-stress tolerance in plants. Mol Breed 5: 493–503.

Bartels D, Sunkar R (2005) Drought and salt tolerance in plants. Crit Rev Plant Sci 24: 23–58.

Blum A (1985) Breeding for crop varieties for stress environments. Crit Rev Plant Sci 2: 199–238.

Bohnert HJ, Gong Q, Li P, Ma S (2006) Unraveling abiotic stress tolerance mechanisms - getting genomics going. Curr Opin Plant Biol 9: 180–188.

Borlaug NE, Dowswell CR (2005) Feeding a world of ten billion people: a 21st century challenge. In: Tuberosa T, Phillips RL, Gale M (eds) Proceedings of "In the Wake of the Double Helix: From the Green Revolution to the Gene Revolution", 27–31 May 2003, at Bologna, Italy. Avenue Media, Bologna, Italy pp 3–24.

Boyer JS (1982) Plant productivity and environment. Science 218: 443–448.

Bray EA, Bailey-Serres J, Weretilnyk E (2000) Responses to abiotic stresses. In: Gruissem W, Buchannan B, Jones R (eds) Biochemistry and Molecular Biology of Plants. American Society of Plant Physiologists, Baltimore, MD, pp 1158–1249.

Cattivelli L, Rizza F, Badeck FW, Mazzucotelli E, Mastrangelo AM, Francia E, Mare C, Tondelli A, Stanca AM (2008) Drought tolerance improvement in crop plants: an integrated view from breeding to genomics. Field Crop Res 105: 1–14.

Chinnusamy V, Jagendorf A, Zhu JK (2005) Understanding and improving salt tolerance in plants. Crop Sci 45: 437–448.

Colmer TD, Munns R, Flowers TJ (2005) Improving salt tolerance of wheat and barley: future prospects. Aust J Exp Agr 45: 1425–1443.

Cuartero J, Bolarín MC, Asíns MJ, Moreno V (2006) Increasing salt tolerance in the tomato. J Exp Bot 57: 1045–1058.

Dasgan HY, Aktas H, Abak K, Cakmak I (2002) Determination of screening techniques to salinity tolerance in tomatoes and investigation of genotype responses. Plant Sci 163: 695–703.

De Block M, Verduyn C, De Brouwer D, Conelissen M (2005) Poly(ADP-ribose) polymerase in plants affects energy homeostasis, cell death and stress tolerance. Plant J 41: 95–106.

Diab AA, Teulat B, This D, Ozturk NZ, Benscher D, Sorrells ME (2004) Identification of drought-inducible genes and differentially expressed sequence tags in barley. Theor Appl Genet 109: 1417–1425.

Dubouzet JG, Sakuma Y, Ito Y, Kasuga M, Dubouzet EG, Miura S, Seki M, Shinozaki K, Yamaguchi-Shinozaki K (2003) OsDREB genes in rice, Oryza sativa L., encode transcription activators that function in drought-, high salt and cold responsive gene expression. Plant J 33: 751–763.

Epstein E, Norlyn JD, Rush DW, Kingsbury R, Kelley DB, Wrana AF (1980) Saline culture of crops: a genetic approach. Science 210: 399–404.

FAO (Food and Agriculture Organization, United Nations) (2003) Unlocking the water potential of agriculture. www.fao.org. Accessed 7 April 2006.

FAO (2008) FAO Land and Plant Nutrition Management Service. http://www.fao.org/ag/agl/agll/spush

Flexas J, Bota J, Cifre J (2004) Understanding down regulation of photosynthesis under water stress, future prospects and searching for physiological tools for irrigation management. Ann Appl Biol 144: 273–283.

Flowers TJ (2004) Improving crop salt tolerance. J Exp Bot 55: 307–319.

Foolad MR, Lin GY, Chen FQ (1999) Comparison of QTLs for seed germination under non-stress, cold stress and salt stress in tomato. Plant Breed 118: 167–173.

Garg AK, Kim JK, Owens TG, Ranwala AP, Choi YD, Kochian LV, Wu RJ (2002) Trehalose accumulation in rice plants confers high tolerance levels to different abiotic stresses. PNAS USA 99: 15898–15903.

Gaxiola RA, Li J, Undurraga S, Dang LM, Allen GJ, Alper SL, Fink GR (2001) Drought- and salt-tolerant plants result from overexpression of the *AVP1* H&pron;pump. PNAS USA 98: 11444–11449.

Gelburd DE (1985) Managing salinity, lesson from the past. J Soil Water Conserv 40: 329–331.

Genc Y, McDonald GK, Tester M (2007) Reassessment of tissue Na^+ concentration as a criterion for salinity tolerance in bread wheat. Plant Cell Environ 30: 1486–1498.

Greenway H, Munns R (1980) Mechanisms of salt tolerance in non-halophytes. Annu Rev Plant Physiol 312: 149–190.

Harris K, Subudhi PK, Borrel A, Jordan D, Rosenow D, Nguyen H, Klein P, Klein R, Mullet J (2007) Sorghum stay-green QTL individually reduce post-flowering drought-induced leaf senescence. J Exp Bot 58: 327–338.

Hasegawa PM, Bressan RA, Zhu J-K, Bohnert HJ (2000) Plant cellular and molecular responses to high salinity. Annu Rev Plant Physiol Plant Mol Biol 51: 463–499.

Hsieh TH, Lee JT, Charng YY, Chan MT (2002) Tomato plants ectopically expressing Arabidopsis CBF1 show enhanced resistance to water deficit stress. Plant Physiol 130: 618–626.

Hussain SS (2006) Molecular breeding for abiotic stress tolerance: drought perspective. Proc Pakistan Acad Sci 43(3): 189–210.

Ito Y, Katsura K, Maruyama K, Taji T, Kobayashi M, Seki M, Shinozaki K, Yamaguchi-Shinozaki K (2006) Functional analysis of rice DREB1/CBF-type transcription factors involved in cold-responsive gene expression in transgenic rice. Plant Cell Physiol 47: 141–153.

Kauser R, Athar HR, Ashraf M (2006) Chlorophyll fluorescence: a potential indicator for rapid assessment of water stress tolerance in Canola (*Brassica napus* L.). Pak J Bot 38(5): 1501–1509.

Kavi Kishore PB, Hong Z, Miao G-H, Hu CAA, Verma DPS (1995) Overexpression of Δ-pyrroline-5-carboxylate synthetase increase proline production and confers osmoltolerance in transgenic plants. Plant Physiol 108: 1387–1394.

Krishnamurthy L, Serraj R, Hash CT, Dakheel AJ, Reddy BVS (2007) Screening sorghum genotypes for salinity tolerant biomass production. Euphytica 156: 15–24.

Kumar D (2005) Breeding for drought resistance. In: Ashraf M, Harris PJC (eds) Abiotic Stress: Plant Resistance Through Breeding and Molecular Approaches. Haworth Press, New York, pp 145–175.

Lanceras JC, Pantuwan G, Jongdee B, Toojinda T (2004) Quantitative trait loci associated with drought tolerance at reproductive stage in rice. Plant Physiol 135: 384–399.

Laporte MM, Shen B, Tarczynski MC (2002) Engineering for drought avoidance: expression of maize NADP-malic enzyme in tobacco results in altered stomatal function. J Exp Bot 53: 699–705.

Lewis LN (1984) A vital resource in danger. Calif Agr 38: 2.

Liang ZS, Ding ZR, Wang STR (1992) Study on type of water stress adaptation in both *Brassica napus* and *B. juncea* L. species. Acta Botanika 12: 38–45.

Maccaferri M, Sanguineti MC, Corneti S, Araus-Ortega JL, BenSalem M, Bort J, DeAmbrogio E, del Moral LFG, Demontis A, El-Ahmed A, Maalouf F, Machlab H, Martos V, Moragues M, Motawaj J, Nachit M, Nserallah N, Ouabbou H, Royo C, Slama A, Tuberosa R (2008) Quantitative trait loci for grain yield and adaptation of durum wheat (*Triticum durum* Desf.) across a wide range of water availability. Genetics 178: 489–511. DOI: 10.1534/genetics.107.077297

Mano Y, Takeda K (1997) Mapping quantitative trait loci for salt tolerance at germination and the seedling stage in barley (*Hordeum vulgare* L.). Euphytica 94: 263–272.

Mano Y, Takeda K (2001) Genetic resources of salt tolerance at germination and seedling stage in wheat. Jpn J Crop Sci 70: 215–220.

Masle J, Gilmore SR, Farquhar GD (2005) The ERECTA gene regulates plant transpiration efficiency in Arabidopsis. Nature 436: 866–870.

Moreno LS, Maiti RK, Gonzales AN, Star JV, Foroughbakhch R, Gonzales HG (2000) Genotypic variability in bean cultivars (*Phaseolus vulgaris* L.) for resistance to salinity at the seedling stage. Indian Agr 44: 1–12.

Morgan JM (2000) Increases in grain yield of wheat by breeding for an osmoregulation gene: relationship to water supply and evaporative demand. Aust J Agr Res 51: 971–978.

Morgan JM, Tan MK (1996) Chromosomal location of a wheat osmoregulation gene using RFLP analysis. Aust J Plant Physiol 23: 803–806.

Munns R (2002) Comparative physiology of salt and water stress. Plant Cell Environ 25: 239–250.

Munns R (2005) Genes and salt tolerance: bringing them together. New Phytol 167(3): 645–663.

Munns R (2007) Utilizing genetic resources to enhance productivity of salt-prone land. CAB Rev: Perspect Agr Vet Sci Nutr Natl Res 2: (009).

Munns R (2008) Strategies for crop improvement in saline soils. In: Ashraf M, Ozturk M, Athar HR (eds) Salinity and Water Stress: Improving Crop Efficiency. Springer, The Netherlands, pp 99–110.

Munns R, Tester M (2008) Mechanisms of salinity tolerance. Annu Rev Plant Biol 59: 651–681.

Munns R, Husain S, Rivelli AR, James RA, Condon AG, Lindsay MP, Lagudah ES, Schachtman D, Hare RA (2002) Avenues for increasing salt tolerance of crops, and the role of physiologically based selection traits. Plant Soil 247: 93–105.

Munns R, James RA, Laüchli A (2006) Approaches to increasing the salt tolerance of wheat and other cereals. J Exp Bot 57: 1025–1043.
Neumann PM (2008) Coping mechanisms for crop plants in drought-prone environments. Ann Bot 101: 901–907.
Nguyen TT, Klueva N, Chamareck V, Aarti A, Magpantay G, Millena AC, Pathan MS, Nguyen HT (2004) Saturation mapping of QTL regions and identification of putative candidate genes for drought tolerance in rice. Mol Gen Genomics 272: 35–46.
Park BJ, Liu Z, Kanno A, Kameya T (2005) Genetic improvement of Chinese cabbage for salt and drought tolerance by constitutive expression of a *B. napus* LEA gene. Plant Sci 169: 553–558.
Parry MAJ, Flexas J, Medrano H (2005) Prospects for crop production under drought: research priorities and future directions. Ann Appl Biol 147: 211–226.
Pellegrineschi A, Reynolds M, Pacheco M, Brito BM, Almeraya R, Yamaguchi-Shinozaki K, Hoisington D (2004) Stress induced expression in wheat of the Arabidopsis thaliana DREB1A gene delays water stress symptoms under greenhoude conditions. Genome 47: 493–500.
Peng S, Ismail AM (2004) Physiological basis of yield and environmental adaptation in rice. In: Blum A, Nguyen H (eds) Physiology and Biotechnology Integration for Plant Breeding. Marcel Dekker, New York, pp 83–140.
Quesada V, García-Martínez S, Piqueras P, Ponce MR, Mico JL (2002) Plant Physiol 130: 951–963.
Rehman S, Harris PJC, Ashraf M (2005) Stress environments and their impact on crop production. In: Ashraf M, Harris PJC (eds) Abiotic Stresses: Plant Resistance Through Breeding and Molecular Approaches. Haworth Press, New York, pp 3–18.
Rengasamy P (2006) World salinization with emphasis on Australia. J Exp Bot 57(5): 1017–1023.
Reynolds M, Tuberosa R (2008) Translational research impacting on crop productivity in drought-prone environments. Curr Opin Plant Biol 11: 171–179.
Reynolds MP, Mujeeb-Kazi A, Sawkins M (2005) Prospects for utilizing plant-adaptive mechanisms to improve wheat and other crops in drought and salinity-prone environments. Ann Appl Biol 146: 239–259.
Ribaut J-M, Ragot M (2007) Marker-assisted selection to improve drought adaptation in maize: the backcross approach, perspectives, limitations, and alternatives. J Exp Bot 58(2): 351–360.
Richards RA, Rebetzke GJ, Condon AG, Herwaarden AF (2002) Breeding opportunities for increasing the efficiency of water use and crop yield in temperate cereals. Crop Sci 42: 111–121.
Rogers ME, Noble CL (1992) Variation in growth and ion accumulation between two selected populations of *Trifolium repens* L differing in salt tolerance. Plant Soil 146: 131–136.
Rogers ME, Noble CL, Halloran GM, Nicolas ME (1997) Selecting for salt tolerance in white clover (*Trifolium repens*): chloride ion exclusion and its heritability. New Phytol 135: 645–654.
Rogers ME, Craig AD, Munns R, Colmer TD, Nichols PGH, Malcolm CV, Barrett-Lennard EG, Brown AJ, Semple WS, Evans PM, Cowley K, Hughes SJ, Snowball R, Bennett SJ, Sweeney GC, Dear BS, Ewing MA (2005) The potential for developing fodder plants for the salt-affected areas of southern and eastern Australia: an overview. Aust J Exp Agr 45: 301–329.
Sakuma Y, Maruyama K, Osakabe Y, Qin F, Seki M, Shinozaki K, Yamaguchi-Shinozaki K (2006) Functional analysis of Arabidopsis transcription factor, DREB2A, involved in drought-responsive gene expression. Plant Cell 18: 1292–1309.
Salvi S, Tuberosa R (2005) To clone or not to clone plant QTLs: present and future challenges. Trends Plant Sci 10: 297–304.
Saranga Y, Cahaner A, Zamir D, Marani A, Rudich J (1992) Breeding tomatoes for salt tolerance inheritance of salt tolerance and related traits in inter-specific populations. Theor Appl Genet 84: 390–396.
Schwabe KA, Iddo K, Knap KC (2006) Drain water management for salinity mitigation in irrigated agriculture. Am J Agr Ecol 88: 133–140.
Serraj R, Sinclair TR (2002) Osmolyte accumulation: can it really help increase crop yield under drought conditions? Plant Cell Environ 25: 333–341.
Serraj R, Hash TC, Buhariwalla HK, Bidinger FR, Folkertsma RT, Chandra S, Gaur PM, Kashiwagi J, Nigam SN, Rupakula A, Crouch JH (2005a) Marker-assisted breeding for crop drought tolerance at ICRISAT: achievements and prospects. In: Tuberosa R, Phillips RL, Gale M (eds) Proceedings of the International Congress "In the Wake of the Double Helix: From the Green Revolution to the Gene Revolution". Avenue Media, Bologna, Italy, pp 217–238.
Serraj R, Hash CT, Rizvi SM, Sharma A, Yadav RS, Bidinger FR (2005b) Recent advances in marker assisted selection for drought tolerance in pearl millet. Plant Prod Sci 8: 334–337.
Serrano R, Mulet JM, Rios G, Marquez JA, de Larrinoa IF, Leube MP, Mendizabal I, Pascual-Ahuir A, Proft M, Ros R, Montesinos C (1999) A glimpse of mechanisms of ion homeostasis during salt stress. J Exp Bot 50: 1023–1036.
Shinozaki K, Yamaguchi-Shinozaki K (2007) Gene networks involved in drought stress response and tolerance. J Exp Bot 58: 221–227.
Slafer GA, Satorre EH (1999) An introduction to the physiological-ecological analysis of wheat yield. In: Satorre EH, Slafer GA (eds) Wheat: Ecology and Physiology of Yield Determination. Food Product Press, New York, pp 3–12.
Subbarao GV, Ito O, Serraj R, Crouch JJ, Tobita S, Okada K, Hash CT, Ortiz R, Berry WL (2005) Physiological perspectives on improving crop adaptation to drought –justification for a systematic component-based approach. In: Pessarakli M (ed) Handbook of Photosynthesis, 2nd edn. Marcel and Dekker, New York, pp 577–594.
Syvänen AC (2005) Toward genome-wide SNP genotyping. Nat Genet 37: S5–S10.
Tambussi EA, Bort J, Araus JL (2007) Water use efficiency in C_3 cereals under Mediterranean conditions: a review of physiological aspects. Ann Appl Biol doi:10.1111/j.1744-7348.2007.00143.x.
Tondelli A, Francia E, Barabaschi D, Aprile A, Skinner JS, Stockinger EJ, Stanca AM, Pecchioni N (2006) Mapping regulatory genes as candidates for cold and drought stress tolerance in barley. Theor Appl Genet 112: 445–455.

Tuberosa R, Giuliani S, Parry MAJ, Araus JL (2007a) Improving water use efficiency in Mediterranean agriculture: what limits the adoption of new technologies? Ann Appl Biol 150: 157–162.

Tuberosa R, Salvi S, Giuliani S, Sanguineti MC, Bellotti M, Conti S, Landi P (2007b) Genome-wide approaches to investigate and improve maize response to drought. Crop Sci 47(S3): S120–S141.

UN Millennium Project (2003) Halving global hunger. Background Paper of the Millennium Task Force on Hunger. UNDP, New York.

Vinocur B, Altman A (2005) Recent advances in engineering plant tolerance to abiotic stress: achievements and limitations. Curr Opin Biotechnol 16: 123–132.

Wang W, Vinocur B, Altman A (2003) Plant responses to drought, salinity and extreme temperatures: towards genetic engineering for stress tolerance. Planta 218: 1–14.

Yamaguchi T, Blumwald E (2005) Developing salt-tolerant crop plants: challenges and opportunities. Trends Plant Sci 10: 615–620.

Yeo A (1998) Molecular biology of salt tolerance in the context of whole-plant physiology. J Exp Bot 49: 915–929.

Yeo AR, Yeo ME, Flowers TJ (1988) Selection of lines with high and low sodium transport from within varieties of an inbreeding species: rice (*Oryza sativa* L). New Phytol 110: 13–19.

Zhang HX, Hodson JN, Williams JP, Blumwald E (2001) Engineering salt-tolerant Brassica plants: characterization of yield and seed oil quality in transgenic plants with increased vacuolar sodium accumulation. PNAS USA 98: 12832–12836.

Zhao XQ, Xu J-L, Zhao M, Lafitte R, Zhu L-H, Fu B-Y, Gao Y-M, Li Z-K (2008) QTLs affecting morph-physiological traits related to drought tolerance detected in overlapping introgression lines of rice (*Oryza sativa* L.). Plant Sci 174: 618–625.

Zheng BS, Yang L, Zhang WP, Mao CZ, Wu YR, Yi KK, Liu FY, Wu P (2003) Mapping QTLs and candidate genes for rice root traits under different water supply conditions and comparative analysis across three populations. Theor Appl Genet 107: 1505–1515.

Zhu BC, Su J, Chan MC, Verma DPS, Fan YL, Wu R (1998) Over expression of a –pyrroline-5-carboxylate synthetase gene and analysis of tolerance to water stress and salt stress in transgenic rice. Plant Sci 139: 41–48.

Part I
Salt and Water Stress

Chapter 2
Prediction of Salinity Tolerance Based on Biological and Chemical Properties of *Acacia* Seeds

S. Rehman, A. Khatoon, Z. Iqbal, M. Jamil, M. Ashraf, and P.J.C. Harris

Abstract Multiple regression equations have been developed to predict the salinity tolerance of *Acacia* seeds, expressed as the I_{50} (the concentration of NaCl required to reduce final germination to 50% of the control value in distilled water). Accurate predictions can be made using one or more chemical and biological seed parameters. In this study relationships were drawn among final germination percentage and rate of germination in distilled water, Ca^{2+} or K^+ contents and their ratios to predicted salinity tolerance (I_{50}) of *Acacia* species. Simulation of the effects of changing final germination, calcium and potassium suggest the possibility of practical application of these results to modify the salinity tolerance of seeds. The predicted I_{50} increased with increase final germination percentage. Similarly, the higher the rate of germination was the higher the predicted salt tolerance of *Acacia* species. The Ca^{2+} content of seeds was found to be positively correlated with I_{50}. Species with higher Ca^{2+} contents had a higher I_{50}. This suggests that that I_{50} might be increased by increasing the Ca^{2+} contents of seeds by pretreatment with calcium salts.

Keywords *Acacia* • calcium • germination • potassium • prediction • salinity

S. Rehman (✉), A. Khatoon, and Z. Iqbal
Botany Department, Kohat University of Science & Technology, Kohat, NWFP, Pakistan
e-mail: drshafiq@yahoo.com

M. Jamil
Biotechnology Department, Kohat University of Science & Technology, Kohat, NWFP, Pakistan

M. Ashraf
Botany Department, University of Agriculture, Faisalabad, Pakistan

P.J.C. Harris
Biosciences, School of Science and Environment, Coventry University, Coventry, CV1 5FB, UK

1 Introduction

Seed germination is a major factor in the establishment of plants under saline conditions. Salinity stress caused various physiological and biochemical disorders which prevent or delay germination (Rehman et al. 1996; Ungar 1996). There are also reports suggesting that salt may affect germination rate to a greater extent than germination percentage (Lovato et al. 1994). Several reports have confirmed that the detrimental effects of salinity on plant growth were due to creating an ionic imbalance, particularly of Ca^{2+} and K^+ (Cerda et al. 1995; Ashraf 2004). For example, it has been shown that essential plant mineral nutrients including Ca^{2+} and K^+ leached from the seed in response to soaking in water or NaCl solution. However, extent of leaching of mineral nutrients varies from species to species among *Acacia* species (Rehman et al. 1996). It is generally accepted that plants must maintain relatively high concentrations of Ca^{2+} and K^+ if they are to grow successfully in a saline environment (Greenway and Munns 1980; Ashraf 2004).

Previously Rehman et al. (2000) found a considerable variation in the germination, germination rate and salinity tolerance (I_{50}) in *Acacia*. The *Acacia* species also varied significantly in the Ca^{2+} and K^+ concentration of their seeds and in the loss of these ions when soaked in DW or in NaCl solution. Furthermore, by applying multiple regression equations on various physiological parameters of 13 species and/or accessions of *Acacia*, it was obvious that salinity tolerance of *Acacia* seeds could be predicted by one or more seed parameters (Rehman et al. 2000). From these results, they suggested that accurate predictions can be made from a combination of the ion concentration and ratios of the untreated seeds, the leaching of ions into

M. Ashraf et al. (eds.), *Salinity and Water Stress*,

DW and NaCl solutions, and the germination of the seeds in DW. There appears to be a real possibility of developing a screening test for salt tolerance of seed germination in *Acacia* species, provenances and individual seed accessions based on simple and rapid laboratory chemical analyses, without lengthy germination trials in saline conditions.

In this study, the predicted effect of final germination percentage in distilled water, germination rate, and seed Ca^{2+} or K^+ concentrations and their ratios, as independent or interrelated factors, on the salinity tolerance of *Acacia* species was investigated.

2 Materials and Methods

Acacia tortilis and *A. coriacea* seeds were manually scarified to overcome hard seed coat dormancy by removing a small portion of the testa at the cotyledon end with nail clippers. Seeds were germinated in incubator at constant temperature 25°C, in Petri dishes (9cm) with two Whatman No. 1 filter papers soaked with 10ml of distilled water (DW) or a range of NaCl concentrations (0–400mol m^{-3} with 25mol m^{-3} increments). The final germination (FG) percentage and germination rate (Rate) were determined as described by Rehman et al. (1996).

It is a well-known phenomenon that natural populations of plants grow well in different soils and vary inherently in their ability to absorb and/or utilize mineral nutrients. The leakage of Ca^{2+} and K from plant species under saline conditions has been reported by many workers. Ca^{2+} and K leakage during water treatment is generally considered not to be an inherited character and, presumably, depends on conditions under which the seeds develop on the parent plant, and are harvested and stored (Simon 1974). The leaching (Ca^{2+} and K^+) of the seeds was evaluated by soaking scarified seeds in 10ml of DW or 250mol m^{-3} NaCl for 24h. The Ca^{2+} and K^+ concentrations, and Ca^{2+}/K^+ and K^+/Ca^{2+} ratios of seeds initially, and after soaking in DW or NaCl were determined as described by Rehman et al. (1996).

The salinity tolerance (I_{50} – the concentration of NaCl required to reduce final germination to 50% of the control value in DW) of *Acacia tortilis* and *A. coriacea* was predicted using a multiple regression equation developed by Rehman et al. (2000).

$$I_{50}\ (\text{mol m}^{-3}) = 108 + 422\ \text{Rate}_{\text{unt}} - 1.32\ K^+_{\text{unt}} + 16.3\ Ca^{2+}_{DW} - 5.4\ K^+_{DW} + 71.6\ K^+/Ca^{2+}_{DW}$$

I_{50} = the concentration of NaCl required to reduce final germination to 50% of the control value in distilled water (DW). Rate = rate of germination ($1/t_{50}$) in distilled water (t_{50} = the time to 50% of final germination in DW). Ca_{unt}, K_{unt} = Ca and K concentration in dry seeds (in µmol g^{-1} dry weight (dw)). K_{DW} = leakage of Ca and K when soaked for 24h in DW. FG = final germination percentage in distilled water.

The interrelationships between parameters in the equation were established from linear regressions fitted to the data for 13 accessions from 10 *Acacia* species (Rehman et al. 2000).

$$\Delta\text{Rate} = 0.0051\Delta\text{FG};\ \Delta\text{Rate} = 0.0018\Delta Ca^{2+}_{\text{unt}}$$

$$\Delta\text{Rate} = -0.0008\Delta K^+;\ \Delta Ca^{2+}_{DW} = 0.1208\Delta Ca^{2+}_{\text{unt}}$$

$$\Delta Ca^{2+} = -0.03058\Delta K^+;\ \Delta K^+_{DW} = 0.0089\Delta K^+_{\text{unt}}$$

$$\Delta K^+ = -0.07267\Delta Ca^{2+}_{\text{unt}}$$

The effect of the following on the I_{50} were tested:

1. Germination rate from 0.1 to 2 at 0.1 intervals
2. Final germination in distilled water from 5% to 100% at 5% intervals
3. Ca^{2+} concentration of seeds from 10 to 300µmol g^{-1} dw at 10µmol g^{-1} dw intervals
4. K^+ concentration of seeds from 10 to 590µmol g^{-1} dw with 20µmol g^{-1} dw intervals

The predicted effect of parameters in the equation on salinity tolerance of *Acacia* seeds was performed by using the Microsoft Excel (Middleton 1995) and Minitab statistical software package (Ryan and Joiner 1994).

3 Results

3.1 Germination Rate (Rate) and Final Germination (FG)

Figure 2.1a shows the predicted I_{50} with different germination rates varied independently of other factors tested. The I_{50} predicted for both species increased linearly with the increase in the germination Rate. Similarly, Fig. 2.1b shows the I_{50} predicted with different values of FG assuming that a change in FG will involve a change in Rate. An increase in FG increased the I_{50} predicted.

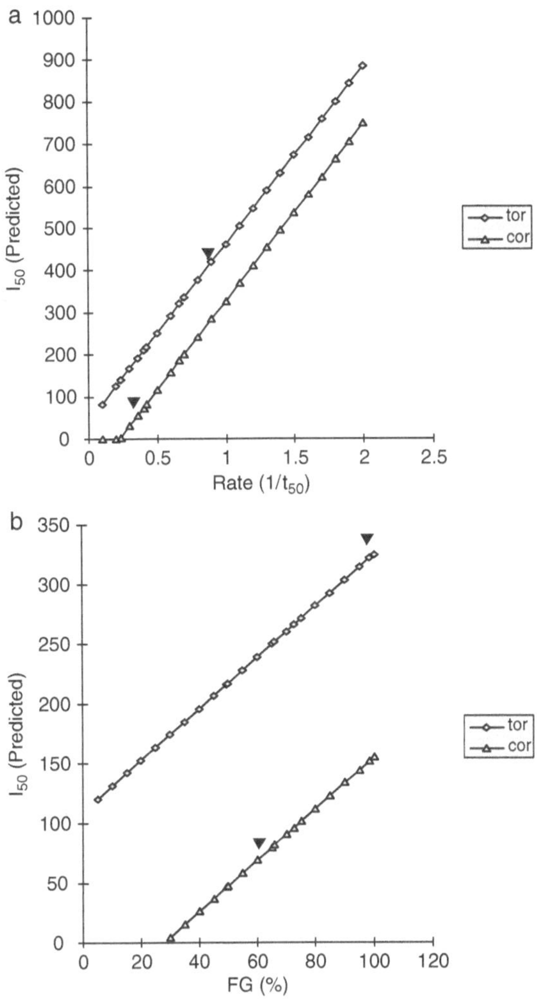

Fig. 2.1 Predicted effect on the I_{50} mol m^{-3} NaCl of *Acacia* species of (a) varying germination rate (Rate) independently and (b) varying final germination (FG), assuming a correlation between FG and Rate. tor (*A. tortilis*) and cor (*A. coriacea*). ▾ indicates the actual (a) Rate and (b) FG

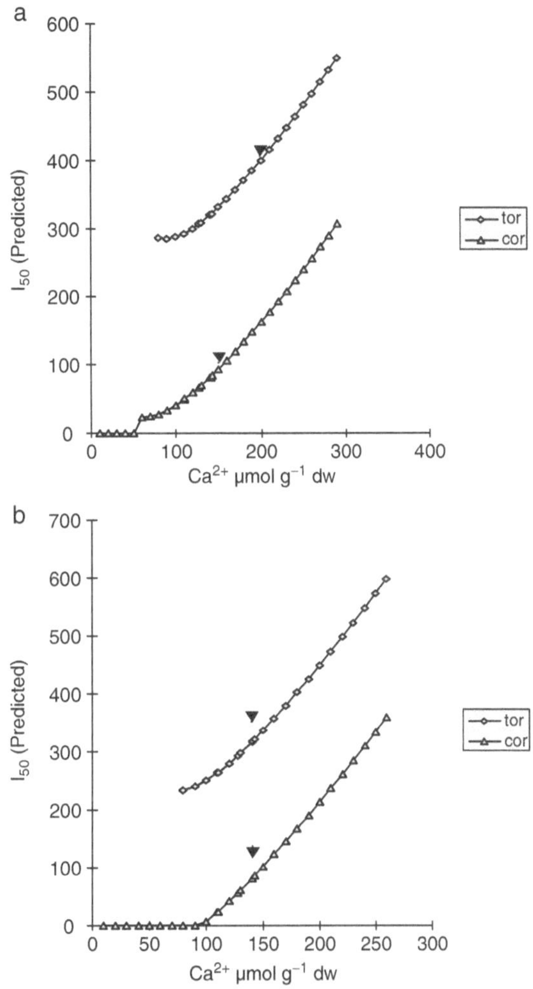

Fig. 2.2 Predicted effect on the I_{50} mol m^{-3} NaCl of *Acacia* species of varying Ca^{2+} (a) independently and (b) assuming a correlation between Ca^{2+} and K^+, Rate, K^+ Ca^{2+} DW, and Ca^{2+} and K^+ loss due to DW. tor (*A. tortilis*) and cor (*A. coriacea*). ▾ indicates the actual Ca^{2+} concentration

A higher I_{50} was predicted for *A. tortilis* than for *A. coriacea*. This equation predicted that *A. coriacea* seeds with FG below 30% would not germinate in NaCl.

3.2 The Predicted Effect of Changing Seed Calcium (Ca^{2+}) Concentration on the Salinity Tolerance of Acacia Species

Figure 2.2a shows the I_{50} predicted with values of Ca^{2+} varied independently of other factors, but assuming that Ca^{2+} is related to Ca^{2+}_{DW} also with a consequent effect on the K^+/Ca^{2+}_{DW} ratio. The predicted I_{50} of both species increased with an increase in Ca^{2+} levels. Figure 2.2b shows the predicted I_{50} with different Ca^{2+} values assuming that change in Ca^{2+} will also change the values of Rate, K^+, Ca^{2+}_{DW} and again the K^+/Ca^{2+}_{DW} ratio. An increase in Ca^{2+} level increased the predicted I_{50} of both species. I_{50} predicted shows that *A. coriacea* with Ca^{2+} contents below 100 µmol g^{-1} dw respectively would not germinate in NaCl. *A. tortilis* was predicted to have a higher I_{50} than the *A. coriacea* species at all Ca^{2+} levels.

3.3 The Predicted Effect of Changing Seed Potassium (K^+) Concentration on the Salinity Tolerance of Acacia Species

Figure 2.3a shows the predicted I_{50} with changing K^+ values varied independently of other factors except assuming that K^+ is related to K^+_{DW} with a consequent effect on the K^+/Ca^{2+}_{DW} ratio. The predicted I_{50} decreased with increasing K^+ levels in both species. Neither species would germinate in NaCl if the seeds contained more than 490 µmol g^{-1} dw K^+. Figure 2.3b shows the predicted I_{50} with varied values of K^+ assuming that change in K^+ will also involve change in Rate, Ca^{2+} and K^+/Ca^{2+} ratio. Similar to Fig. 2.3a, the I_{50} predicted decreased with increasing K^+ level.

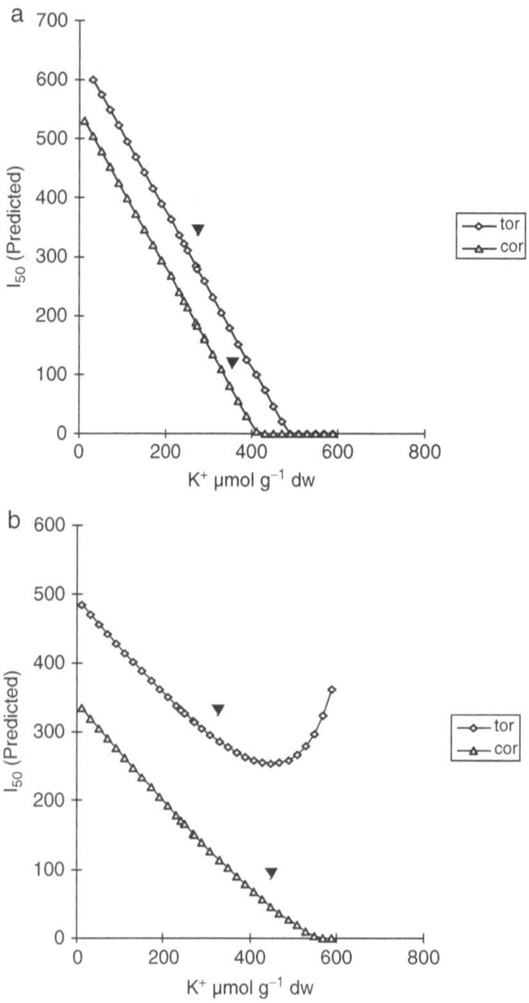

Fig. 2.3 Predicted effect on the I_{50} mol m^{-3} NaCl of *Acacia* species of varying K^+ (a) independently and (b) assuming a correlation between K^+ and Ca^{2+}, Rate, K^+ Ca^{2+} DW, and Ca^{2+} and K^+ loss due to DW. tor (*A. tortilis*) and cor (*A. coriacea*). ▾ indicates the actual K^+ concentration

4 Discussion

Simulation of the effects of changing final germination, germination rate, and calcium and potassium concentrations of seeds suggests possible practical application of these results to modify salinity tolerance. For example, the predicted I_{50} increased with increasing final germination and germination rate. Practically, hardening or osmoconditioning is used to increase germination and germination rate. Therefore, hardening or osmoconditioning may increase salinity tolerance. This hypothesis was partially supported by Rehman et al. (1998a) who showed that hardening the seeds of *Acacia nilotica* and *A. tortilis* increased both germination rate and salinity tolerance, while hardening *A. elata* seeds decreased both germination rate and salinity tolerance. However, seeds of four other *Acacia* species showed increased or decreased germination rate while their salinity tolerance was unaffected. Seed ageing is associated with a loss of vigour (Rehman et al. 1999). Rehman et al. (1999) showed that artificially ageing of *Acacia* seeds at 55°C for 24 h reduced both final germination and germination rate. As predicted from the above results, the salinity tolerance of *A. tortilis* and *A. coriacea* was also reduced by ageing.

The above results predict that salinity tolerance can be increased by increasing the Ca^{2+} contents of seeds and this has been reported for wheat by Idris and Aslam (1975). The Ca^{2+} contents of seeds may be increased by pre-treating seeds with Ca^{2+} salts, or possibly by supplying Ca^{2+} to the mother plants. Although increasing the Ca^{2+} content of *Acacia* seeds by pre-treating them with dilute $CaCl_2$ or $Ca(NO_3)_2$ increased salinity tolerance (Rehman et al. 1998b), this was mainly attributed to the hardening effect of soaking and drying on germination rather than to the effects of altered Ca^{2+} content *per se*.

References

Cerda A, Pardines J, Botella MA, Martinez V (1995). Effect of potassium on growth, water relations and the inorganic and organic solute contents for two maize cultivars grown under saline conditions. J Plant Nutr 18: 839–851.

Greenway H, Munns R (1980). Mechanism of salt tolerance in non-halophytes. Ann Rev Plant Physiol 31: 149–190.

Idris M, Aslam M (1975). The effect of soaking and drying seeds before planting on the germination and growth of *Triticum vulgare* under normal and saline conditions. Can J Bot 53: 1328–1332.

Lovato MB, Martins PS, Lemos-Filho JP (1994). Germination in *Stylosanthes humilis* population in the presence of NaCl. Aust J Bot 42: 717–723.

Middleton MR (1995). Data Analysis Using Microsoft Excel 5.0. Duxbury Press/ Wadsworth Publishing, Belomont, CA.

Rehman S, Harris PJC, Bourne WF, Wilkin J (1996). The effect of sodium chloride on germination and the potassium and calcium contents of *Acacia* seeds. Seed Sci Technol 25: 45–57.

Rehman S, Harris PJC, Bourne WF (1998a). Effects of presowing treatment with calcium salts, potassium salts or water on germination and salt tolerance of *Acacia* seeds. J Plant Nutr 21: 277–285.

Rehman S, Harris PJC, Bourne WF (1998b). The effect of hardening on the salinity tolerance of *Acacia* seeds. Seed Sci Technol 26: 743–754.

Rehman S, Harris PJC, Bourne WF (1999). Effect of ageing on the germination and ion leakage of *A. tortilis* and *A. coriacea* seeds in distilled water and sodium chloride. Seed Sci Technol 27: 141–150.

Rehman S, Harris PJC, Bourne WF, Wilkin J (2000). The relationship between ions, vigour and salinity tolerance of *Acacia* seeds. Plant Soil 220: 229–233.

Ryan BF, Joiner BL (1994). Minitab Handbook. Duxbury Press, Wadsworth Publishing, Belomont, CA.

Simon ED (1974). Phospholipids and plant membrane permeability. New Phytol 73: 377–420.

Ungar IA (1996). Effect of salinity on seed germination, growth and ion accumulation of *Atriplex patula* (Chenopodiaceae). Am J Bot 83: 604–607.

Chapter 3
Antioxidant-Enzyme System as Selection Criteria for Salt Tolerance in Forage Sorghum Genotypes (*Sorghum bicolor* L. Moench)

M. Hefny and D.Z. Abdel-Kader

Abstract The involvement of antioxidant enzyme activities in mitigating the damage of NaCl stress was studied in 26 genotypes of forage sorghum exhibiting different responses to salinity, including a local hybrid with unknown performance under salinity stress. The 2-week old sorghum seedlings were subjected to 0, 50 and 100 mM NaCl for 4 weeks, which correspond to 0.7, 8.2 and 15.11 dS m^{-1} salinity levels. Plants were sampled for enzyme analyses and dry weight determinations 4 weeks after starting salt treatments. Salt stress resulted in significant reduction of dry weight of both tolerant and sensitive genotypes. The reduction was stronger in the later group compared with the former one at 8.2 dS m^{-1}. In contrast, at the highest salinity level, there was sever reduction in plant dry weights for both groups, meanwhile the highest value was recorded by the local genotype. Five out of the 21 salt tolerant genotypes and the local hybrid produced the highest dry weights at 50 and 100 mM NaCl. The effect of salinity levels on antioxidant enzymes and lipid peroxidation was examined. Both salinity levels induced significant increase in superoxide dismutase (SOD) activity, glutathione (GSH) levels and carotenoid concentrations in all tolerant genotypes and the local genotype compared to sensitive group. Moreover, the activities of peroxidase and glutathione reductase (GR) have increased at 8.2 dS m^{-1} NaCl for most of tolerant genotypes, then the activity was declined at 15.11 dS m^{-1} salinity level for the second enzyme and was somehow constant for the first enzyme. There was a common trend in increasing lipid peroxidation activity for sensitive genotypes at both levels of stresses, and reducing the activity for some tolerant genotypes. Five salt tolerant genotypes and the local hybrid maintained beside high SOD, GSH, GR activities, reasonable lipid peroxidation and pigment contents. It could be concluded that the local genotype could be considered as salinity tolerant genotype as it exhibited the same trend of tolerant genotypes. Moreover, antioxidant system SOD, GSH, GR, ASPX and carotenoids could be considered as selection criteria for salt tolerance in sorghum species.

Keywords Antioxidant enzymes • oxidative stress • salinity • *Sorghum bicolor* L

1 Introduction

Sorghum (*Sorghum bicolour* L.) is one of the stable crops grown in arid and semi-arid countries. It is the fifth most important cereal crops grown on 44 million hectares in 99 countries in Africa, Asia and the Americans. The majority of sorghum plantings are concentrated in poor countries where it constitutes a valuable source of grains for human consumption. In addition, characterized by its high nutritional as animal feeding source.

Sorghum exhibit excellent tolerance and yield potential to environmental stresses such as water shortage and salinity compared to millet (Boursier and Läuchli 1990). In such regions, salinity is impose a limiting factor for crop production, where osmotic stress, ion toxicity and mineral deficiencies are all considered as consequence of the effect of salt stress on plant growth and performance. Abiotic stresses lead to oxidative stress through increase in the production of reactive oxygen species (ROS). These species are toxic

M. Hefny (✉)
Agronomy Department, Faculty of Agriculture, Suez Canal University, Ismailia 41522, Egypt

D.Z. Abdel-Kader
Botany Department, Faculty of Science, Suez Canal University, Ismailia 41522, Egypt

M. Ashraf et al. (eds.), *Salinity and Water Stress*,

and cause damage to DNA, proteins, lipids, chlorophyll, and almost every other organic constituent of the living cells (Davies 1987; Imlay and Linn 1988).

In this regard, there are many important adaptive mechanisms that plants use to cope with the adverse effects of salinity. Synthesis of compatible solutes such as: amino acid (proline), sugar alcohols (mannitol) and quaternary ammonium (glycinebetaine) that retain water within cells to combat from dehydration is one of these mechanisms (Nuccio et al. 1999). Lacerda et al. (2003, 2005) subjected the seedlings of two forage sorghum to 0 and 100 mM of NaCl, and suggested that proline accumulation is an expression of the plant reaction to the stress damage and not a salt tolerance factors. On the other hand, Mickelbart et al. (1999) stated the role of glycinebetaine (GB) under a variety of unfavorable conditions. It has been shown that high concentration alleviate salt-induced destabilization of DNA helices and maintain the activity of enzymes when plants experiencing extremes of pH, high temperature and salt concentration.

Other mechanism plants use to alleviate the effect of oxidative stress is evolving antioxidant systems such as: superoxide dismutase (SOD), catalase (CAT), ascorbate peroxidase (APX), glutathione reductase (GR) (Alscher et al. 1997; Apel and Hirt 2004). The response of field crops to salt stress through increasing activity of responsible enzymes were studied in cotton (Gosset et al. 1994), wheat (Sairam et al. 2002), rice (Vaidyanathan et al. 2003), sugar beet (Bor et al. 2003) and maize (Azevedo Neto et al. 2005). Most of the previous studies confirmed a correlation between salinity tolerance and the activity of the enzymes. On the other hand, results presented by Costa (2005), on two forage genotypes of sorghum differ in their salt tolerance disagreed with previous data. Lipid peroxidation activity in leaves showed no difference between salt-sensitive and salt tolerant genotypes, this in turn may suggested that lipid peroxidation is not a useful marker for salt tolerance discrimination for all plant species.

Many valuable literatures are published on ion accumulation and compartmentation in salt-stressed grain and forage sorghum (Lacerda et al. 2005; Netondo et al. 2004; Sunseri et al. 1998)

Abundant information are reported on the capacity of antioxidant systems in conferring tolerance to salinity in many field crops, only one reference (Costa et al. 2005), presented a detailed study on the pattern of activity of most related antioxidant enzymes in this species.

Identification of salt-tolerant lines to design a breeding programs aim at tolerance for stress environments is a great challenge because of defining the suitable criterion associated with the stress and the complexity of the inheritance to stress environments as well. It is crucial to characterize and identify those criteria and determine their relative importance and contribution to the imposed stress. Then, characterization of the individuals under investigation in relation to these criteria is the next step to study the inheritance of these traits.

The present study was performed to understand the activity of antioxidant enzymes and their role in protecting against salt-induces oxidative damage in 26 forage sorghum genotypes during seedling stage. The experiment is a preliminary study to screen the genetic materials based on the activity of the anti-oxidant enzymes and select the promising genotypes for crossing and hybrid evaluation.

2 Materials and Methods

2.1 *Plant Materials, Salinity Treatments and Growth Conditions:*

The present study was conducted for two summer seasons, 2005 and 2006 under greenhouse conditions where no light or temperature supplements. Twenty five genotypes of forage sorghum were introduced from ICRISAT (International Crops research Institute for the Semi-Arid Tropics, India), and used for this investigation, in addition to one local genotype (hybrid 101). The exotic materials consist of 21 salt-tolerant and 4 salt susceptible control lines as certified by ICRISAT. No information is available on the performance of the local hybrid under salinity stress. Table 3.1 lists origin and classification of the studied germplasm based on ICRISAT notification.

Five seeds from each genotype were planted in plastic pot, filled with a mixture of sand, peatmoss and vermiculite (1:1:1). Pots were watered regularly using the irrigation source for 14 days, then thinned to three plants per pot and salt treatments were applied. Salinization was induced by adding sodium chloride to the one-fourth strength commercial nutrient solution free from Na^+ and Cl^- salts in three concentrations; control (0), 50 mM and 100 mM NaCl l^{-1}. These corresponded to

Table 3.1 Origin and classification of 26 forage sorghum genotypes according to their salt tolerance

Genotype	Origin	Classification
1	ICSV 93046	T
2	ICSV 745	T
3	SP 47513	T
4	SP 39262	T
5	SP 47529	T
6	S 35	T
7	ICSV 93048	T
8	ICSV 112	T
9	SP 47519	T
10	SP 39105	T
11	ICSR 93034	T
12	SP 39053	T
13	SP 40567	T
14	SP 47503	T
15	SP 39007	T
16	ICSR 170	T
17	A 2267-2	T
18	ICSB 707	T
19	ICSV 96020	T
20	NTJ 2	T
21	GD 65008	T
22	ICSB 406	S
23	ICSB 676	S
24	ICSV 21029	S
25	SP 36257	S
26	Hybrid 101	Local

T: Tolerant; S: susceptible

electrical conductivities of 0.7, 8.2 and 15.11 dS m^{-1}. The electric conductivity of the soil was estimated after the experiment termination in 1:1 soil extract and found to be: 0.86, 5.12 and 13.23 dS m^{-1} for the three NaCl levels, respectively. Salt treatments lasted for 4 weeks, and then the experiment was ended.

Dry weight of plants was determined 4 weeks from the start of salt treatments, by uprooting fresh plants, drying at 70°C for 48 h and weighed.

For enzyme assays, samples were taken 6 weeks after planting and 4 weeks from application of stress treatments, the following enzymes and pigments were determined.

3 Enzyme Extraction

Enzyme extracts were prepared by homogenizing plant tissue in a pre-chilled mortar in 20 ml chilled extraction buffer (pH 7.5). Extracts were then centrifuged at 6,000 rpm for 20 min at 5°C. Enzyme assays were conducted immediately following extraction.

Super Oxide dismutase (SOD) was measured by the photochemical method described by Giannopolitis and Ries (1977). Assays were carried out under illumination. One unit of SOD activity was defined as the amount of enzyme required to cause 50% inhibition of the rate of ρ–nitro blue tetrazolium chloride reduction at 560 nm.

Peroxidase activity (Per) was determined by following the dehydrogenation of guaicol at 436 nm (Malik and Singh 1980).

Catalase (CAT) activity was assayed in a method following Aebi (1983). Activity was determined by following the decomposition of H_2O_2 at 240 nm.

Ascorbate peroxidase (ASPX) activity was determined using the method of Nakano and Asada (1987). Activity was determined by following the H_2O_2-dependent decomposition of ascorbate at 290 nm.

Glutathione reductase (GR) activity was determined as the oxidation of NADPH at 340 nm (extinction coefficient 6.2 mM cm^{-1}) according to Donahue et al. (1997). Leaf samples (1 g) were homogenized in 5 ml phosphate buffer (pH 7.6), 2 mM EDTA. The homogenate was centrifuged at 15,000 g for 10 min, and supernatant was used for analyses. The assay mixture contained 0.1 mM buffer (pH 7.6), 2 mM EDTA, 50 nicotinamioleaolenine-dinucleotide phosphate (NADPH), 0.5 mM glutathione oxidised (GSSG) and 500 μl of the extract.

The reaction was initiated by addition of NADPH and followed for 5 min at 25°C.

Total glutathione content (GSH) was determined spectrophotometrically following the method described by Griffith (1980).

Lipid peroxidation (MDA) was assayed spectrophotometrically using TBA-MDA assay. Lipid peroxides were extracted with 5 ml of 5% (w/v) metaphosphoric acid and 100 μl of 2% (w/v in ethanol) butyle hydroxytoluene. An aliquot of the supernatant was reacted with thiobarbituric acid 95°C and cooled to room temperature. The resulting thiobarbituric acid malondialdehyde adduct was extracted with 1-butanol (Hodges et al. 1999).

Determination of pigments: The pigments were extracted in 80% chilled acetone. The amount of total chlorophyll a & b and carotenoids were estimated spectrophotometrically according to Lichtenthaler (1987).

4 Statistical Analysis

The experimental design was split-plot with four replicates per treatment and three plants/replicate. Main plots represented salt concentrations, whereas the 26 sorghum genotypes constituted the split-plots. Data were combined over the two seasons for statistical analyses.

For each enzyme assays, three replicate extracts were used per treatment. All statistical analyses and least significant differences (LSD) were performed by GenStat Software statistical Program, release 4.24.

5 Results

Analysis of variance revealed significant effect of genotypes, salinity levels and the interaction between both factors on plant dry weight. Growth of sorghum plants was sensitive to salinity, as indicated by the reduction of dry weight of all genotypes compared to control treatment. The reduction was severe at 100 mM NaCl since tolerant, sensitive and local genotypes were strongly affected, although the local genotype showed the highest dry weight per plant (Table 3.2). The tolerant genotype 1 and the sensitive genotypes, 23 and 25 had a dry weight record above 1.00 g and presented the least reduction at 100 mM salinity concentration. The reduction percent ranged from 29.4–59.0% for the abovementioned genotypes. At 50 mM, the reduction in dry weight was less compared to the highest salinity level, the majority of tolerant genotypes maintained high dry weight relative to sensitive group. The genotypes; 11, 16, 3, 12, 20 and 26 maintained high dry weight at 50 mM.

Table 3.2 Response of plant dry matter (g plant^{-1}) averaged over two seasons for 26 genotypes of forage sorghum to increasing NaCl concentration during seedling stage

	NaCl concentrations (mM)			
Genotypes	0	50	100	Mean
1	5.22	2.62	1.62	3.15
2	3.34	2.59	0.49	2.14
3	3.07	3.90	0.54	2.50
4	3.06	2.89	0.52	2.16
5	4.24	1.96	0.39	2.20
6	3.31	2.12	0.65	2.03
7	2.25	2.49	0.59	1.78
8	2.22	2.02	0.56	1.60
9	3.65	2.46	0.53	2.21
19	5.67	2.55	0.59	2.94
11	9.50	4.99	0.73	5.07
12	5.10	3.78	0.48	3.12
13	4.81	2.50	0.84	2.72
14	4.66	2.48	0.56	2.57
15	4.45	3.01	0.51	2.66
16	4.20	4.09	0.49	2.93
17	3.60	2.75	0.82	2.39
18	2.91	1.93	0.72	1.85
19	2.16	1.74	0.27	1.39
20	6.07	3.34	1.13	3.51
21	3.43	2.15	0.70	2.09
22	2.71	1.11	0.51	1.44
23	2.54	1.87	1.32	1.91
24	3.86	2.04	0.55	2.15
25	3.00	0.95	1.23	1.73
26	6.46	3.32	2.15	3.98
Mean	4.06	2.60	0.75	
$LSD_{(0.05)}$				
Salt levels (S)	0.13			
Genotypes (G)	0.31			
S * G	0.53			

The activities of anti-oxidant enzymes were statistically analyzed for sorghum plants at different salinity levels and presented as group mean values (T, S, and Local) and means of individual genotypes. The profile of enzymes activities has differed among the tested groups at control treatment, indicating genetic differences for the enzyme contents in this species. On the other hand, when the plants were subjected to NaCl, salinity induced oxidative stress in plant tissues and the groups were differentiated accordingly. The mean activities, as shown in Fig. 3.1, of the enzymes; SOD, GR, ASPX, and GSH were increased markedly in tolerant group (T). The activities of SOD and GSH levels were increased in both tolerant and sensitive groups at 50 mM, then decreased in S group at higher salinity, whereas increased in T group. On the other hand, GR and ASPX activities were increased in both groups up to 50 mM, and then declined for both groups at 100 mM. The increasing in activity was 345.67%, 120.10%, 94.6%, and 503.82% for the previous enzymes in tolerant group in response to 50 mM NaCl compared to control treatment. The sensitive group, in contrast, exhibited the lowest activities for the same enzymes at the same salinity level (229.69%, 148.6%, 61.0% and, 286.78%, respectively). Interestingly, the local genotype surpassed the tolerant group in the mean activity of these enzymes at 100 mM, but showed close values for GR activity at 50 mM relative to control treatment (114.60%).

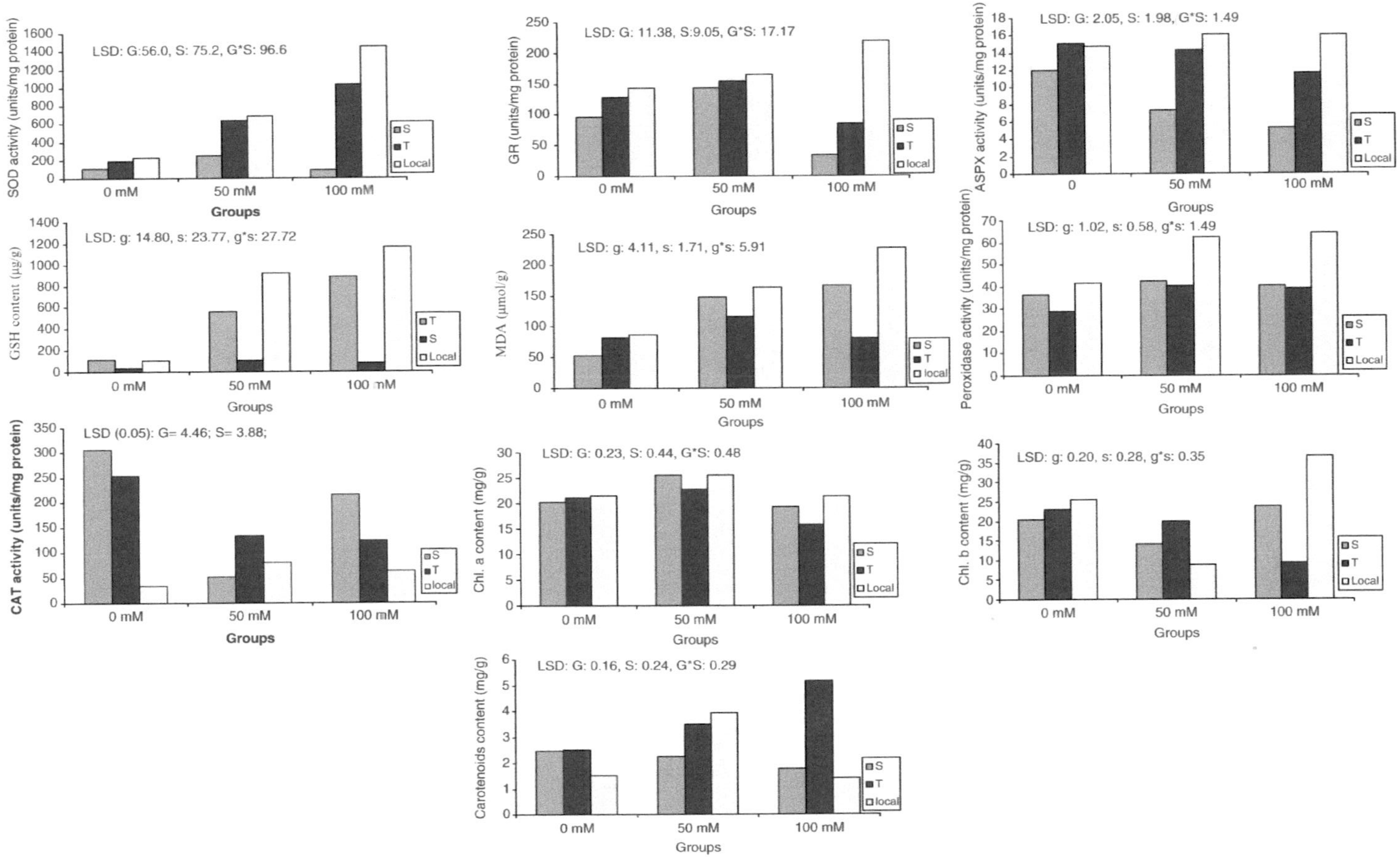

ig. 3.1 Activity of antioxidant enzymes measurec in seedlings of 26 forage sorghum genotypes, irrigated with 0, 50 and 100 mM NaCl. The assayed enzymes are: SOD, GR, ASPX, GSH level, MDA content, Peroxidase and CAT. The determined pigments are: chlorophyll a, chlorophyll b, and carotenoid contents. T, S and local represents: tolerant, sensitive and local groups, respectively

Lipid peroxidation, assessed through malondialdehyde (MDA) content, was higher in plants from sensitive group comparing with plants of tolerant group when salt-stressed and with elevating salinity level (Fig. 1). The MDA contents decreased for the tolerant group at 100 mM to values equal to control treatment. Increase in MDA content was 279.47%, 139.4% and 188.24% for the sensitive, tolerant and local groups at 50 mM NaCl, respectively.

Chlorophyll a content (Fig. 3.1) was increased due to salinity effect for all the three groups at 50 mM, and then decreased at 100 mM. Carotenoid contents, in contrast, increased with increasing salinity stress for T at both levels of stress (Fig. 3.1), in contrast, sensitive group showed some decline in contents toward stress intensity. The local genotype still has the highest content of pigments at both salinity concentrations and the lowest carotenoids at 100 mM compared with tolerant group.

The pattern of peroxidase and CAT activity values (Fig. 1) were different from other enzymes. Both T and S groups showed slight increase in the activity of Peroxidase when exposed to both levels of salinity relative to unstressed treatment. Whereas the local genotype, hybrid 101, showed the highest activity at all levels of salinity. The view of CAT activity was differed from other enzymes, it showed reduced activity at 50 mM for all groups compared to control treatment, then the activity was increased at 100 mM but still lower than unstressed condition. At 50 mM NaCl, the plants of T group demonstrated the highest activity, followed by the local genotype. In contrast, the mean activity of the sensitive group has exceeded those of tolerant and local genotypes at the highest level of salinization.

Data on the activities of different anti-oxidant systems of individual genotypes treated with different concentrations of NaCl are presented in Figs. 3.2 and 3.3 The enzymes activity were affected significantly by the genotypes, salt application levels and the interaction of both factors. At both levels of NaCl (Fig. 3.2) the genotypes from the tolerant group: 20, 12, 17, 21, 18, 15 and 5 showed the highest SOD activity. The local genotype showed close value of SOD activity to genotypes from T group at 100 mM. All tested genotypes showed reduction in ASPX activity (Fig. 3.2), except, 19 and 4 which recorded the highest values at 50 mM. On the other hand, the genotypes 4, 17 and 18 maintained high and close values by increasing NaCl levels when compared with other genotypes. There was significant increase in GR activity for the genotypes: 19 and 25 at 50 mM NaCl, in contrast other genotypes showed weak activities at both levels of stress (Fig. 3.2). The genotypes: 15, 13, 18, 21, 20, 26 and 19 were apparently distinguished from other genotypes for its high GSH content at 50 mM (Fig. 3.2), however the highest salinity level caused clear increase in GSH production for the same genotypes but lower than 50 mM concentration. The activity of CAT was low for all genotypes under stress conditions, although its activity was somewhat high under control (Fig. 3.2). The only three genotypes: 1, 19 and 18 exhibited the highest activity at both levels of stress. Lipid peroxidation, measured in the form of its degradative product malondialdehyde, increased under salt stress. There was a general tendency in the direction of increasing MDA content under stress conditions for all tested genotypes (Fig. 3.3). There was a constant increase in MDA contents of S genotypes at both levels of NaCl, in contrast T genotypes fluctuated in these contents. The lowest content was observed in the genotypes: 11, 21, 13 and 14 at both NaCl concentrations. Exposing plants to NaCl decreased peroxidase activity in tested genotypes, furthermore, only three genotypes: 25 (sensitive group); and 8, 18 and 3 (tolerant group) and the local one demonstrated high peroxidase activity under salt conditions (Fig. 3.3). In general, carotenoids showed increased contents at 100 mM compared to control and the lowest level of stress (Fig. 3.3), in contrast were chlorophyll a and b which represented reduction with increasing NaCl level. There was much reduction in chlorophyll a pigment for the majority of genotypes as a result of stress except for: 16, 18, 17, 19, 20, 26, 22, 23 and 24 which gave high contents at 50 mM, whereas at the highest concentration the Chlorophyll content was adversely affected in all genotypes with no exception.

6 Discussion

The sensitivity of crop growth to environmental stresses is well-recognized and investigated by many researchers. In the present study growth of sorghum genotypes was severely affected by increasing salinization, as indicated by the reduction in dry weight of plants. The reduction percent reached a maximum values of 68% and 90% at 50 (8.2 dS m^{-1}) and 100 mM (15.11 dS m^{-1}), respectively. According to Maas (1990)

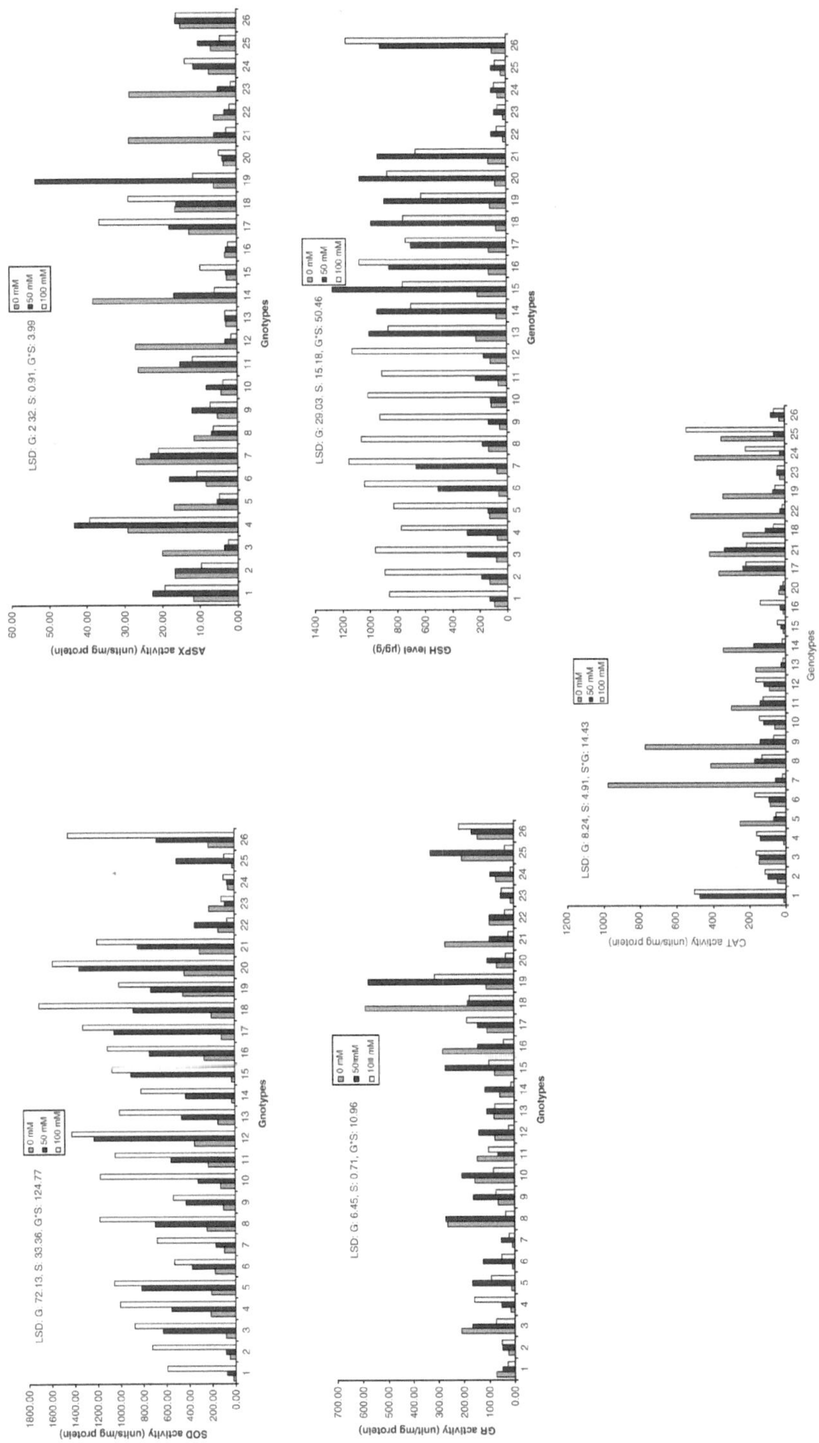

Fig. 3.2 Activity of antioxidant enzymes measured in seedlings of 26 forage sorghum genotypes, irrigated with 0, 50 and 100 mM NaCl. The assayed enzymes are: SOD, ASPX, GR, CAT and GSH level

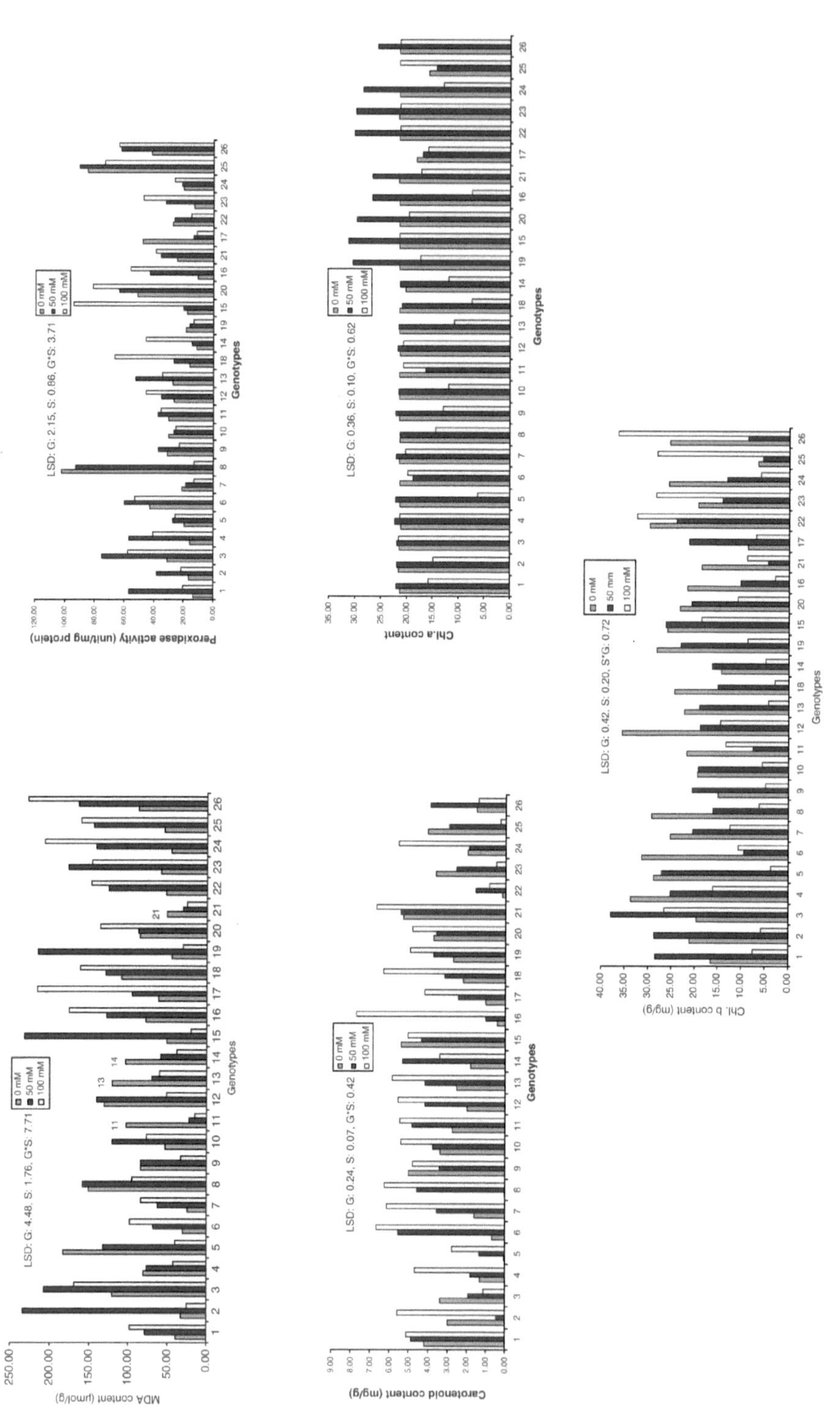

Fig. 3.3 Activity of antioxidant enzymes and pigment contents measured in seedlings of 26 forage sorghum genotypes, irrigated with 0, 50 and 100 mM NaCl. The determined enzymes are: MDA, Peroxidase, in addition to carotenoids, chlorophyll a and chlorophyll b pigments

sorghum is classified as moderately tolerant with threshold of 6.8 dS m^{-1} and slope percentage of 16 dS m^{-1}. Similar data were obtained by Netondo et al. (2004) who recorded a reduction of 75 and 53% in stem dry weight for two sorghum genotypes exposed to 15.01 dS m^{-1} salinity level. They commented that, salinity increases energy cost and carbon gain and reduce photosynthetic rates/unit leaf area.

The effects of various environmental stresses on plants are known to be mediated, at least partially, by an enhanced generation of ROS (Able et al. 2003; Hernandez et al. 2001). Salinity causes oxidative stress by inhibition the CO_2 assimilation, exposing chloroplasts to excessive excitation energy, which in turn increases the generation of ROS from triplet chlorophyll (Asada 1994; Gosset et al. 1994). As soon as the carbon fixation inside chloroplasts decreases, there is also a lower NADP availability to accept electrons from PSI, thus initiating O_2 reduction resulting in the ROS generation (Sudhakar et al. 2001). In addition, considering the fact that Cl^- is involved in electron flux during the H_2O oxidation, the Cl^- toxicity is likely to disrupt the normal electron flow to PSII, which in turn leads to excess electron leakage and increased production of ROS (Gosset et al. 1994). Plants with higher levels of antioxidants, whether constitutive or induced, have been reported to possess greater resistance to different types of environmental stress conditions (Dionisio-Sese and Tobita 1998; Young and Jung 1999).

There are different arrays of mechanisms that plant breeders use to study the adaptive response of different genotypes, and as criteria for selection under unfavorable conditions. Among those, is generation of reactive oxygen intermediate scavenging systems (ROS) (Able et al. 2003). The significant increase in the antioxidative enzymes: SOD, GR, ASPX, and GSH at 50 mM for the three groups indicate the role of those enzymes in protection and tolerance against salinity damage at this level and genetic differences for enzyme production. Further increase in stress resulted in increasing activity of SOD and GSH level for the tolerant group and the local one.

The general comparison of the examined antioxidants in sorghum genotypes revealed that SOD activity and GSH levels were significantly higher under the two salinity levels when compared to control plants. The increase in these two parameters was highly pronounced in tolerant genotypes and local genotypes than in sensitive genotypes.

The diverse responses of the SOD enzyme activities in the plants subjected to saline conditions suggest that oxidative stress is an important component of salt stress. SOD is reported to play an important role in cellular defense against oxidative stress, as its activity directly modulates the amount of O_2^- and H_2O_2, the two substrates of the metal-catalyzed site-specific Haber–Weiss reaction resulting in generation of the high-reactive OH (Sudhakar et al. 2001). The higher SOD level observed in tolerant and local sorghum genotypes could be considered as an advantage that allows the plants to resist the potential oxidative damages. These results are in good agreement with that obtained by Acar et al. (2001) and Bor et al. (2003), who found a higher constitutive and induced level of SOD activity in more tolerant barley and sugar beet cultivars under drought and salt stresses.

Glutathione is widely used as a marker of oxidative stress to plants, although its part in plant metabolism is a multifaceted one (Grill et al. 2001). As it is a nonprotein sulphur-containing tripeptide, glutathione acts as a storage and transport form of reduced sulphur. Glutathione is related to the sequestration of xenobiotics and heavy metals and is also an essential component of the cellular antioxidative defence system, which keeps reactive oxygen species (ROS) under control (Noctor and Foyer 1998). Antioxidative defense and redox reactions play a central role in the acclimation of plants to their environment, which made glutathione a suitable candidate as a stress marker. In view of the stress-response concept of the glutathione system, higher concentrations of glutathione would confer better antioxidative protection and would be considered as an acclimation. An increase in the GSH/GSSG ratio (more reduced would indicate an 'overcompensation' by intensified recycling of glutathione to keep it in its active, reduced form.

The present study revealed also that GR activity increased significantly in most tolerant and local genotypes at 50 mM NaCl. The highest NaCl concentration, however, induced a fluctuating response of the two enzymes activity in all genotypes under investigation. The role of GR and glutathione in the H_2O_2 scavenging in plant cells has been well established in Halliwell–Asada pathway (Bray et al. 2000). GR catalyses the last rate-limiting step of the ascorbate–glutathione cycle. This enzyme maintained high ratio of GSH/GSSG which is required for the regeneration of ascorbate and for the activation of several chloroplastic CO_2-fixing enzymes. The GSH and GR action suggests that the more active ascorbate–glutathione cycle may be related to the development of relatively higher salt tolerance in sorghum.

The results for CAT and peroxidase activities were varied and did not follow the other enzymes pattern, the mean of CAT activity decreased at 50 mM NaCl for all groups, then increased for sensitive group at 100 mM concentration. Peroxidase did not discriminate both groups from each other as the values were nearly unchanged, but the local genotype was distinguished by possessing the highest activity. Both CAT and peroxidase are not considered a distinguished marker for selection for salinity in the present materials of sorghum. Its activity was generally low and only three genotypes namely: 1, 19 and 18 recorded the highest activity at both salinity levels, although the values are low. Costa et al. (2005), found increase in CAT activity when sorghum plants subjected to 75 mM NaCl and the increase was more conspicuous in tolerant than in sensitive genotype. Sairam and Srivastava (2001) stated that scavenging of H_2O_2 as represented by GR and CAT is limited and less efficient in susceptible wheat genotypes leading to higher H_2O_2 accumulation and increasing in lipid peroxidation under water limited environments. ASPX uses ascorbate as the electron donor for the H_2O_2 reduction and is known to be a major enzyme in the detoxification of H_2O_2 (Asada 1992; Sairam and Saxena 2000). The increase in ASPX activity (at 50 mM NaCl) observed in the present study was in agreement with gradual application of salinity. The increase in enzyme activity may be due to increasing the synthesis of the enzyme or an increased activation of constitutive enzyme pool. An increase in the transcription of genes involved in the synthesis of various stress metabolites, including antioxidant enzymes, has been reported (Scandalios 1994). Malondialdehyde as the main decomposition product of polyunsaturated fatty acids in biomembranes is known to show greater accumulation under salt stress (Gosset et al. 1994; Meloni et al. 2003; Sudhakar et al. 2001). Such results are consistent with those in present investigation. A significant increase in the malondialdehyde level (used as an indicator of the extent of membrane damage) was observed in all sensitive genotypes and the local genotype. In contrast there was a progressive increase in MDA content in tolerant group. The study of Sairam and Srivastava (2001) conducted on wheat cultivars revealed a lower MDA content in tolerant genotypes compared to susceptible one under water stress conditions. As a result of the greater antioxidant defense in tole-rant sorghum genotypes, the malondialdehyde content did not raise the high level that of sensitive genotypes. It is presumably due to the high constitutive level of the antioxidant enzymes activities in tolerant genotypes which is sufficient to avoid a substantial elevation in the lipid peroxidation. Moreover, the increase in antioxidants activities was negatively associated with the level of lipid peroxidation. Cell membrane stability has been widely used to differentiate stress-tolerant and susceptible cultivars of many crops, and it could be correlated with better field performance (Premachandra et al. 1991). The chlorophylls compared to carotenoids have markedly decreased in most sorghum genotypes under investigation; this gives the appearance of senescing plant. The reduction in chlorophylls was reduced with elevating stress while the opposite was true for carotenoids. The decrease in chlorophyll a & b contents could be attributed to the increase their degradation (Abdel-Kader 2000). Moreover, the damage caused by ROS may also affect macromolecules as mentioned by Pastori and Foyer (2002) and Costa et al. (2005). On the other hand, carotenoids concentration was significantly increased in most tolerant and local sorghum genotype under salinity treatments. The most important role of carotenoids is preventing the formation of singlet oxygen and protecting chlorophylls by quenching their triplet states via thermal dissipation of energy. Additionally, carotenoids play a central structural role for chlorophyll-binding proteins of both the antenna system and the reaction center (Paulsen 1997). Recent reports have shown that *b*-Car is essential for the assembly of D1 protein during its turnover in the formation of functional PS2 complexes in *Chlamydomonas reinhardtii* under high light conditions (Trebst and Depka 1997; Depka et al. 1998). The increase in carotenoid concentration in most tolerant sorghum genotypes may be due to a shift in the synthesis of carotenoids to protect chloroplast from oxidative damage. Carotenoids could be considered as salinity tolerant marker or criteria.

7 Conclusion

Although the exotic genotypes are classified as tolerant and sensitive (based on ICRISAT confirmation), there were significant differences in genotypic responses within each group in the pattern of anti-oxidative mechanisms represented by the activity of antioxidant enzymes. The H_2O_2 scavenging system: SOD, ASPX, GR, and GSH in addition to carotenoid contents are more important in imparting salt stress in the present materials. MDA content, showed decreased level in tolerant compared to sensitive group, however some genotypes have shown some high values. Although most of the previous enzymes involved in the amelioration of oxidative stress, they do not show uniform increase/decrease in activity in a given genotype. The tolerance of the following genotypes: 11, 16, 12, 20, 15 and 26 seems to be related to the production of high dry weight per plant, efficiency of the anti-oxidant enzymatic systems: SOD, GSH and MDA content. Other genotypes which represented high enzymes activity although did not produce reasonable dry weight are: 18, and 21. The local genotype, hybrid 101, has some degree of salt tolerance as it showed similar enzyme activities as some tolerant genotypes and reasonable dry weight. Sorghum plants can tolerate a maximum level of 8.0 dS m^{-1}, then obvious reduction in plant performance and productivity occur. Genotypes which produce high dry weight do not necessarily exhibit high enzyme activities, as there may be no correlation between high production and intensity of oxidation-protecting enzymes. The antioxidative systems, Sod, ASPX, GSH and the MDA and carotenoids content are considered good physiological markers to distinguish between tolerant and sensitive genotypes; consequently it becomes a selection criteria for breeding for saline environments.

References

Abdel-Kader DZ (2000). Salinity and adaptation effects on lipid peroxidation and antioxidant enzymes in Raphanus sativus and Eruca sativa seedlings. J Union Arab Biol (8B): 59–71.

Able AJ, Sutherland MW, Guest DI (2003). Production of reactive oxygen species during non-specific elicitation, non-host resistance and field resistance expression in cultures of tobacco cells. Funct Plant Biol. 30: 91–99.

Acar O, Urkan IT, Özdemir F (2001). Superoxide dismutase and peroxidase activities in drought sensitive and resistant barley (Hordeum vulgare L.) varieties. Acta Physiol Plant 3: 351–356.

Aebi HE (1983). Catalase. In: Bergmeyer HU, ed. Methods of Enzymatic Analysis, Vol. 3. Verlag Chemie, Weinhem, pp. 273–286.

Alscher RG, Donahue JL, Cramer CL (1997). Reactive oxygen species and antioxidants: relationship in green cells. Physiol Plant 100: 224–233.

Apel K, Hirt H (2004). Rea ctive oxygen species: metabolism, oxidative stress, and signal transduction. Ann Rev Plant Biol 55: 373–399.

Asada K (1992). Ascorbate peroxidase – a hydrogen proxide scavenging enzyme in plants. Physiol Plant 55: 235–241.

Asada K (1994). Production of active oxygene species in photosynthetic tissue. In: Foyer CH, Mullineaux PM, eds. Causes of Photooxidative Stress and Amelioration of Defense Systems in Plants, 1st Ed. CRC Press, Boca Roton, FL, pp. 77–104.

Azevedo Neto AD, Prisco JT, Eneas-Filho J, Medeiros JR, Gomes-Filho E (2005). Hydrogen peroxide pre-treatment induces salt-stress acclimation in maize plants. J Plant Physiol 162: 1114–1122.

Bor M, Özdemir F, Türkan I (2003). The effect of salt stress on lipid peroxidation and antioxidants in leaves of sugar beet *Beta vulgaris* L. and wild beet *Beta maritime* L. Plant Sci 164: 74–77.

Boursier P, Läuchli A (1990). Growth responses and mineral nutrient relations of salt-stressed sorghum. Crop Sci 30: 1226–1233.

Bray EA, Bailey-Serres J, Weretilnyk E (2000). Responses to abiotic stresses. In: Buchanan BB, Gruissem W, Jones RL, eds. Biochemistry & Molecular Biology of Plants. American Society of Plant Physiologists, Rockville, MD, pp. 1158–1203.

Costa PHA, Neto A, Bezerra M, Prisco J, Filho E (2005). Antioxidant-enzymatic system of two sorghum genotypes differing in salt tolerance. Braz J Plant Physiol 17: 353–361.

Depka B, Jahans P, Trebst A (1998). *b*-Carotene to zeaxanthin conversion in the rapid turnover of the D1 protein of photosystems. FEBS Lett 424: 267–270.

Dionisio-Sese ML, Tobita S (1998). Antioxidant responses of rice seedlings to salinity stress. Plant Sci 135: 1–9.

Donahue JL, Okpodu CM, Cramer CL, Grabau EA, Alscher RG (1997). Responses of antioxidants to paraquat in Pea leaves. Plant Physiol 113: 249–257.

Giannopolitis N, Ries SK (1977). Superoxide dismutase. 1. Occurrence in higher plants. Plant Physiol 59: 309–314.

Gosset DR, Millhollon EP, Lucas MC (1994).Antioxidant response to NaCl stress in salt- tolerant and salt-sensitive cultivars of cotton. Crop Sci 34. 706–714.

Griffith OW (1980). Potent and specific inhibition of glutathione synthesis by buthionine sulfoximine (S-n butyl homocysteine sulfoximine). J Biol Chem 254: 7558–7560.

Grill D, Tausz M, De Kok LJ (2001). Significance of glutathione in plant adaptation to the environment. In: De Kok LJ, ed. Handbook of Plant Ecophysiology, Vol. 2. Kluwer, Dordrecht.

Herna'ndez JA, Ferrer MA, Jime'nez A, Ros-Barcelo'A, Sevilla F (2001). Antioxidant system and O_2/H_2O_2 production in the apoplast of Pisum sativum L. Leaves: its relation with NaCl-induced necrotic lesions in minor veins. Plant Physiol 127: 817–831.

Hodges DM, DeLong JM, Forney C, Prange RK (1999). Improving the thiobarbituric acid-reactive-substances assay for estima-ting lipid peroxidation in plant tissues containing anthocyanin and other interfering compounds. Planta 207: 604–611.

Imlay J, Linn S (1988). DNA damage and oxygen radical toxicity. Science 240: 1302–1309.

Lacerda C, Cambraia J, Oliva M, Ruiz H (2005). Changes in growth and solute concentrations in sorghum leaves and roots during salt stress recovery. Environ Exp Bot 54: 69–76.

Lacerda C, Cambraia J, Oliva M, Ruiz H, Prisco J (2003). Solute accumulation and distribution during shoot and leaf development in two sorghum genotypes under salt stress. Environ Exp Bot 49: 107–120.

Lichtenthaler HK (1987). Chlorophylls and carotenoids – pigments of photosynthetic membranes. Method Enzymol 48: 350–382.

Malik CP, Singh MB (1980) In Plant Enzymology and Histoenzymology. Kalyani Publishers, New Delhi, 53 pp.

Meloni DA, Oliva MA, Martinez CA, Cambraia J (2003). Photosynthesis and activity of superoxide dismutase, peroxidase and glutathione reductase in cotton under salt stress. Environ Exp Bot 49: 69–76.

Mickelbart MV, Ejeta G, Rhodes D, Joly RJ, Goldsbrough PB (1999). Assessing the contribution of glycinebetain to environmental stress: tolerance in sorghum. In Ribaut JM, Poland D, eds. Molecular Approaches for the Genetic Improvement of Cereals for Stable Production in Water-Limited Environments, 21–25 June. Cimmyt, El Batan, Mexico.

Nakano A, Asada K (1987) Purification of ascorbate peroxidase in spinach chloroplasts; its inactivation in ascorbate-depleted medium and reactivation by monodehydroascorbate radical. Plant Cell Physiol 28: 131–140.

Netondo G, Onyango JC, Beck E (2004). Sorghum ans salinity: I. Response of growth, water relations, and ion accumulation to NaCl salinity. Crop Sci 44: 797–805.

Noctor G and Foyer CH (1998). Ascorbate and glutathione: keeping active oxygen under control. Ann Rev Plant Physiol Plant Mol Biol 49: 249–279.

Nuccio ML, Rhodes D, McNeil SD, Hanson AD (1999). Metabolic engineering of plants for osmotic stress resistance. Curr Opin Plant Biol 2: 128–134.

Pastori GM, Foyer CH (2002). Common components, networks, and pathways of cross-tolerance to stress. The central role of "redox" and abscisic acid-mediated controls. Plant Physiol 129: 7460–7468.

Paulsen H (1997). Pigment ligation to proteins of the photosynthetic apparatus in higher plants. Physiol Plant 100: 760–768.

Premachandra GS, Soneoka H, Kanya M, Ogata S (1991). Cell membrane stability and leaf surface wax content as affected by increasing water deficits in maize. J Exp Bot 42: 167–171.

Sairam RK, Saxena DC (2000). Oxidative stress and antioxidants in wheat genotypes: possible mechanism of water stress tolerance. J Agron Crop Sci 184: 55–61.

Sairam RK, Srivastava GC (2001). Water stress tolerance of wheat (*Triticum aestivium* L.): variation in hydrogen peroxide accumulation and antioxidant activity in tolerant and susceptible genotypea. J Agron Crop Sci 186: 63–70.

Sairam RK, Rao KV, Srivastava GC (2002). Differential response of wheat genotypes to long-term salinity stress in relation to oxidative stress, antioxidant activity and osmolyte concentration. Plant Sci 163: 1037–1046.

Scandalios JG (1994). Regulation and properities of plant catalases. In Foyer CH, Mullineaux PM, eds., Causes of Photooxidative Stress and Amelioration of Defense Systems in Plants. CRC Press, Boca Raton, FL, pp. 275–316.

Sudhakar C, Lakshmi A, Giridarakumar S (2001). Changes in the antioxidant enzymes efficacy in two high yielding genotypes of mulberry (Morus alba L.) under NaCl salinity. Plant Sci 161: 613–619.

Sunseri F, Palazzo D, Montemurro N, Montemurro F (1998). Salinity tolerance in sweet sorghum (Sorghum bicolorL. Moench): field performance under salt stress. Italian J Agron 2: 111–116.

Trebst A, Depka B (1997). Role of carotene in the rapid turnover and assembly of photosystem II in *Chamydomonas reinhardtiis*. FEBS Lett 400: 59–362.

Vaidyanathan H, Sivakumar P, Ghakrabarty R, Thmas G (2003). Scavenging of reactive oxygen species in NaCl-stressed rice (Oriza sativa L.)-differential response in salt-tolerant and sensitive varieties. Plant Sci 165: 1411–1418.

Young CB, Jung J (1999). Water-induced oxidative stress and antioxidant defenses in rice plants. J Plant Physiol 155: 255–261.

Chapter 4
Genetic Variation in Wheat (*Triticum aestivum* L.) Seedlings for Nutrient Uptake at Different Salinity and Temperature Regimes

E.V. Divakara Sastry and M. Gupta

Abstract In the present study, 20 genetically diverse genotypes of wheat were evaluated for salt (0.0% and 0.3%; EC 2.8 and 11.4 mS/cm, respectively) and heat stress (15°C and 25°C) tolerance. The petridishes were irrigated with 5 ml of test solution after draining the previous day's solution for the first 5 days which were later increased to 10 ml. On the 11th day, the experiment was terminated and the observations were recorded on germination percentage, fresh weight of shoot/seedling (mg), fresh weight of root/seedling (mg), dry weight of shoot/seedling (mg), dry weight of root/seedling (mg). The data for Na^+, K^+, Ca^{2+}, Na^+/K^+ ratio, Na^+/Ca^{2+} ratio, Cu^{2+}, Zn^{2+}, Mn^{2+} contents in roots and shoots were also recorded. Salt and high temperature stress reduced the growth of all genotypes of wheat. However, cultivars differed significantly to both salt and temperature stress. Mukta, Raj-3765, Sonalika, Kharchia-65 were found to be best suited to salinity, while PBW-226 and Raj-2535 were very sensitive to salinity and higher temperature. Accumulation of Cu^{2+}, Zn^{2+} and Mn^{2+} was positively correlated with all other attributes. However, Na^+ was negatively correlated with K^+ and Ca^{2+}. Therefore, in order to breed efficient genotypes which can withstand the effects of salinity the positive association between Na^+ and other contents will have to be broken. This can be done by biparental mating design or recurrent selections.

Keywords Salinity • temperature • wheat • Na • K • variation • correlation

E.V. Divakara Sastry (✉) and M. Gupta
Department of Plant Breeding and Genetics, SKN College of Agriculture, Rajasthan Agricultural University, Jobner 303329, Rajasthan, India

1 Introduction

Wheat is an important cereal crop in India and there is great variation among the temperature and environmental conditions throughout India where wheat is cultivated. Its growth, development and yield are influenced by a range environmental factors such as temperature, radiation, photoperiod etc. Among the various abiotic stresses, temperature (both low and high) and salinity affect wheat crop productivity considerably (Langridge et al. 2006; Rosegrant et al. 2007). The high temperature cause various biochemical and physiological changes in plants. The duration of high temperature stress is of fundamental importance in the growth and survival of a plant. Not only does the effect of killing temperature vary inversely with the exposure time, but relationship to the time is actually exponential. Similarly, salinity stress remains one of the oldest environmental problems (Downton 1985), which is caused by presence of soluble salts. It has often been referred as physiological drought and the possible effects of salinity are to be mediated via disturbed water balance. Recently with the accumulation of information on ion excesses it has been questioned whether the whole plant responses to salinity are the result of water deficiency or ion toxicity. Due to high concentration of salts, the metabolism and uptake of minerals are also affected, the contents of nitrogen, calcium and sodium increased while the content of P, K, Mg and B decreased with increasing levels of EC of irrigation water. In order to verify the effect of the combined effect of salinity and temperature on the uptake of minerals the present experiment was conducted.

M. Ashraf et al. (eds.), *Salinity and Water Stress*,

2 Material and Methods

The experimental material for the present study consisted of 20 genotypes of wheat (*Triticum aestivum* L.) as listed in Table 4.3. The genotypes were selected at random except for Kharchia 65 and Raj 3077 which are supposed to be tolerant to salinity.

The method reported by Sharma (1995) has been used for the evaluation of genotypes in an environment. Petri dishes of 10 cm diameter were used for evaluation of genotypes. Each Petri dish were layered with blotting paper and sterilized at 120°C temperature, 20 lbs/in.^{2} for 30 min in an autoclave. Ten seeds were sown in each Petri dish after surface sterilization by 0.1% $HgCl_2$ solution for 2 min. Two such dishes were prepared for each replication/genotype/salinity levels/temperature. The entire experiment was replicated three times.

The experiment was maintained for 10 days in an environmental chamber. Every day each Petri dish was irrigated with 5 ml of sterilized test solution after draining out the previous day's left over to maintain the uniform salinity, during the first 5 days. After 5th day each Petri dish was irrigated with 10 ml of test solution. All the Petri dish were kept in darkness for 72 h, later dishes were exposed to artificial light (10 h/day) achieved by the use of fluorescent lamps and incandescent bulbs. The temperature and salinity levels used in the experiment have been selected based on an earlier study (Sharma 1995).

Two test solutions were prepared by supplementing Hogland solution with 0, and 300 mg sodium chloride (NaCl) per 100 ml. The salinity levels are represented as 0.0% (control), and 0.3% respectively. The corresponding ECe values were 2.08 and 11.4 dS m^{-1} respectively. The whole set (20 genotypes × 2 salinity × 3 replication × 2 dishes per replicate) was evaluated at two temperatures viz. 15°C and 25°C in an environmental chamber (Make-Khera).

The experiment was terminated after 10 days. On the 11th day, the seedling, were harvested from each of the Petri dish/replication/genotype/salinity/temperature level. The dry matter of shoot and root obtained from each Petri dish/replicate/genotypes was used for the estimation of nutrient contents such as Na, K, Ca, Cu, Zn and Mn. The data recorded so far was subjected to statistical analysis. The character associations represented by simple correlation coefficients were also estimated between different pairs of characters.

3 Results and Discussion

The analysis of variance of all growth attributes suggested existence of significant differences between the treatments for all the traits studied. Partitioning of treatment sum of squares into various components revealed significant differences between genotypes, salinities and temperatures indicating that inherent differences between genotypes existed, also, the genotypes responded to temperature and salinity gradients, but significant interaction sum of squares suggested that the genotypes responded non-linearly to temperature and salinity gradients (Table 4.1). The higher order interaction namely Temperature × Salinity × Genotypes was significant in only few cases, therefore further analyses were all based on pooled analysis. Sastry and Prakash (1993) reported significant differences between varieties for Na^+ and K^+ contents, although the interaction between salinity levels and varieties was non-significant which is in contrast to the present observations. Muralia (1989) reported significant differences between genotypes for K^+, Ca^{2+} and Na^+ in the roots, the present investigations supports this observation but the contents of Na^+, Ca^{2+} were found non-significant in shoot (Muralia 1989) which is in contrast to the present investigations. Differences in the reports might be because of the differences in the experimental material.

Comparison of pooled means (Table 4.2) at different temperatures indicated that the values of Na^+ were higher at 15°C than at 25°C (Fig. 4.1). Similarly this value was generally higher in shoot than in root while for the rest of characters, the values were higher in roots than in shoots. Also the values were generally higher at 25°C than at 15°C, Similarly the comparison of mean values of other traits indicated that the values were higher in control (0.0% salinity) than at 0.3%, which is expected for most of the traits in shoot as well as root, except for Na^+ content or Na^+ related ratios namely Na^+/K^+ and Na^+/Ca^{2+} which is again expected. Increased uptake of Na^+ at higher salinities is a common phenomenon (Singh 1993) and most of the deleterious effects of salinity on plant growth is due to this increased uptake Na^+ (Levitt 1972). This may also be the reason why the contents of K^+ and Ca^{2+} are less in salinity which is a commonly reported observation (Sastry and Prakash 1993). However, Sastry and Prakash (1993) reported that K^+ content does not show any specific trend along

Table 4.1 Pooled ANOVA for fresh weight, dry weight and various nutrients content in shoot and root

		Mean squares															
		Na^+		K^+		Ca^{2+}		Cu^{2+}		Zn^{2+}		Mn^{2+}		Na^+/K^+		Na^+/Ca^{2+}	
Source	d.f.	Shoot	Root	Shoot	Root	Shoot	Root	Shoot	Root	Shoot	Root	Shoot	Root	Shoot	Root	Shoot	Root
Replications	2	117.18**	343.78**	75.91**	46.01**	2.64**	3.58**	0.01	0.01	0.19	1.17**	0.00	0.02	0.27*	13.35	172.78	880.87
Treatments	79	157.37**	205.32**	163.27**	70.24**	7.10**	7.50**	0.07**	0.07**	3.18**	4.74**	0.01**	0.01**	1.32**	52.74**	1028.32**	1137.22
Temperature	1	328.62**	487.06**	380.53**	127.59**	41.81**	69.32**	0.61**	1.61**	17.73**	18.90**	0.04**	0.02*	0.04	12.61	480.80*	1651.0701
Salinity	1	3897.01**	5860.21**	1387.12**	695.29**	44.35**	1.80**	2.50**	1.97**	2.19**	43.38**	0.00	0.11**	29.15**	787.31**	9252.14**	3267.93*
Genotypes	19	326.99**	355.52**	470.33**	181.00**	13.11**	24.06**	0.07**	0.05**	7.70**	9.87**	0.01**	0.01**	3.18**	116.08**	2004.60**	2497.39**
Temp. × salinity	1	20.22*	0.11	8.06*	1.22	2.64**	0.36	0.23**	0.11**	0.06	1.17**	0.00	0.00	0.00	13.65	50.81	1010.28
Temp. × genotypes	19	6.92	12.73**	1.52	12.47**	0.9349**	0.7995**	0.0063	0.0043	0.3504**	0.1422	0.0013	0.0058	0.0325	4.6748	111.16	702.05
Salinity × genotypes	19	92.25**	134.92**	113.22**	46.09**	10.1284**	2.5088**	0.0629**	0.0490**	3.8688**	6.1582	0.0070**	0.0183**	0.7057**	50.8360**	1529.56**	554.52
Temp. × salinity × genotypes	19	4.72	16.46**	0.34	9.14**	0.7157**	0.0808	0.0051	0.0053	0.2761*	0.2204**	0.0033	0.0055	0.0479	4.8747	115.40	662.385
Error	158	4.64	3.95	1.31	2.9748	0.1087	0.1284	0.0046	0.0087	0.1547	0.11	0.0031	0.0071	0.0648	5.6503	75.73	605.2362

*Significant at 5%

**Significant at 1%

Table 4.2 Means of different characters at different temperature and salinity levels

	15°C		25°C	
Character	0% salinity	0.30% salinity	0% salinity	0.30% salinity
Shoot Na^+	5.73	13.21	7.49	16.1
Root Na^+	13.12	23.04	16	25.85
Shoot K^+	18.21	13.77	21.09	15.85
Root K^+	9.03	5.77	10.63	7.08
Shoot Ca^{2+}	2.05	1.39	3.09	2.05
Root Ca^{2+}	2.28	2.37	3.27	3.53
Shoot Na^+/K^+	0.39	1.09	0.42	1.11
Root Na^+/K^+	2.03	6.13	2.05	5.19
Shoot Na^+/Ca^{2+}	5.51	18.85	3.6	15.09
Root Na^+/Ca^{2+}	8.19	19.67	7.05	10.32
Shoot Cu^{2+}	0.28	0.14	0.44	0.17
Root Cu^{2+}	0.22	0.38	0.36	0.59
Shoot Zn^{2+}	1.42	1.26	1.95	1.77
Root Zn^{2+}	1.99	2.69	1.7	1.28
Shoot Mn^{2+}	0.13	0.12	0.16	0.15
Root Mn^{2+}	0.16	0.11	0.18	0.13

the salinity gradient. But the observations in the present investigation support the report of Muralia (1989).

The comparison of mean values of genotypes for different nutrient contents did not indicate any specific trend. In such a situation it is ideal to look at ranking of genotypes for each attribute and comparing the relative ranking. Perusal of Table 4.3 reveals that Mukta with lowest rank total is the ideal genotype with high K^+, Ca^{2+}, Cu^{2+}, Zn^{2+}, Mn^{2+} contents as well as low Na^+ contents. This is followed by Kharchia-65 Raj-3765, Sonalika, and Raj-3856. Most of the "Raj" lines thus have shown tolerance to salinity. This is expected as these lines have specially been bred for salinity tolerance (Dr. C.P. Nagpal, personal communication, 2007). C306 and Sonalika were reported to be salinity tolerant by Gupta and Tyagi (1973). Prakash and Sastry (1992) and Sastry and Prakash (1993) also found Sonalika to be moderately tolerant. High tolerance observed in the present investigation is thus in line with the earlier reports. This can be ascribed to the fact that the study is based on the nutrient content/uptake and that too at the seedling stage. It is a common knowledge that the tolerance changes with age, stage, temperature and other cultural conditions. In order to get more meaningful conclusions, evaluations at adult stage is warranted.

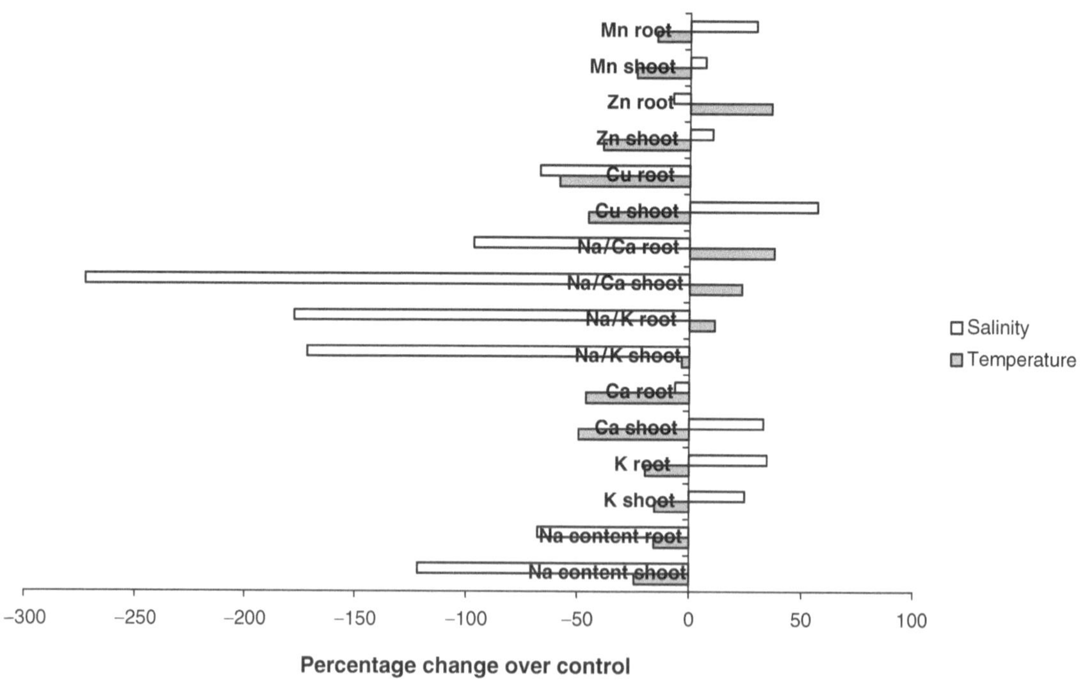

Fig. 4.1 Percent change over control in salinity and temperature in different characters

Table 4.3 The rank totals[1] of different genotypes over different environments

		Shoot								Root									
S. no.	Genotypes	Na^+	K^+	Ca^{2+}	Na/K	Na/Ca	Cu^{2+}	Zn^{2+}	Mn^{2+}	Na^+	K^+	Ca^{2+}	Na^+/K^+	Na^+/Ca^{2+}	Cu^{2+}	Zn^{2+}	Mn^{2+}	Total	Rank
1	Raj-3077	35	40	47	39	43	54	11	16	20	69	62	45	47	32	62	27	649	13
2	Raj-3856	39	39	13	32	20	48	24	20	50	25	29	32	35	39	50	19	514	5
3	Raj-1972	17	61	27	26	18	38	41	27	24	60	55	50	42	34	49	27	596	10
4	Raj-3765	6	31	53	6	25	46	56	31	16	27	46	19	19	31	28	18	458	3
5	Raj-1482	26	68	23	42	19	41	43	40	32	47	36	46	33	40	79	29	644	12
6	Raj-2535	73	14	72	57	77	28	9	16	63	11	76	18	70	41	46	32	703	15
7	Raj-2184	62	60	41	70	61	25	31	8	62	45	32	60	49	30	40	32	708	16
8	Job-601	28	56	24	38	28	20	29	38	25	64	19	59	18	32	49	37	564	7
9	Job-2030	59	55	45	62	56	22	30	38	66	63	41	68	60	26	58	32	781	19
10	Job-1002	37	42	37	61	33	35	54	39	41	45	41	43	45	15	60	90	718	17
11	Job-984	51	41	25	56	43	32	74	29	58	45	28	50	50	24	18	25	649	13
12	Kalyansona	78	54	23	77	67	22	46	14	68	68	11	76	31	27	37	43	742	18
13	Kharchia-65	43	20	22	23	19	23	54	15	27	25	8	22	9	34	16	31	391	2
14	Mukta	15	23	13	18	5	24	39	31	15	16	6	6	5	40	18	35	309	1
15	Sonalika	45	10	39	15	43	9	41	28	39	40	40	41	35	13	50	13	501	4
16	WL-711	48	13	53	27	48	30	31	12	64	34	47	52	67	15	6	29	576	9
17	HD-2285	49	22	61	31	70	21	72	27	30	11	66	14	55	13	49	19	610	11
18	C-306	29	18	45	19	40	20	57	29	43	14	48	17	53	18	43	37	530	6
19	UP-2338	32	41	52	32	50	22	22	10	34	37	70	35	48	17	42	25	569	8
20	PBW-226	43	68	41	58	67	47	42	21	73	58	86	77	79	37	32	16	845	20

[1]This table is prepared by pooling the rank of each genotypes over different environments for a character

Correlation analysis indicated positive and significant correlation of Na content with all the attributes in both root and shoot except with K^+ and Ca^{2+} contents. Negative associations of Na^+ content with K^+ and Ca^{2+} has been reported by Muralia (1989). Reduced uptake of K^+ and Ca^{2+} in the presence of higher concentrations of Na^+ is commonly reported. The reduced growth of seedlings in salinity is basically because of this reason (Singh 1994). Interestingly K^+ and Ca^{2+} had positive and significant associations with Cu^{2+}, Zn^{2+}, and Mn^{2+} in both roots as well as shoots. This indicates that the uptake of Na^+, K^+, Ca^{2+} may have a correlated uptake of Cu^{2+}, Zn^{2+} and Mn^{2+}. This indicates that the genes governing the uptake of all these elements might have related with each other and this will have to be explored further.

Before the value of the results obtained from the present investigations are discussed for their manipulation through breeding, the following conclusions can be drawn

1. As the nutrients supplied to the plant being similar, the contents were supposed to reflect their uptake. Variations exists among the genotypes with regard to the contents of various micronutrients namely K^+, Ca^{2+}, Na^+, Zn^+, Cu^+ and Mn^+.
2. The uptake of elements is affected by temperature and salinity, the uptake was lower at 15°C than at 25°C and in general, the contents of various minerals excepting Na^+ were high in control than in salinity. This is expected as Na^+ uptake increases in salinity while the uptake of K^+ and Ca^{2+} relatively decreased. Interestingly the uptake of other minerals namely Cu^{2+}, Zn^{2+}, Mn^{2+} was also lower in salinity.
3. The correlation studies indicated that the contents of Cu^{2+}, Zn^{2+}, Mn^{2+} were positively correlated with all the other attributes in both root and shoot. But the Na^+ was negatively correlated with K^+ and Ca^{2+}.

In the light of above conclusions, it may be suggested that ample scope for improvement of uptake of various minerals exists. As the contents of Na^+, Cu^{2+}, Zn^{2+}, Mn^{2+} and K^+, Ca^{2+}, Cu^{2+}, Zn^{2+}, Mn^{2+} were all positively correlated, selection based on any one of these minerals is expected to yield concurrent improvements in the uptake of other minerals. This also suggests that probably the genes governing their uptake are related, either through linkage or pleiotrophy. Positive associations also indicate selection based on any one of these attributes is expected to increase the growth as well. However, beyond a threshold limit, the higher uptake may prove toxic for certain of these ions namely Na^+. Therefore, in order to breed efficient genotypes which can withstand the effects of salinity the positive association between Na^+ and other contents will have to be broken. This can be done by biparental mating design or recurrent selections, alternatively, an evaluation of tribe *Triticeae* for above attributes is also suggested. It is suggested that D genome of tribe *Triticeae* has genes conferring salt tolerance to wheat (Shah et al. 1987; Gorham et al. 1985). This points to the possibility of widening the gene pool using *Aegilops squarossa* the donor of D genome and other D genome carrying *Aegilopes* species (Forster et al. 1988; Shah et al. 1987). However, more distantly related species may offer even more potent salt tolerant genes.

Agropyron junceum (*Thinopyrum bessarabicum*) a littoral diploid grass native to China, USSR has been shown to tolerate salinity levels lethal to wheat. Efforts made by Forster et al. (1988) and William and Mujeeb-Kazi (1995) met with success to transfer this ability into wheat.

References

Downton WJS (1985) Salt tolerance of food crops: Prospectives for improvements. CRC Crit Rev Plant Sci 1: 183–201.

Forster, B.P., Millar, T.E. and Law, C.N. 1988. Salt tolerance of wheat *Agropyron junceum* disomic addition lines. Genome 30: 559–564.

Gorham J, Forster BP, Budrewicz E, Wyn Jones RG, Miller TE, Law CN (1985) Salt tolerance in *Triticeae*. Salt accumulation and distribution in an amphiploid derived from *T. aestivum* cv. Chinese spring and *Thinopyrum bossarabicum*. J Exp Bot 347: 1435–1449.

Gupta US, Tyagi AP (1973) Number of noringenes and early salt tolerance in bread wheat. Biochem Physiol Pflanzen 164: 349–356.

Langridge P, Paltridge N, Fincher G (2006) Functional genomics of abiotic stress tolerance in cereals. http://bfgp.oxfordjournals.org/cgi/content/full/4/4/343. Accessed 27 March 2008.

Levitt J (1972) Responses of plants to environmental stresses. Vol. I, Academic, New York.

Muralia S (1989) Studies on effect of salinity on the variation and stability of seedling emergence and establishment in wheat. M.Sc. (Ag.) thesis, Rajasthan Agricultural University, Bikaner, Rajasthan (unpublished).

Prakash V, Sastry EVD (1992) Effects of salinity on germination and seedling growth in wheat. Ann Arid Zone 31: 71–72.

Rosegrant M, Ringler C, Msangi S (2007) Food security in Asia: the role of agricultural research and knowledge in a changing environment. Sat ejournal http://www.icrisat.org/Journal/SpecialProject/sp6.pdf. Accessed 27 March, 2007.

Sastry EVD, Prakash V (1993) Effect of salinity on variation in Na and K contents in wheat seedlings. Ann Arid Zone 32: 257–259.

Shah SH, Gorham J, Forster BP, Wyn Jones RG (1987) Salt tolerance in *Triticeae*. The contribution of D Genome to cation selectivity in hexaploid wheat. J Exp Bot 38: 254–269.

Sharma H (1995) Genetic variation in germination and seedling establishment traits in wheat (*Triticum aestivum* L.) seedling at different salinity and temperature levels. M.Sc. (Ag.) thesis, Rajasthan Agricultural University, Bikaner, Rajasthan (unpublished).

Singh U (1993) Variation in germination and seedling characteristics in certain *Triticum* species and selected progenies of their crosses with *Triticum aestivum* L. grown at different levels of salinity. M.Sc. (Ag.) thesis, Rajasthan Agricultural University, Bikaner, Rajasthan (unpublished).

William MDHM, Mujeeb-Kazi A (1995) Biochemical and molecular diagnostics of *Thinopyrum bessarabicum* chromosomes in *Triticum aestivum* germplasm. TAG 90(7–8): 952–956.

Chapter 5
The Role of Plant Hormones in Plants Under Salinity Stress

C. Kaya, A.L. Tuna, and I. Yokaş

Abstract Plant hormones can be defined as organic substances that are produced in one part of plant and translocated to another parts, where at very low concentration, They stimulate physiological response. Plant hormones are natural products and when they are synthesized chemically they are called plant growth regulators. Plants are usually subjected to environmental factors such as drought or high soil and water salinity. The reduction in plant growth exposed to saline environments could be due to either the effects of specific ions on metabolism, or adverse water relations. Different strategies are being employed to maximize plant growth under saline conditions. One of them is to produce salt tolerant genotypes of different crops. Attempts to improve tolerance to salinity through conventional plant breeding methods are time consuming and laborious, and rely on existing genetic variability. In addition, many other attempts have been made to overcome this disorder, including proper management and exogenous application of plant growth regulators. In this context, the levels of gibberellic acid (GA), abscisic acid (ABA), indoleacetic acid (IAA), cytokinins (CK), jasmonates (JA) and triazoles under salt stress are being discussed in this review.

Keywords Cytokinins • plant growth regulators • hormones

C. Kaya (✉)
Harran University, Agriculture Faculty, Soil Science and Plant Nutrition Department, Sanliurfa, Turkey

A.L Tuna
Mugla University, Science and Art Faculty, Biology Department, Mugla, Turkey

I. Yokaş
Mugla University, Ortaca Polytechnic, Mugla, Turkey

1 Introduction

Crops are usually subjected to environmental factors such as drought or high soil and water salinity. The reduction in plant growth exposed to saline environments could be due to either adverse water relations or the effects of ions on metabolism. Different strategies are being employed to optimise growth under saline conditions. One of them is to produce salt tolerant genotypes in different crops (Ashraf 1994; Kingsbury and Epstein 1984; Shannon and Grieve 1999). In addition, many other attempts have been made to overcome this disorder, including proper management and exogenous application of plant growth regulators.

Plant hormones are active members of the signal compounds involved in the induction of plant stress responses (Pedranzani et al. 2003). Abiotic stresses result in both altered levels of phytohormones and decreased plant growth (Morgan 1990). The decreased cytokinin and gibberellic acid and increased abscisic acid contents reported in salt stressed plants (Boucaud and Ungar 1976; Itai et al. 1968; Mizrahi et al. 1971) and this had led to the suggestion that salt stress induces changes in water relations and membrane permeability (Ilan 1971; Karmoker and Van Steveninck 1979). In this review, levels of some plant growth regulators in crop plants grown under salinity stress have been discussed.

1.1 Gibberellic Acid (GA)

Gibberellic acid (GA) accumulates rapidly when plants are exposed to both biotic (Farmer and Ryan 1992; McConn et al. 1997) and abiotic stresses (Xu et al.

M. Ashraf et al. (eds.), *Salinity and Water Stress*,

1994; Lehmann et al. 1995). In order to alleviate deleterious effects of salinity, different types of phytohormones have been used. Of these, gibberellins have been the main focus of some plant scientists (Munjal and Goswami 1995; Basalah and Mohammad 1999; Hisamatsu et al. 2000). For instance, gibberellic acid (GA_3) has been reported to be helpful in enhancing wheat and rice growth under saline conditions (Parasher and Varma 1988; Prakash and Prathapasenan 1990). Under saline conditions, seed germination has been improved by application of GA_3 and in this experiment, growth and grain yield of wheat were decreased with increasing salinity levels, but were also increased relatively by seed treatment with GA_3 (Kumar and Singh 1996). In another study, wheat seeds, after treatment with various growth regulators including GA_3, showed highest percent germination when treated with 20 mg/l GA_3 (Nayyar et al. 1995). Free radicals-induced lipid peroxidation are inhibited by GA (Choudhuri 1988). Those results show that GA_3 application could improve salinity tolerance in crop plants grown under saline condition.

1.2 Abscisic Acid (ABA)

Considerable researches are available on the effects of salinity on ABA production in plants. Increases in the endogenous ABA concentration of leaf tissue for salt stressed *Brassica* (He and Cramer 1996), *Phaseolus* (Montero et al. 1998) and *Zea mays* (Cramer and Quarrie 2002) correlated strongly with growth inhibition. Concentrations of ABA can also be increased in the roots (Lachno and Baker 1986; He and Cramer 1996) whilst root growth is maintained suggesting that these tissues may have different sensitivities to the localised concentration of ABA whether applied or endogenous in origin (Creelman et al. 1990). Stress responses of the roots and shoot tissues appear also to be coordinated by increased amounts of hormones moving in the xylem sap by 'root-to-shoot communication (Davies et al. 1994).

However, some doubt remains concerning the ability of ABA to act as a signal that mediates the effects of root-zone stress (Munns and King 1988). There is a significant evidence that ABA acts as the root-to-shoot stress signal. Jeschke et al. (1997) reported that increased ABA concentration in the xylem is correlated with reduced leaf conductance and a general inhibition of leaf growth. They also reported that salt stress caused an increase in ABA concentrations of mature *Ricinus* leaves by a factor of 18 at 128 mM NaCl concentration. Salt stress stimulated synthesis in roots and xylem transport of ABA was well correlated to stomatal reactions. This may be explained by the fact that when roots are directly exposed to the salt, ABA in roots stimulates ion accumulation in vacuoles of barley roots, which may be necessary for adaptation to saline conditions (Jeschke et al. 1997). It has also been reported that ABA improved the adaptation of cultured tobacco cells to salt, but not to equivalent osmotic concentrations of sorbitol and polyethyleneglycol and this suggests that ABA levels change more in response to salt than osmotic stress (LaRosa et al. 1985). Jae-Ung and Youngsook (2001) reported that ABA, a signal for stomatal closure, induces rapid depolymerization of cortical actin filaments and the slower formation of a new type of actin that is randomly oriented throughout the cell. This change in actin organisation has been suggested to be basic in signalling pathways involved in stomatal closing movement, since actin antagonists interfere with normal stomatal closing responses to ABA. Montero et al. (1997) reported that salt-induced ABA mediated the inhibition of leaf expansion and limited the accumulation of Na and Cl in leaves. ABA delayed the deleterious effect of NaCl and improved tolerance of ionic stress in sorghum (Amzallag et al. 1990).

1.3 Indoleacetic Acid (IAA)

IAA has a major role on controlling plant growth. It can control cell elongation, vascular tissue development and apical dominance (Wang et al. 2001). IAA has been reported that it responds to salinity in crop plants. However, little information seems to be available on the relationship between salinity stress and auxin levels in plants. The variations in indol asetic acid (IAA) content under stress conditions appeared to be similar to those of abscisic acid (Ribaut and Pilet 1991), and increased levels of IAA have also been correlated with reduced growth (Ribaut and Pilet 1994). Therefore, reduction in plant growth under stress conditions could be an outcome of altered hormonal balance and, hence, their exogenous application provides an attractive approach to counter the stress conditions. However, Prakash and Prathapasenan (1990) reported that NaCl caused a significant reduction in IAA concentrations in rice leaves. In this experiment, GA_3 application during the salinisation period partly overcome the effect of salinity on reducing IAA levels and this shows that salinity may influence hormone balances by affecting plant growth and development. There was also a significant reduction in IAA levels in rice five days after NaCl treatment (Nilsen and Orcutt 1996) and also salinity caused 75% reduction in IAA levels of tomato (Dunlap and Binzel 1996). As mentioned above further researches should be conducted to understand real mechanism.

1.4 Cytokinins (CKs)

Cytokinins (CKs) are well known in the regulation of many aspects of growth and differentiation, including cell division, apical dominance, nutrient mobilisation, chloroplast development, senescence and flowering (Hare and Van Staden 1997; Van Staden and Davey 1979). CKs retard senescence having effect on membrane permeability to mono and divalent ions and loklised induction of metabolic sinks (Letham 1978). They are generally considered to be antagonists of ABA, with the two hormones having opposing effects in several developmental processes including stomatal opening (Blackman and Davies 1984), cotyledon expansion and seed germination (Thomas 1992). CK levels tend to decrease under adverse environmental conditions. A general view has emerged that during stress a reduction of CK supply from the root alters gene expression in the shoot and thereby elicits appropriate responses to ameliorate the effects of stress (Hare et al. 1997).

Kinetin is capable of breaking stress-induced dormancy during germination of tomato, barley and cotton seeds (Bozcuk 1981). Moreover the observed reduction in endogenous cytokinins under stress conditions points towards the possibility that cytokinin levels could be a limiting factor under stress conditions and thus explain the fact that an exogenous application of kinetin resulted in increased growth of chickpea seedlings (Boucaud and Ungar 1976). It was suggested that a decrease in CK content was an early response to salt stress, but that the effects of NaCl on salt sensitive varieties is not mediated by CKs since a reduction in growth rate preceded any decline in CK levels (Walker and Dumbroff 1981).

However, endogenous levels of zeatin-type CKs remained unaltered in both roots and leaves during salt-stress in the facultative halophyte *Mesembryanthemum crystallinum* (Thomas et al. 1992). Exogenous application of KIN overcame the effects of salinity stress on the growth of wheat seedlings (Naqvi et al. 1982) and treatment of potato plants with KIN prior to salt stress diminished salt-related growth inhibition (Abdullah and Ahmad 1990). However, earlier studies reported that application of KIN to bean plants during salinity stress exacerbated its effects (Kirkham et al. 1974). Addition of benzyl adenin (BA) inhibited growth during stress of a salt-sensitive variety of barley, but overcame the decline in growth rate, shoot:root ratio and internal CK content in a salt-tolerant variety (Kuiper et al. 1990).

Kinetin acts as a direct free radical scavenger or it may involve in antioxidative mechanism related to the protection of purine breakdown (Chakrabarti and Mukherji 2003).

1.5 Jasmonates (JA)

Jasmonates, jasmonic acid in particular, are potent signal compounds not only in host defense but also in a variety of physiological mechanisms (Wasternack and Hause 2002). They accumulate rapidly when plants are grown under biotic (McConn et al. 1997) and abiotic stress (Lehmann et al. 1995). Jasmonic acid (JA) and its methylester (JAME) are involved in the plant signaling response to wounding and pathogen attack (Bohlmann 1994; Pena-Cortes and Willmitzer 1995). Sembdner and Parthier (1993) reported that induction of JA by wounding or pathogens has been attributed to production of fatty acids by damaged cell membranes which then metabolize via lipoxygenase to JA.

Allene oxide synthase (AOS) is an enzyme involved in JA synthesis. AOS was shown to be tightly linked with elevated JA content during the wound response in *Arabidopsis thaliana* (Laudert and Weiler 1998). The expression of AOS genes is also activated by systemin, wounding, 12-oxophytodienoic acid and JAME in tomato plants (Sivasankar et al. 2000). Another enzyme involved in JA synthesis is lipoxygenases (LOXs) which have been identified and localised within the chloroplasts (Feussner et al. 1995). Phospholipase D (PLD) has also been shown to trigger the release of linolenic acid and to stimulate JA biosynthesis (Creelman and Mullet 1997). PLD activity has been linked with stress processes playing a main function in membrane deterioration, although there is evidence for a role in plant signal transduction (Wang 1999).

Jasmonic acid and its derivatives also respond to salinity (Wang et al. 2001). JAME levels in rice roots increased significantly in 200 mM NaCl (Moons et al. 1997). It has been reported that jasmonate treatments or endogenous of these compounds in response to abiotic stress is accompanied by the synthesis of abundant proteins, called JIPs (Sembdner and Parthier 1993). At same time, pretreatment with JA reduced the inhibitory effect of high salt concentration on growth and photosynthesis of barley (Tsonev et al. 1998). It has been reported that JA levels in tomato cultivars changed in response to salt-stress and JA increase was observed in salt tolerant cv. HF (Hellfrucht Fruhstamm) from the beginning of salinisation, while in salt sensitive, cv. Pera, JA level decreased after 24 h of salt treatment (Pedranzani et al. 2003). Kramell et al. (2000) found a rapid increase in endogenous JA content in barley leaf segments subjected to osmotic stress with sorbitol or mannitol; however, endogenous jasmonates did not increase when treated with a high NaCl concentration (Kramell et al. 1995). There seems to be no information about how salinity affects endogenous JA levels in natural plant populations.

1.6 Triazoles

Triazoles are a group of compounds that have been developed for use as either fungicides or plant growth regulators, although in varying degrees they possess both properties (Fletcher et al. 2000). Triazoles can also protect plants from various environmental stresses, including anoxia, air pollutants, drought, extreme temperatures, and ultraviolet light (Davis et al. 1988; Fletcher and Hofstra 1988). Amongst the various triazoles developed as plant growth regulators, uniconazole was the most active stress protectant (Fletcher and Hofstra 1990; Fletcher et al. 1986). However, its commercial use in agriculture is limited by it's residual properties in both soil and plant tissues. Therefore, recent studies have focused on paclobutrazol which is an effective protectant of chilling damage in cucumber seedlings (Whitaker and Wang 1987). It has been shown that triazole compounds may counteract the effect of salinity e.g., in sunflower and mungbean seedlings it has been observed that the pretreatment of seeds with LAB 150978 (a triazole compound) counteracted the inhibitory effect of salinity on root growth, but it inhibited hypocotyls growth (Saha and Gupta 1993). Paclobutrazol treatment reduced shoot elongation and leaf length giving thicker vegetative tissues in wheat (Kraus et al. 1995). Paclobutrazol also protects wheat and corn from extreme temperatures (Pinhero and Fletcher 1994), where these effects have been shown to be associated with enhanced activity of free radical scavenging systems (Kraus and Fletcher 1994).

2 Conclusion

It can be concluded that:

1. Increasing salinity is associated with decreases in auxin, cytokinin and gibberellins in the plant tissues and an increase in ABA.
2. Changes in hormone levels in plant tissue are thought to be an initial process controlling growth reduction due to salinity.
3. NaCl-induced reduction in the plant growth can be mitigated by application of plant growth regulators.

References

Abdullah Z, Ahmad R (1990) Effect of pre- and post-kinetin treatments on salt tolerance of different potato cultivars growing on saline soils. J Agron Crop Sci 165: 94–102.

Amzallag GN, Lerner HR, Poljakoff-Mayber A (1990) Exogenous ABA as a modulator of response of sorghum to high salinity. J Exp Bot 41: 1389–1394.

Ashraf M (1994) Breeding for salinity tolerance in plants. CRC Crit Rev Plant Sci 13: 7–42.

Basalah MO, Mohammad S (1999) Effect of salinity and plant growth regulators on seed germination of *Medicago sativa* L. Pak J Biol Sci 2: 651–653.

Blackman PG, Davies WJ (1984) Modification of the CO2 responses of maize stomata by abscisic acid and by naturally occurring and synthetic cytokinins. J Exp Bot 35: 174–179.

Bohlmann H (1994) The role of thionins in plant protection. Crit Rev Plant Sci 13: 1–16.

Boucaud J, Ungar IA (1976) Hormonal control of germination under saline conditions of three halophyte taxa in genus Suaeda. Physiol Plant 36: 197–200.

Bozcuk S (1981) Effect of kinetin and salinity on germination of tomato, barley and cotton seeds. Ann Bot 48: 81–84.

Chakrabarti N, Mukherji S (2003) Alleviation of NaCl stress by pretreatment with phytohormones in *Vigna radiata*. Biologia Plant 46(4): 589–594.

Choudhuri MA (1988) Free radicals and leaf senescence – a review. Plant Physiol Biochem 15: 18–29.

Cramer GR, Quarrie SA (2002) Abscsic acid is correlated with the leaf growth inhibition of four genotypes of maize differing in their response to salinity. Funct Plant Biol 29: 111–115.

Creelman RA, Mullet JE (1997) Oligosaccharins, brassinolides and jasmonates: nontraditional regulators of plant growth, development, and gene expression. Plant Cell 9: 1211–1223.

Creelman RA, Mason HS, Bensen RJ, Boyer JS, Mullet JE (1990) Water deficit and abscisic acid cause differential inhibition of shoot versus root growth in soybean seedlings. Plant Physiol 92: 205–214.

Davies WJ, Tardieu F, Trejo CL (1994) How do chemical signals work in plants that grow in drying soil. Plant Physiol 104: 309–314.

Davis TD, Steffens GL, Sankhla N (1988) Triazole plant growth regulators. Hortic Rev 10: 63–105.

Dunlap JR, Binzel ML (1996) NaCl reduces indole-3-acetic acid levels in the roots of tomato plants independent of stress-induced abscisic acid. – Plant Physiol. 112: 379–384.

Farmer EE, Ryan CA (1992) Octadecanoid precursors of jasmonic acid activate the synthesis of wound-inducible proteinase inhibitors. Plant Cell 4: 129–134.

Feussner I, Hause B, Voros K, Parthier B, Wasternack C (1995) Jasmonate-induced lipoxygenase forms are localized in chloroplasts of barley (*Hordeum vulgare* cv. Salome) leaves. Plant J 7: 949–957.

Fletcher RA, Hofstra G (1990) Improvement of uniconazole-induced protection in wheat seedlings. J Plant Growth Regul 9: 207–212.

Fletcher RA, Hofstra G, Gao J (1986) Comparative fungitoxic and plant growth regulating properties of triazole derivatives. Plant Cell Physiol 27: 367–371.

Fletcher RA, Hofstra G (1988) Triazoles as potential plant protectants. In: Berg K, Plempel M (eds) Sterol biosynthesis inhibitors: pharmaceutical and agrochemical aspects. Ellis Harwood, Ltd., London

Fletcher RA, Gilley A, Sankhla N, Davis TD (2000). Triazoles as plant growth regulators and stress protectants. Hortic. Rev. 24, 55–138.

Hare PD, Van Staden J (1997) The molecular basis of cytokinin action. Plant Growth Regul 23: 41–78.

Hare PD, Cress WA, van Staden J (1997) The involvement of cytokinins in plant responses to environmental stress. Plant Growth Regul 23: 79–103.

He T, Cramer GR (1996) Abscisic acid concentrations are correlated with leaf area reductions in two salt-stressed rapid-cycling Brassica species. Plant Soil 179: 25–33.

Hisamatsu T, Koshioka M, Kubota S, Fujime Y, King RW, Mander LN (2000) The role of gibberellin in the control of growth and flowering in *Matthiola incana*. Physiol Plant 109: 97–105.

Ilan I (1971) Evidence for hormonal regulation of the selectivity of ion uptake by plant cells. Physiol Plant 25: 230–233.

Itai C, Richmond AE, Vaadia Y (1968) The role of root cytokinins during water and salinity stress. Israel J Bot 17: 187–195.

Jae-Ung H, Youngsook L (2001) Abscisic acid-induced actin reorganization in guard cells of dayflower is mediated by cytosolic calcium levels and by protein kinase and protein phosphatase activities. Plant Physiol 125: 2120–2128.

Jeschke WD, Peuke AD, Pate JS, Hartung W (1997) Transport, synthesis and catabolism of abscisic acid (ABA) in intact plants of castor bean (*Ricinus communis* L.) under phosphate deficiency and moderate salinity. J Exp Bot 48: 1737–1747.

Karmoker JL, Van Steveninck FM (1979) The effect of abscisic acid on the uptake and distribution of ions in intact seedlings of *Phaseolus vulgaris* cv. Redland Pioneer. Physiol Plant 45: 453–459.

Kingsbury RW, Epstein E (1984) Selection for salt resistance in spring wheat. Crop Sci 24: 310–315.

Kirkham MB, Gardner WR, Gerloff GC (1974) Internal water status of kinetin-treated, salt-stressed plants. Plant Physiol 53: 241–243.

Kramell R, Atzorn R, Schneider G, Miersch O, Bruckner C, Schmidt J, Sembdner G, Parthier B (1995) Occurrence and identification of jasmonic acid and its amino acid conjugates induced by osmotic stress in barley leaf tissue. J Plant Growth Regul 14: 29–36.

Kramell R, Miersch O, Atzorn R, Parthier B, Wasternack C (2000) Octadecanoid-derived alteration of gene expression and the 'oxylipin signature'in stressed barley leaves. Implications for different signaling pathways. Plant Physiol 123: 177–187.

Kraus TE, Fletcher RA (1994) Paclobutrazol protectswheat seedlings from heat and paraquat injury; Is detoxification of active oxygen involved? Plant Cell Physiol 35: 45–52.

Kraus TE, McKersie BD, Fletcher RA (1995) Paclobutrazol-induced tolerance of wheat leaves to paraquat may involve increased antioxidant enzyme activity. J Plant Physiol 145: 570–576.

Kuiper D, Schuit J, Kuiper PJC (1990) Actual cytokinin concentrations in plant tissue as an indicator for salt resistance in cereals. Plant Soil 123: 243–250.

Kumar B, Singh B (1996) Effect of plant hormones on growth and yield of wheat irrigated with saline water. Ann Agric Res 17: 209–212.

Lachno DR, Baker DA (1986) Stress induction of abscisic acid in maize roots. Physiol Plant 68: 215–221.

LaRosa PC, Handa AK, Hasgawa PM, Bressan RA (1985) Abscisic acid accelerates adaptation of cultured tobacco cells to salt. Plant Physiol 79: 138–142.

Laudert D, Weiler EW (1998) Allene oxide synthase: a major control point in *Arabidopsis thaliana* octadecanoid signalling. Plant J. 15:675–684.

Lehmann J, Atzorn R, Bruckner C, Reinbothe S, Leopold J, Wasternack C, Parthier B (1995) Accumulation of jasmonate, abscisic acid, specific transcripts and proteins in osmotically stressed barley leaf segments. Planta 197: 156 162.

Letham DS (1978) Cytokinins. In: Letham DS, Goodwin PB, Higgins TJV (eds) Phytohormones and related compounds. Vol 1. Elsevier, Amsterdam, pp. 205–243.

McConn M, Creelman RA, Bell F, Mullet JE, Browse J (1997) Jasmonate is essential for insect defense in *Arabidopsis*. Proc Natl Acad Sci USA 94: 5473–5477.

Mizrahi Y, Blumonfeld A, Bittner S, Richmond AE (1971) Abscisic acid and cytokinin content of leaves in relation to salinity and relative humidity. Plant Physiol 48: 752–755.

Montero E, Cabot C, Barcelo J, Poschenrieder C (1997) Endogenous abscisic acid levels are linked to decreased growth of bush bean plants treated with NaCl. Physiol Plant 101: 17–22.

Montero E, Cabot C, Poschenrieder CH, Barcelo J (1998) Relative importance of osmotic-stress and ion-specific effects on ABA-mediated inhibition of leaf expansion growth in *Phaseolus vulgaris*. Plant Cell Environ 21: 54–62.

Moons A, Prisen E, Bauw G, Montagu MV (1997) Antagonistic effects of abscisic acid and jasmonates on salt-inducible transcripts in rice roots. Plant Cell 92: 243–259.

Morgan PW (1990) Effects of abiotic stresses on plant hormone systems. In: Alscher RG, Cumming JR (eds) Stress responses in plants: adaptation and acclimation mechanism. Wiley-Liss, New York.

Munjal R, Goswami CL (1995) Response of chloroplastic pigments to NaCl and GA_3 during cotton cotyledonary leaf growth and maturity. Agric Sci Digest 15: 146–150.

Munns R, King RW (1988) Abscisic acid is not the only stomatal inhibitor in the transpiration stream of wheat plants. Plant Physiol 88: 703–708.

Naqvi SSM, Ansari R, Kuawada AN (1982) Responses of salt-stressed wheat seedlings to kinetin. Plant Sci Lett 26: 279–283.

Nayyar H, Walia DP, Kaistha BL (1995) Performance of bread wheat (*Triticum aestivum* L.) seed primed with growth regulators and inorganic salts. Indian J Agric Sci 65: 116–122.

Nilsen E, Orcutt DM (1996) The physiology of plants under stress - abiotic factors. Wiley, New York, pp. 118–130.

Parasher A, Varma SK (1988) Effect of pre-sowing seed soaking in gibberellic acid on growth of wheat (*Triticum aestivum* L.) under different saline conditions. Indian J Biol Sci 26: 473–475.

Pinhero RG, Fletcher RA (1994) Paclobutrazol and ancymidol protect corn seedlings from high and low temperatures stresses. Plant Growth Regul 15: 47–53.

Prakash L, Prathapasenan G (1990) NaCl and gibberellic acid-induced changes in the content of auxin, the activity of cellulase and pectin lyase during leaf growth in rice (*Oryza sativa*). Ann Bot 365: 251–257.

Pedranzani H, Racagni G, Alemano S, Miersch O, Ramirez I, Pena-Cortes H, Taleisnik E, Machado-Domenech E, Abdala G (2003) Salt tolerant tomato plants show increased levels of jasmonic acid. Plant Growth Regul 41: 149–158.

Pena-Cortes H, Willmitzer L (1995) The role of hormones in gene activation in response to wounding. In: Davies PJ (ed) Plant hormones: physiology, biochemistry and molecular biology. Kluwer, Dordrecht, pp. 395–414.

Ribaut JM, Pilet PE (1991) Effect of water stress on growth, osmotic potential and abscisic acid content of maize roots. Physiol Plant 81: 156–162.

Ribaut JM, Pilet PE (1994) Water stress and indole-3ylacetic acid content of maize roots. Planta 193: 502–507.

Saha K, Gupta K (1993) Effect of LAB-150978-a plant growth retardant on sunflower and mungbean seedlings under salinity stress. Indian J Plant Physiol 36(3): 151–154.

Sembdner G, Parthier B (1993) The biochemistry and physiology and molecular actions of jasmonates. Ann Rev Plant Physiol Plant Mol Biol 44: 569–586.

Shannon MC, Grieve CM (1999) Tolerance of vegetable crops to salinity. Scientia Hortic 78: 5–38.

Sivasankar S, Sheldrick B, Rothstein S (2000) Expression of allene oxide synthase determines defense gene activation in tomato. Plant Physiol 122: 1335–1342.

Thomas JC, McElwain EF, Bohnert HJ (1992) Convergent induction of osmotic stress-responses: abscisic acid, cytokinin, and the effects of NaCl. Plant Physiol 100: 416–423.

Thomas TH (1992) Some reflections on the relationship between endogenous hormones and light-mediated seed dormancy. Plant Growth Regul 11: 239–248.

Tsonev TD, Lazova GN, Stoinova ZG, Popova LP (1998) A possible role for jasmonic acid in adaptation of barley seedlings to salinity stres. J Plant Growth Regul 17: 153–159.

Van Staden J, Davey JE (1979) The synthesis, transport and metabolism of endogenous cytokinins. Plant Cell Environ 2: 93–106.

Walker MA, Dumbroff EB (1981) Effects of salt stress on abscisic acid and cytokinin levels in tomato. Z Pflanzenphysiol 101: 461–470.

Wang X (1999) The role of phospholipase D in signaling cascade. Plant Physiol 120: 645–651.

Wang Y, Mopper S, Hasentein KH (2001) Effects of salinity on endogenous ABA, IAA, JA, and SA in *Iris hexagona*. J Chem Ecol 27(2): 327–342.

Wasternack C, Hause B (2002) Jasmonates and octadecanoids – signals in plant stress response and development. In: Moldave K (ed) Progress in nucleic acid research and molecular biology. Vol. 72. Academic, New York, pp. 165–221.

Whitaker BD, Wang CY (1987) Effect of paclobutrazol and chilling on leaf membrane lipids in cucumber seedlings. Physiol Plant 70: 404–411.

Xu Y, Chang PL, Liu D, Narasimhan ML, Raghothama KG, Hasegawa PM, Bressan RA (1994) Plant defense genes are synergistically induced by ethylene and methyl jasmonate. Plant Cell 6: 1077–1085.

Chapter 6
Effects of Temperature and Salinity on Germination and Seedling Growth of *Daucus carota* cv. *nantes* and *Capsicum annuum* cv. *sivri* and Flooding on *Capsicum annuum* cv. *sivri*

M. Ozturk, S. Gucel, S. Sakcali, Y. Dogan, and S. Baslar

Abstract The germination and seedling growth of *Daucus carota* cv. *nantes* and *Capsicum annuum* cv. *sivri* were investigated under stress conditions. *D. carrota* seeds germinated well in the dark at 20°C (92%) but in the light germination was only 54%. The germination in dark at 20°C was 92%, 84% and 80% at 0.1%, 0.5% and 1% salt (NaCl) solutions, as the concentration increased germination decreased and length of radicle and plumule also got reduced. Germination was 90% at 0.1% NaCl + 10 ppm GA_3 and 43% at 2% NaCl +10 ppm GA_3. The seeds of *C. annuum* cv. *sivri* germinated well at constant temperatures of 15°C (83%), 20°C (100%), and 30°C (88%). The germination was 100% in the seeds placed in distilled water and left at 20°C, but it was delayed or inhibited when salt solutions were applied to these seeds. An application of growth regulators showed that GA_3 was stimulatory under saline conditions, However plant survival was low at 2% and 3% salt conditions as compared to 0.5% and 1% salt solutions. *C. annuum* cv. *sivri* did not show salt tolerance. When 4 weeks old seedlings of *C. annuum* were subjected to flooding using tap water, the performance of seedlings was better on unflooded soils. The plants survived under short periods of flooding but growth was poor, however, under longer periods of flooding the mortality increased and plants became sensitive.

Keywords *Daucus carota* • *Capsicum annuum* • salt • germination • seedling growth • hormones • flooding

1 Introduction

Nearly 82% of potential yield of crops is lost due to abiotic stress every year, and the amount of available productive arable land continues to decrease worldwide, forcing agriculture to areas where the potential for abiotic stress is even greater (Hirt and Shinozaki 2004). There are a number of abiotic stresses common in nature like salinity, drought, heavy metals, extreme temperatures, moistures, light, mineral deficiencies or toxicities, pH, and pollutants, which can diminish plant yields (Foolad 1996; Ashraf 2004; Öztürk et al. 2006; Munns et al. 2006; Ulfat et al. 2007; Sabir and Ashraf 2008; Chedlly et al. 2008).

Out of these stresses salinity can be disastrous because it causes many direct and indirect harmful effects, inhibits seed germination, induces physiological dysfunctions and often kills nonhalophytes even at low concentrations and limits agricultural development (Shannon 1997; Bartels and Sunkar 2005). More than 80 million hectares are facing this problem globally which accounts for over 6% of the world's total area (FAO 2003; Szabolcs 1994; Ghassemi et al. 1995). Out of 1,500 million hectares of dryland agricultural farming more than 30 million hectares are affected by secondary salinity, whereas more than 40 million hectares are salt affected out of present 230 million hectares of irrigated land (FAO 2008). Salinization transforms fertile and productive land to barren land,

M. Ozturk
Ege University, Botany Department, Bornova-Izmir, Turkey

S. Gucel
Near East University, Institute for Environmental Sciences, Nicosia-North Cyprus

S. Sakcali
Fatih University, Biology Department, İstanbul, Turkey

Y. Dogan
Dokuz Eylul University, Faculty of Education, Biology Department, Buca-Izmir, Turkey

S. Baslar
Dokuz Eylul University, Faculty of Education, Biology Department, Buca-Izmir, Turkey

M. Ashraf et al. (eds.), *Salinity and Water Stress*,

and often leads to loss of habitat and reduction of biodiversity (Ghassemi et al. 1995).These salts have accumulated over time due to natural causes particularly in arid and semiarid zones, or as a result of weathering of parent material or deposition of oceanic salts carried in wind and rain (Szabolcs 1994). Rainwater contains 6–50 mg/kg of sodium chloride; the concentration decreases with distance from the coast. Rain containing 10 mg/kg of sodium chloride would deposit 10 kg/ha of salt for each 100 mm of rainfall per year (Munns and Tester 2008). Susceptibility to salt injury varies with both species and the source of contamination. An electical conductivity is 4 dS/m or more reduces the yield of most crops by aggravating water stress conditions (Chachar et al. 2008). Some plant species and ecotypes have developed numerous and interlinked mechanisms to overcome salinity stress. For example, sugarbeet approximately shows a 20% reduction in dry weight after left for some time in 200 mm NaCl, whereas cotton faces a 60% reduction, and soybean does not live long (Greenway and Munns 1980). Similarly, high salinity causes alfalfa yield to decrease while the leaf/stem ratio increases, influencing forage quality. However, some species of glycophytes tolerate salt by a strategy more typical of halophytes, whereby sodium or chloride, or both, are taken up into leaves and compartmentalized in cell vacuoles, usually with the concomitant production of organic solutes in the cytoplasm for osmotic adjustment (Flowers 2004).

Recent advances in the understanding of these abiotic stress responses provided the impetus for compiling up-to-date reviews (Munns 2002, 2008; Munns et al. 2006; Munns and Tester 2008; Ashraf 2004; Ashraf and Haris 2004). In most plants, sodium and chloride are effectively excluded by roots during water uptake from the soil (USDA-ARS 2008; Munns 2002). The plants in saline habitats are stunted with dark green leaves which, and in many cases are thicker and more succulent than normal. In general the fruits and vegetables seem to be more salt sensitive than forage or field crops, but the sensitivity is higher during seedling stages. It is difficult to quantify the salt tolerance of plants because it varies appreciably with many environmental factors (e.g., soil fertility, soil physical conditions, distribution of salt in the soil profile, irrigation methods, and climate) and plant factors (e.g., stage of growth, variety, and rootstock) (Ghassemi et al. 1995; Juan et al. 2005).

Turkey is one of the eight major gene centres on earth (Öztürk et al. 1998), due to the presence of wild relatives of many domesticated plants. Due to its diversity in geological features, climate, plant cover and topography the country with a total area of approximately 78 million hectares embodies almost all soil groups distributed in the world. More than 90% of the soils are found in dry climatic zone with a poor grassland vegetation resulting in calcareous soils, generally with a clayey and loamy texture, slightly alkaline to alkaline pH, highly saline at places, rich in potassium, poor in nitrogen, very poor in organic matter content, and with low infiltration capacity (Öztürk et al. 2006a, b). The salinity alkalinity problems are threatening our soils, nearly 4.3 million hectares of agricultural land are degraded, out of which 1.5 million hectares are facing aridity and 2.8 million hectares show salinity-alkalinity problems (Öztürk et al. 2006a). The factors responsible for the salinity-alkalinity problems can be summarized as: accumulation of salts in plains due to heavy rains, a long standing high water table, the impact of sea water on the coastal alluvial plains, and geological features of the country, in particular the existence of saline areas as internal seas or sodic lakes, and over irrigation practices (Öztürk et al. 2006a).

For this reason, in order to develop practicable strategies for the evaluation of these areas and other marginal lands it is imperative to gain detailed information on the ecophysiological behaviour of plants of agricultural and economic value.

The focus of the current study was to provide fundamental biological understanding and knowledge on the germination and seedling growth behaviour of two important vegetables carrot and pepper; to different levels of salinity in order to have a knowledge of their salt tolerance. Several studies have shown that flooding in some areas of Turkey is also producing detrimental effects. Another objective of this study was therefore to determine the seedling growth response of locally grown conical pepper variety to flooding.

2 Material and Methods

2.1 Germination and Seedling Growth

The seeds of carrot (*Daucus carota* L. cv. *nantes*) and conical pepper (*Capsicum annuum* L. cv. *sivri*) were purchased locally. In all experiments fresh seeds were used because seeds lose viability with age, and develop dormancy (Bosland and Votava 2000; Özçoban and Demir 2002). The seeds of both species were placed in 100 ml beakers and 20 ml of 1% sodium hypocholorite was added for sterilization. These were left in the solution for 5 min followed by washing under running tap water and deionized water. Sterilized seeds of both species (50 seeds/dish) were placed in petri dishes with a double layer of Watman No l filter paper. The dishes (two replicates) were left in growth chamber at 15°C, 20°C and 25°C under continuous light (40 W flourescent tubes) and continuous dark. For darkness dishes were covered with aluminum foils and counts made under 25 W green lamp in the evening.

In the second set of experiments 10 ml of distill water, salt solution (0.1%, 0.5%, 1%, 2%, 3%, 4% NaCl) or salt solution + growth regulators (10, 50,100 ppm of GA_3, KIN, IAA) were added to the petri dishes left under 12 + 12 h light/dark condition in growth chamber at 20°C. The solutions were changed daily. Germination was followed for 14 days. Length of hypocotyle and radicle was measured on 14th day by using a ruler. For fresh weight (Wf) determination, the seedlings were weighed on a high-precision electrical balance and then kept in an oven at 80°C for 48 h and reweighed to determine their dry weight (Wd). The root moisture content (M), expressed as a percentage, was calculated from Wf and Wd values: M = 100(Wf – Wd)/ Wf.

2.2 Flooding

5 mm long, 4 mm wide and 1 mm thick seeds (Demir and Ellis 1992; Chen and Lott 1992) were sown in wide glass pans containing sandy-loam soil and after 2 weeks seedlings were transferred randomly to plastic cans. These were allowed to grow for 2 weeks. Four weeks old seedlings (5/can) in 5 cans were subjected to flooding by leaving water to stand 1 cm above the soil throughout the experiment. Other 5 cans received normal watering. Morphological features (root/shoot length; number of leaves, flowers, fruits) were noted at the start and harvest in both flooded and unflooded series, and root sections were taken at the end of experiment. Moisture of the soil samples taken 2.5 cm below the upper surface and pH of the samples taken 5 cm below the soil surface were recorded (Öztürk et al. 1983).

In a separate experiment 4 weeks old seedlings of *Capsicum annuum* cv. *sivri* were transplanted to small plastic containers with control, 0.5%, 1%, 2% and 3% levels of NaCl and allowed to grow for 6 weeks. During transplantation seedlings were uniform in size. Each treatment had two replicates. The experiment was terminated after 25 days. Following parameters were recorded: root/shoot length, number of leaves and roots, leaf area ratio, root/shoot ratio, specific leaf area, leaf/shoot ratio, root/leaf area ratio, and root/leaf ratio.

3 Results and Discussion

The wild carrot *D. carota* is said to have originated from Afghanistan, which is still the centre of its diversity. The garden vegetable we have today is a naturally-occurring subspecies *Daucus carota* subsp. *sativus*. In early days only aromatic leaves and seeds were used, and use of roots started much afterwards. It was introduced to Europe only few centuries back, and orange-coloured carrots are said to have appeared in the Netherlands in the 17th century (Öztürk 1996).

D. carota cv. *nantes* belongs to the family Apiaceae. It is a high quality coreless variety with reddish-orange color, sweet flavour, smooth, cylindrical sides, blunt tip and a fine-grain crispy texture, and is preferred in fresh markets. The crop is cultivated over an area of 7,575,030,030 Da in Turkey with a production of a 232,000,232,000 t. The highest production is seen in Marmara region (128.785 t), followed by Ege (60.022 t), Mediterranean (21.114 t) and Central and East Anatolia (22.079 t). A major part of this production is exported and is thus very important for the economy of the country.

3.1 Germination and Seedling Growth of D. carota cv. nantes

A lot of literature is available on the germination, temperature and salinity interactions of different plant species notable among these are Garcia et al. (1995), Baskin and Baskin (1998), Bell et al. (1999), Aiazzi et al. (2002), Khan and Gulzar (2003), Al-Khateeb (2006), Noreen et al. (2007), Nisa et al. (2007) and Nasim et al. (2008). However, not many papers have been published on *D. carota* cv. *nantes* (Szafirowska et al. 1981; Szafirowska 1984; Dearman et al.1987; Murray 1989; Yanmaz and Özdil 1992; Demiray and

Eşiz Dereboylu 2005; Eraslan et al. 2007). In the present study the germination behaviour of this economically important crop of Turkey was investigated, because it faces the greatest problem during germination and seedling emergence (Murray 1989; Duman and Eşiyok 1998). The seeds of this plant are included among the late germinating group (Duman and Eşiyok 1998) and these develop secondary dormancy. The results obtained by us revealed that, at 15°C (under light) germination is 74% but in the dark it goes up to 83%. At 20°C germination is 54% under light but reaches a value of 92% in the dark. At 25°C germination is 82% and 88% under light and dark conditions respectively (Fig. 6.1). Salinity reduces the total number of seeds germinating and postpones initiation of germination processes. Seed germination of many glycophytes may be inhibited by 0.5% salt. In *D. carota* cv. *nantes* germination was 92%, 84% and 82% respectively in 0.1%, 0.5% and 1% NaCl at 20°C. As the salt concentration increased germination decreased and length of radicle and plumule got reduced too. The seeds of *D. carota* cv. *nantes* did not get effected at lower salt concentrations, even at 1% they germinated well, but at 2%, 3% and 4% NaCl germination was inhibited.

Salt-induced inhibition of germination can sometimes be partially alleviated by exogenous application of growth regulators (Ashraf et al. 2002). GA_3, IAA and KIN applied at 10, 50 and 100ppm concentration to *D. carota* cv. *nantes* revealed that germination was 90% at 0.1% NaCl + 10ppm GA_3 and 43% at 2% NaCl + 10ppm GA_3. An alleviation of the salt stress was thus achieved to some extent at 2% salinity level. Similar results were obtained for Kinetin and IAA. GA_3 proved more effective than KIN and IAA. Germination got slightly reduced in 0.1% NaCl + 10ppm GA_3 and started on 4th day and ended on 12th day. Similarly in 10ppm of Kinetin and IAA same results were obtained but germination started on 6th and 7th day. Germination percentage also decreased. Even seedling growth is effected, radicle as well as hypocotyle lengths got reduced. In other studies the effect of higher concentrations of growth regulators on flowering, number of umbels, seed yield and seed quality of carrot has been investigated (Öztürk et al. 1995). Bud application of 1,000ppm Gibberellic acid (GA_3), 2,000 or 5,000ppm Daminozide, 250 or 500ppm phosphon-D, drench application of Ancymidol at 100 or 250ppm or chlormequat at 2,000 or 5,000ppm revealed that all chemicals had no significant effect on the number of days to flowering. Gibberellic acid increased seed stalk height, while Ancymidol and phosphon-D at 500ppm reduced it. Ancymidol and Daminozide reduced the number of umbels per plant at both concentrations of chemical used, however, Ancymidol reduced seed yield. The percentage of seed from the primary and secondary umbels was increased by Daminozide (Öztürk et al. 1995).

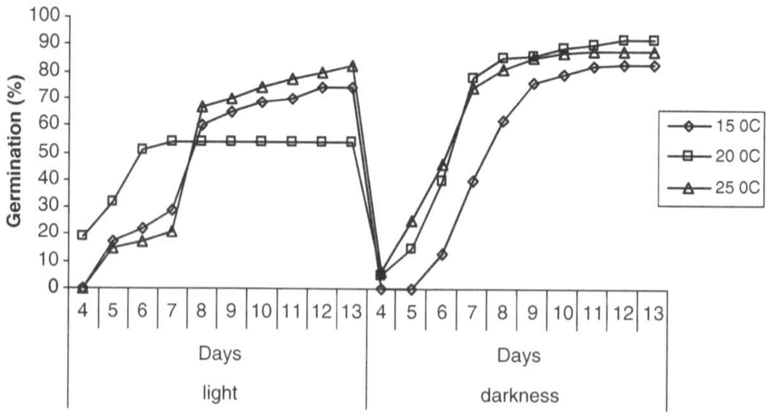

Fig. 6.1 Germination of *Daucus carota* cv. *Nantes* seeds at different temperatures under continuous dark and light

3.2 *Germination of* Capsicum annuum *cv.* sivri

The archaeological evidence has revealed that *C. annuum* has been used and domesticated in central-eastern and south-central Mexico as early as 9000 BP (Whitmore and Turner 2002; Smith 2005). The domesticated species of peppers are; *Capsicum annuum*, *C. frutescens*, *C. chinense*, *C. baccatum* and *C. pubescens* (Heiser 1985; DeWitt and Bosland 1996; Yoon et al.; 2004; Yamamoto and Nawata 2005). *C. annuum*, *C. frutescens* and *C. chinense* are grouped in a taxonomic complex, with the three clusters of domesticated plants appearing to be more divergent than their wild progenitors (Jarret and Dang 2004). The remaining two domesticated species are in other taxonomic complexes of the genus (Eshbaugh et al. 1983; Eshbaugh 1993). The name *C. frutescens* instead of *C. annuum* has been used for the domesticated chili peppers, so in some literature caution is needed to ascertain whether the plants discussed are actually *C. annuum*, or *C. frutescens* (Heiser 1985). *C. annuum* is usually grown as a herbaceous annual in temperate areas. It shows diversity in plant habit, shape, size, colour, pungency, and other qualities of the fruit (Idu and Ogbe 1997; Andrews 1999; Aleemullah et al. 2000; Dabauza and Peña 2001; Dağ and Kamer 2001; Estrada et al. 2002; Geleta and Labuschagne 2004; Shirai and Hagimori 2004; Derera et al. 2005). It provides the ingredient for a non-lethal deterrent or repellent to some human and animal behaviours (Cichewicz and Thorpe 1996; Blum et al. 2002; Cronin 2002; Krishna De 2003). These features have helped to make *C. annuum* globally important as a fresh as well as cooked vegetable, as a source of food ingredients, as a colourant and medicinal importance (Andrews 1999; Bosland and Votava 2000). The seed is covered by a parchment-like seed coat. Seed size is somewhat dependent on the variety and growing conditions. *C. annuum* cv. sivri is a warm-season crop, and highly susceptible to frost. Cultivated *C. annuum* is very diverse worldwide and has many varieties like Cerasiforme Group (cherry peppers), Conoides Group (conical peppers), Longum Group (e.g. Cayenne peppers) and Grossum Group (blocky sweet or bell peppers) (Eshbaugh 1993). *C. annuum* is one of the most important crops in Turkey and several varieties are grown with a total export of more than 250,000 t, out of which 100,000 t are fresh.

The seeds of *C. annuum* cv. *sivri* germinated well at constant temperatures of 15°C (83%), 20°C (100%), and 30°C (88%), but germination was very poor at 10°C (23%), and no germination took place at 5°C. There was no effect of light and darkness on the germination. These findings coincide with those reported by Choi (1985), Choi et al. (1999), Hernández-Verdugo et al. (2001) and Dell'Aquila (2004).

Generally salinity reduces seed germination and postpones initiation of germination processes in nonhalophytes, however the responses are variable and species specific (Ozturk et al. 1997; Khan and Ungar 1998; El-Keblawy 2004). Very few studies have been undertaken on the salt tolerance of pepper (Khan and Sheikh 1976; Chung and Choi 2002). In the present study we found out that the seeds of *C. annuum* cv. *sivri* showed a 100% germination at 20°C in distilled water, but an inhibition or delay was observed in the salt, but no germination occurred at 4% salt. In 1% and 2% treatments the percentage germination was relatively better than 3% (Fig. 6.2). Salt-induced inhibition of germination can

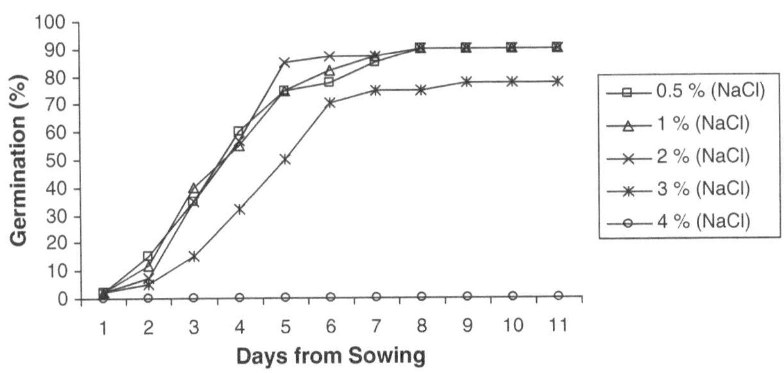

Fig. 6.2 Seed germination of *Capsicum annuum* cv. *sivri* in different concentrations of NaCl at 20°C

sometimes be partially alleviated by exogenous application of growth regulators like kinetin and gibberellin, but no effect on germination is seen at high salt concentrations (>400 mM NaCl) (Watkins et al. 1985; Öztürk et al. 1993,1994,1995). 10, 50 and 100 ppm concentrations of growth regulators (GA_3, IAA, Kinetin, GA_3 + KI, GA_3 + IAA and GA_3 + KIN + IAA) were applied to overcome the effects of salt. These regulators in particular GA_3 stimulated the germination of seeds. In 0.5% and 1% germination was better than 2% and 3%.

3.3 *Seedling Growth of* C. annuum cv. sivri

Plants differ greatly in their tolerance of salinity, as reflected in their different growth responses. The variation in salinity tolerance in dicotyledonous species is even greater than in monocotyledonous species. Some legumes are very sensitive, even more sensitive than rice (Läuchli 1984, 2002) and alfalfa or lucerne (*Medicago sativa*) is very tolerant. The mechanisms of salinity tolerance fall into three categories: tolerance to osmotic stress, sodium exclusion from leaf blades and tissue tolerance (Munns and Tester 2008b). The relative importance of these various processes clearly varies with the species (i.e., the strategy a particular plant species has evolved for responding to the salinity stress), but probably also varies with the length of exposure to the salinity, the concentration of the salt, and possibly the local environmental conditions, notably soil water supply and air humidity.

The studies on the growth behaviour of seedlings transplanted to the pots with 0.5%, 1%, 2% and 3% levels of soil salinity revealed that plant survival was low at 2% and 3% as compared to 0.5% and 1%. The growth of the plants was best in the control followed by 0.5% NaCl. In all other treatments both shoot as well as root growth was reduced. There was a lesser production of roots than shoots at higher levels of salinity. Seedlings in 0.5% had significantly longer radicle/hypocotyle than other concentrations. Seedlings of 1% were significantly longer than 2% and 3%. Seedling fresh weight in 0.5% and 1% differed significantly from 2% and 3%. In general the seedling growth was better in the control than salt. However, seedling dry weight in 2% and 3% was significantly greater than 0.5% and 1% (Figs. 6.3, 6.4). Watkins and Cantliffe (1983a, b) showed that at 25°C radicle emergence required 3.5 days, whereas at 15°C, 9 days were required. Lowest seedling fresh weight in 3% was accompanied by highest dry weight. This might be due to the toxicity of high concentrations of chloride ions.

The survival of plants is maximum in 0.5% (66%), double than other treatments (31%). The leaves in 0.5% and 1% were normal green, but turned yellow in 2% after 7 days, dropped after few days in 3%. Greater leaf fresh and dry weight, stem dry weight and shoot fresh/dry weight in 0.5% and 1%. Longer roots in 0.5% than 1%, higher stem dry weight in 1% than 0.5%. In 3% roots smaller, number of leaves less, fresh weight and dry weight of leaves/shoots less than 2%. A progressive increase in leaf area ratio and specific leaf area with increase in salinity from 0.5% to 3% (Figs. 6.5–6.8). This may be an adaptive response by the plants

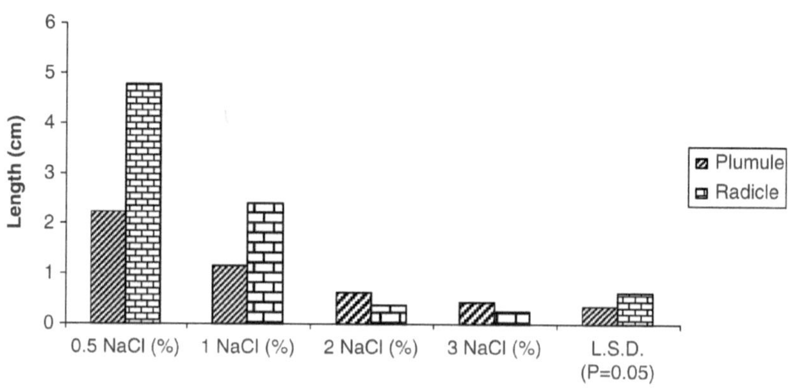

Fig. 6.3 Measurements of length of radicle and plumule of *Capsicum annuum* cv. *sivri* in different NaCl concentrations

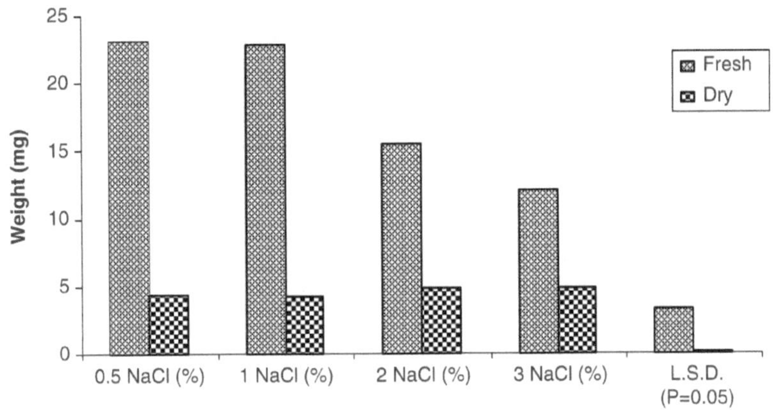

Fig. 6.4 Measurements of seedling fresh and dry weight of *Capsicum annuum* cv. *sivri* in different NaCl concentrations

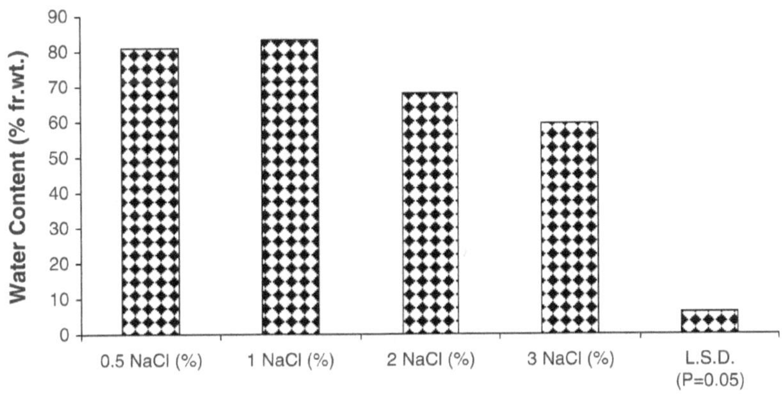

Fig. 6.5 Measurements of water content in the seedling of *Capsicum annuum* cv. *sivri* in different NaCl concentrations

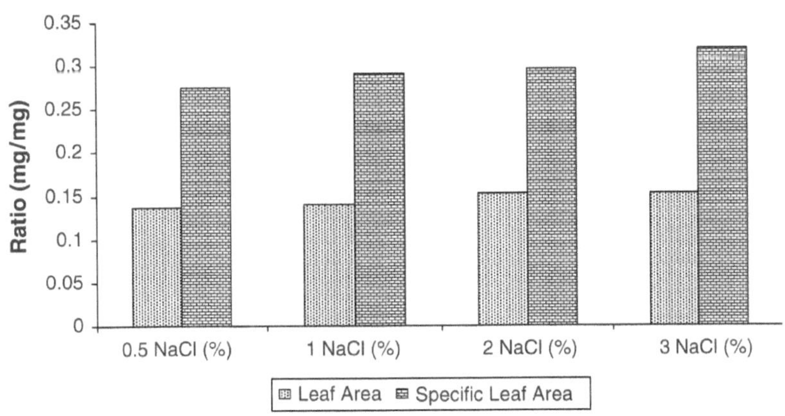

Fig. 6.6 Measurements of the ratio of leaf area and specific leaf area in the seedling of *Capsicum annuum* cv. *sivri* in different NaCl concentrations

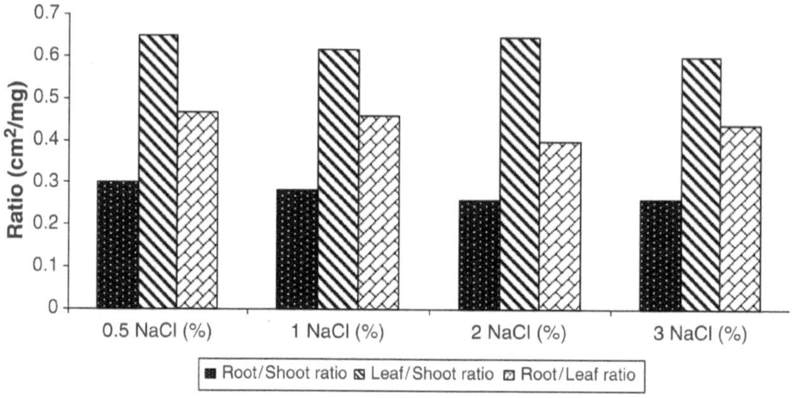

Fig. 6.7 Measurements of the ratio of root/shoot, leaf/shoot and root/leaf in the seedling of *Capsicum annuum* cv. *sivri* in different NaCl concentrations

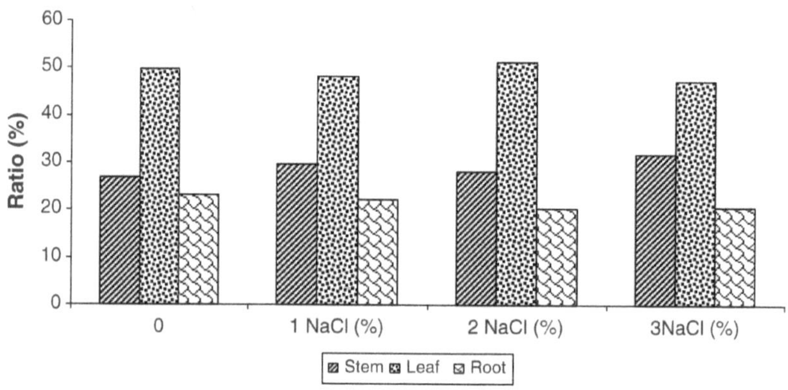

Fig. 6.8 Measurements of the ratio of root, stem and leaf in the seedling of *Capsicum annuum* cv. *sivri* in different NaCl concentrations

when subjected to a harsh environment. The time taken and ability for roots to recover may depend on whether or not plasmolysis has occurred (Munns 2002).The decreased rate of leaf growth after an increase in soil salinity is primarily due to the osmotic effect of the salt around the roots.

Root growth is usually less affected than leaf growth, and root elongation rate recovers remarkably well after exposure to sodium chloride or other osmotica (Munns 2002). For example, sodium does not increase in the leaf blade of grapevines until after several years of exposure to saline soil, then the exclusion within the root, stem, and petiole breaks down, and Na^+ starts to accumulate in the leaf blade, whereas leaf blade chloride concentrations increase progressively (Prior et al. 2007).

In cereals, the major effect of salinity on total leaf area is a reduction in the number of tillers; in dicotyledonous species, the major effect is the dramatic curtailing of the size of individual leaves or the numbers of branches. Curiously, shoot growth is more sensitive than root growth, a phenomenon that also occurs in drying soils and for which there is as yet no mechanistic explanation.

3.4 *Effects of Flooding on* Capsicum annuum *cv.* sivri

Flooding is an environmental stress for many natural and man-made ecosystems worldwide. Flooding during the growing season adversely affects all developmental stages of flood-intolerant plants, whereas flooding during the dormant season generally has little effect in the short term (Kozlowski 1984; Bailey-Serres and Voesenek 2008). Flooding of soil with nonsaline or saline water adversely affects the distribution of many plants (Boland et al. 1996; Butsan et al. 2004). Plant responses to flooding during the growing season include suppression of leaf formation and expansion of leaves and internodes, premature leaf abscission and senescence, shoot dieback, generally decreased cambial growth, injury to vegetative and reproductive growth, fruit quality, smaller fruit size, altered chemical composition, appearance of fruits, changes in plant anatomy, and ethylene production (Abbott and Gough 1987; Casenave et al. 1999). Injury and growth inhibition typically are preludes to plant mortality. Root growth typically is reduced more than shoot growth.

It affects the soils as well, by altering soil structure, depleting O_2, accumulating CO_2, inducing anaerobic decomposition of organic matter, and reducing iron and manganese. However, the specific plant responses vary with many factors including plant species and genotype, age of plants, properties of the floodwater, site characteristics and time and duration of flooding (Kozlowski 1984). The mechanisms by which flood-tolerant plants survive waterlogging are complex (Pezeshki 1994). Important adaptations include production of hypertrophied lenticels, aerenchyma tissue, and adventitious roots. Ion imbalances due to flooding indicate a breakdown in root membrane integrity, which would affect passive uptake of ions.

The 4 weeks old seedlings of *C. annuum* cv. *sivri* were transplanted to the cans containing soils with sandy loam texture. At the start of experiment (PI-Period) the plant height was 13 cm in unflooded and 13.3 cm in flooded series. During the PII-Period it was 26.55 cm in the unflooded set but only 14.60 cm in the flooded set. In the final period the height in unflooded set was 59.70 cm. In the unflooded (UF) soils root/shoot growth increased twice in each period, but in flooded soils this increase was very low. Lower root/shoot ratio in unflooded set in period II can be attributed to the an increase in root length. In the unflooded set number of leaves at the start was 6 but was 19 at the harvest. In the flooded soils number of leaves at the start was 6, but only 4 at the harvest. The flowers and fruits were present in unflooded series, but none in the flooded one. The roots in the unflooded set were whitish-brown but orange brown with black dots at the root tips in the flooded series, with a poorer root development. Root functions were severely impaired due to ion imbalances occurring in the roots. In flooded series upper 1 cm part of stem in water swollen and soft. As compared to root biomass leaf biomass declined substantially in flooded series (Figs. 6.9–6.11). Some studies have been carried out on the effects of flooding on pepper (Khan and Sheikh 1976; Suh et al. 1987).

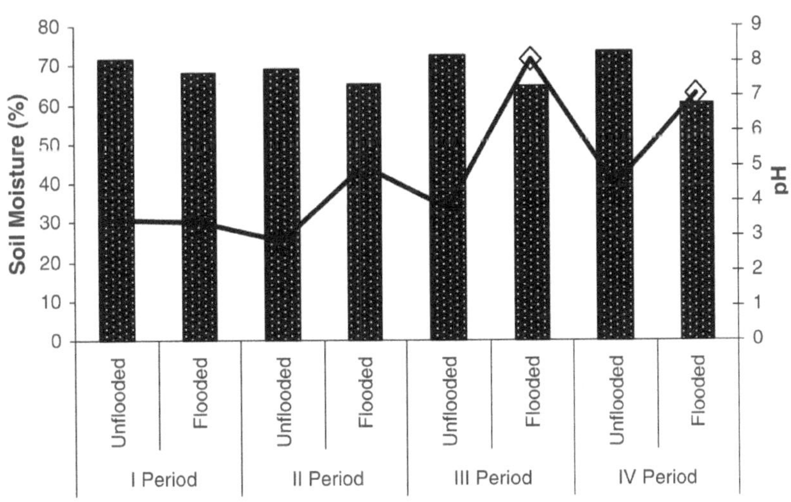

Fig. 6.9 Moisture content and ph of the soil samples from unflooded and flooded *Capsicum annuum* cv. *sivri* cans at the end end of experiment

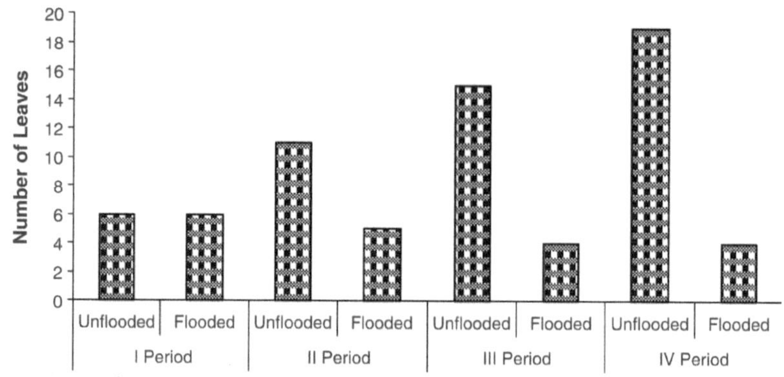

Fig. 6.10 Effect of flooding on the number of leaves in *Capsicum annuum* cv. *sivri*

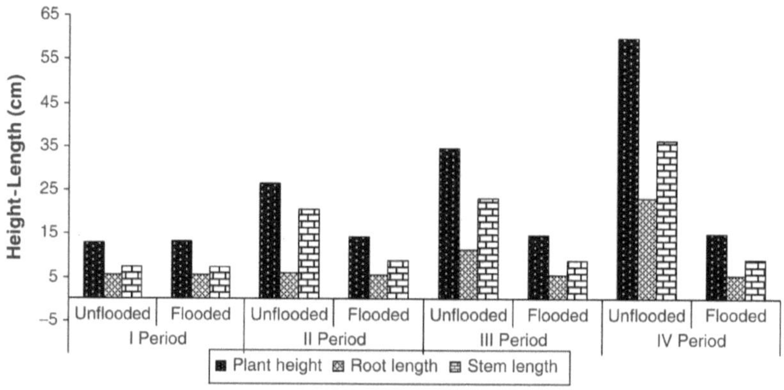

Fig. 6.11 Effect of flooding on the plant height, root and stem length of *Capsicum annuum* cv. *sivri*

In the unflooded series the diameter of roots, vascular bundles and cortex at the start of experiment was 1,840, 344 and 416∝, but at the end of experiment the values were 3,680, 2,484 and 579∝ respectively. In the flooded series these values were 2,769, 368, and 522µ.

The organic matter in unflooded soils was higher, but lower in flooded due to root rottening. $CaCO_3$ in both sets was higher at the start due to the water given. Soil pH was low in flooded but higher in drained soils. Soil moisture increased in flooded soils. In flooded series flagellar algae and *Chara* covered the surface.

Many studies have been undertaken on the salinity effects on *Olea europaea* (Benlloch et al. 1991; Cresti et al. 1994; Al-Absi et al. 2003). In contrast to our studies growth of salt-tolerant *Olea europaea* plants flooded with saline water for 4 weeks recovered readily when salinization was relieved. The rate of recovery depended on the salt concentration to which the plants had been exposed (0, 50, 100, or 200 mM NaCl). Growth was inhibited by all salt solutions but growth rates of plants treated with 50 or 100 mM NaCl returned to the rates of control plants within 4 weeks of relief from flooding. Plants exposed to 200 mM NaCl recovered to only 60% of the growth rate of control plants after 4 weeks. Similarly in *Citrus*, the exclusion of sodium and chloride ions occurs continuously and progenies separate widely on the basis of their capacity to restrict foliar accumulation of these ions (Cole 1985).

4 Conclusion

Several studies have shown that a combination of flooding and salinity is considerably more detrimental to seedlings than the effect of either stress alone, and the detrimental effects of a combination of flooding and salinity increase with increasing salinity (Pezeshki 1994; Allen et al. 1996; Bosland and Votava 2000). Salinity adversely affects nonhalophytes in several ways (Bernstein 1980; Waisel 1991). Injury is more severe when salts absorbed from the soil are augmented by salts deposited on leaves. The evidence for nonosmotic effects of salinity on injury to plants can be summarised as follows: organic solutes do not injure plants at osmolalities higher than the critical concentrations for salt injury, individual salts have different critical concentrations for inducing injury, certain organic solutes increase the critical salt concentration for injury, and injurious effects of salts are antagonized by calcium (Munns 2008).

This work shows that *D. carota* cv. *nantes* and *C. annuum* cv. *sivri* did not show tolerance to the salt during their later growth as such, the use of germination trials only as a method of testing the salt tolerance of a plant is not a sound approach. Studies on the establishment of seedlings and their subsequent growth to maturity must be carried out in order to draw valid conclusions regarding the salt tolerance of a species or a variety. It did not show tolerance to the continuous flooding as well.

References

Abbott JD, Gough RE (1987) Prolonged flooding effects on anatomy of highbush blueberry. J Hort Sci 22: 622–625.

Aiazzi MT, Carpane PD, Rienzo JA, Arguello JA (2002) Effects of salinity and temperature on the germination and early seedling growth of *Atriplex cordobensis* Gandoger et Stuckert (Chenopodiaceae). Seed Sci Technol 30: 329–338.

Al-Absi K, Qrunfleh M, Abu-Sharar T (2003) Mechanism of salt tolerance of two olive (*Olive europaea* L.) cultivars as related to electrolyte concentration and toxicity. Acta Hort 618: 281–290.

Al-Khateeb SA (2006) Effect of salinity and temperature on germination and ion relations of *Panicum turgidum* Forssk. Bioresource Technol 97: 292–298.

Allen JA, Pezeshki SR, Chambers JL (1996) Interaction of flooding and salinity stress on baldcypress (*Taxodium distichum*). Tree Physiol 16: 307–313.

Aleemullah M, Haigh AM, Holford P (2000) Anthesis, anther dehiscence, pistil receptivity and fruit development in the Longum Group of *Capsicum annuum*. Aust J Exp Agr 40: 755–762.

Andrews J (1999) The Pepper Trail: History and Recipes from Around the World. University of North Texas Press, Denton, TX, pp. 261.

Ashraf M (2002) Salt tolerance of cotton: some new advances. Crit Rev Plant Sci 21(1): 1–30.

Ashraf M (2004) Some important physiological selection criteria for salt tolerance in plants. Flora 199: 361–376.

Ashraf M, Karim F, Rasul E (2002) Interactive effects of gibberellic acid (GA_3) and salt stress on growth, ion accumulation and photosynthetic capacity of two spring wheat (*Triticum aestivum* L.) cultivars differing in salt tolerance. Plant Growth Regul 36(1): 49–59.

Ashraf M, Haris PJC (2004) Potential biochemical indicators of salinity tolerance in plants. Plant Sci 166: 3–16.

Benlloch M, Arboleda F, Barranco D, Fernandez-Escobar R (1991) Response of young olive trees to sodium and boron excess in irrigation water. J Hort Sci 26: 867–870.

Blum E, Liu K, Mazourek M, Yoo EY, Jahn MM, Paran I (2002) Molecular mapping of the C locus for presence of pungency in *Capsicum*. Genome 45: 702–705.

Bosland PW, Votava EJ (2000) Peppers: Vegetable and Spice *Capsicums*. Crop Production Science in Horticulture 12. CAB International, Wallingford, England, pp. 204.

Bailey-Serres J, Voesenek LACJ (2008) Flooding stress: acclimations and genetic diversity. Annu Rev Plant Biol 59: 313–339.

Bartels D, Sunkar R (2005) Drought and salt tolerance in plants. Crit Rev Plant Sci 24: 23–58.

Baskin CC, Baskin JM (1998) Seeds: Ecology, Biogeography, and Evolution of Dormancy and Germination. Academic, San Diego, CA, pp. 666.

Bell DT, King LA, Plummer JA (1999) Ecophusiological effects of light quality and nitrate on seed germination in species from Western Australia. Aust J Ecol 24: 2–10.

Bernstein L (1980) Salt tolerance of fruit crops. Agr Inf Bull 292: USDA, Riverside, CA.

Boland AM, Jerie PH, Mitchell PD, Nordella JL, Nordella IN (1996) The effect of a saline and non-saline water table on peach tree water use, growth, productivity and ion uptake. Aust J Agr Res 47: 121–139.

Butsan A, Sagi M, Malach YD, Pasternak D (2004) Effects of saline irrigation water and heat waves on potato production in an arid environment. Field Crop Res 90: 275–285.

Casenave EC, Degano CAM, Toselli ME, Catan EAA (1999) Statistical studies on anatomical modifications in the radicle and hypocotyl of cotton induced by NaCl. Biol Res 32(4): 289–295.

Chachar QI, Solangi AG, Verhoef A (2008) Influence of sodium chloride on seed germination and seedling root growth of cotton (*Gossypium hirsutum* L.). Pak J Bot 40(1): 183–197.

Chedlly A,Ozturk M, Ashraf M,Grignon C (2008) (eds) Biosaline Agriculture and High Salinity Tolerance. Birkhauser Verlag-AG (Springer Science), Basel, pp. 367.

Chen P, Lott JN (1992) Studies of *Capsicum annuum* seeds: structure, storage reserves, and mineral nutrients. Can J Bot 70: 518–529.

Choi KS (1985) Physiological and ecological studies on the germinability of pepper (*Capsicum annuum* L.) seeds at low temperature. Research Report of RDA (Hort) 27: 46–87.

Choi WY, Oh YB, Choi IH, Nam SH (1999) Effect of rooting in different kinds of bed soil for *Capsicum annuum* L. cuttings propagation. Kor J Hort Sci Technol 17: 628.

Chung HD, Choi YJ (2002) Enhancement of salt tolerance of pepper plants (*Capsicum annuum* L.) by grafting. J Kor Soc Hort Sci 43: 556–564.

Cichewicz RH, Thorpe PA (1996) The antimicrobial properties of chile peppers (*Capsicum* species) and their uses in Mayan medicine. J Ethnopharm 52: 61–70.

Cole PJ (1985) Chloride toxicity in citrus. Irrigation Sci 6: 63–71.

Cresti M, Ciampolini F, Tattini M, Cimato A (1994) Effect of salinity on productivity and oil quality of olive (*Olea europaea* L.) plants. Adv Hort Sci 8: 211–214.

Cronin JR (2002) The chili pepper's pungent principle: capsaicin delivers diverse health benefits. Altern Complement Ther 8: 110–113.

Dabauza M, Peña L (2001) High efficiency organogenesis in sweet pepper (*Capsicum annuum* L.) tissues from different seedling explants. Plant Growth Regul 33: 221–229.

Dağ A, Kamer Y (2001) Comparison between the effectiveness of honeybee (*Apis mellifera*) and bumblebee (*Bombus terrestris*) as pollinators of greenhouse sweet pepper (*Capsicum annuum*). Am Bee J 141: 447–448.

Dearman J, Brocklehurst PA, Drew RLK (1987) Effects of Osmotic Priming and Ageing on the Germination and Emergence of Carrot and leek Seed. Ann Appl Biol 111: 717–722.

Dell'Aquila A (2004) Cabbage, lentil, pepper and tomato seed germination monitored by an image analysis system. Seed Sci Technol 32: 225–229.

Demiray H, Eşiz Dereboylu A (2005) The effects of boron and niacin on the growth of nantes carrot (*Daucus carota* L.). Ege Univ Ziraat Fakultesi Dergisi 42(1): 191–201.

Demir I, Ellis RH (1992) Development of pepper (*Capsicum annuum*) seed quality. Ann Appl Biol 121: 385–399.

Derera NF, Nagy N, Hoxha A (2005) Condiment paprika research in Australia. J Bus Chem 2: 4–18.

DeWitt D, Bosland PW (1996) Peppers of the World: An Identification Guide. Ten Speed Press, Berkeley, CA, pp. 219.

Duman I, Eşiyok D (1998) Effects of presowing PEG and KH_2PO_4 treatments on germination emergence and yield of carrot. Turk J Agric For 22: 445–450.

El-Keblawy A (2004) Salinity effects on seed germination of the common desert range grass,*Panicum turgidum*. Seed Sci Technol 32: 943–948.

Eraslan F, Inal A, Gunes A, Alpaslan M (2007) Impact of exogenous salicylic acid on the growth, antioxidant activity and physiology of carrot plants subjected to combined salinity and boron toxicity. Sci Hort 113(2): 120–128.

Eshbaugh WH (1993) Peppers: History and exploitation of a serendipitous new crop discovery. In: Janick J, Simon JE (eds) New Crops. Wiley, New York, pp. 132–139.

Eshbaugh WH,Guttman SI, McLeod MJ (1983) The origin and evolution of domesticated *Capsicum* species. J Ethnobiol 3: 49–54.

Estrada B, Bernal MA, Díaz J, Pomar F, Merino F (2002) Capsaicinoids in vegetative organs of *Capsicum annuum* L. in relation to fruiting. J Agr Food Chem 50: 1188–1191.

FAO (2003) Food and Agriculture Organization of the United Nations Production Yearbook 2001, Vol. 55, Statistics Series No. 170. FAO, Rome, 333 pp.

FAO (2008) FAO Land and Plant Nutrition Management Service. http://www.fao.org/ag/agl/agll/spush

Flowers TJ (2004) Improving crop salt tolerance. J Exp Bot 55: 307–319.

Foolad MR (1996) Response to selection for salt tolerance during germination in tomato seed derived from PI174263. J Am Soc Hort Sci 121: 1001–1006.

Garcia FP, Pita JM, Benito ME, Iriondo G (1995) Effect of light, temperature and seed priming on germination of celery seeds (*Apium graveolens* L.). Seed Sci Technol 23: 377–383.

Ghassemi F, Jakeman AJ, Nix HA (1995) Salinisation of Land and Water Resources: Human Causes, Extent, Management and Case Studies. UNSW Press, Sydney-Australia and CAB International, Wallingford.

Geleta LF, Labuschagne MT (2004) Hybrid performance for yield and other characteristics in peppers (*Capsicum annuum* L.). J Agr Sci (Cambridge) 142: 411–419.

Greenway H, Munns R (1980) Mechanisms of salt tolerance in nonhalophytes. Ann Rev Plant Physiol 31: 149–90.

Heiser Jr CB (1985) How many kinds of peppers are there? In: Heiser CB Jr (ed) Of Plants and People. University of Oklahoma Press, Norman and London, pp. 142–154.

Hernández-Verdugo S, Oyama K, Vázquez-Yanes C (2001) Differentiation in seed germination among populations of *Capsicum annuum* along a latitudinal gradient in Mexico. Plant Ecol 155: 245–257.

Hirt H, Shinozaki K (2004) Plant Responses to Abiotic Stress. Springer, Netherlands, Vol. 4, 300 pp.

Idu M, Ogbe FM (1997) Studies on the morphology of *Capsicum* species. J Plant Anat Morphol 7: 114–123.

Jarret RL, Dang P (2004) Revisiting the waxy locus and the *Capsicum annuum* L. complex. Georgia J Sci 62: 117–133.

Juan M, Rivero RM, Romero L, Ruíz JM (2005) Evaluation of some nutritional and biochemical indicators in selecting salt-resistant tomato cultivars. Env Exp Bot 54: 193–201.

Khan SS, Sheikh KH (1976) Effects of different levels of salinity on seed germination and growth of *Capsicum annuum* L. Biologia 22: 15–25.

Khan MA, Gulzar S (2003) Light, salinity, and temperature effects on the seed germination of perennial grasses. Am J Bot 90: 131–134.

Khan MA, Ungar IA (1998) Seed germination and dormancy of *Polygonum aviculare* L. as influenced by salinity, temperature and gibberellic acid. Seed Sci Technol 26: 107–117.

Kozlowski TT (1984) Plant responses to flooding of soil. Bioscience 34: 162–167.

Krishna De A (2003) The Genus *Capsicum*. Medicinal and Aromatic Plants - Industrial Profiles. Taylor & Francis, London and New York, Vol. 33, 275 pp.

Läuchli A (1984) Salt exclusion: an adaptation of legumes for crops and pastures under saline conditions. In: Staples RC (ed) Salinity Tolerance in Plants: Strategies for Crop Improvement. Wiley, New York, pp. 171–187.

Läuchli A (2002) Introduction to salinity. In: Läuchli A, Lüttge U (eds) Environment-Plants-Molecules. Kluwer, Dordrecht, The Netherlands.

Munns R (2002) Comparative physiology of salt and water stress. Plant Cell Environ 25: 239–250.

Munns R, James RA, Läuchli A (2006) Approaches to increasing the salt tolerance of wheat and other cereals. J Exp Bot 5(57): 1025–1043.

Munns R, Tester M (2008) Mechanisms of salinity tolerance. Ann Rev Plant Biol 59: 651–681.

Munns R (2008) Strategies for crop improvement in saline soils. In: Ashraf M, Ozturk M, Athar HR (eds) Salinity and Water Stress: Improving Crop Efficiency. Springer (this volume), Dodrecht, The Netherlands (in press).

Murray GA (1989) Osmoconditioning carrot seed for improved emergence. Hortsci 24(4): 701.

Nasim M, Qureshi RH, Aziz T, Saqib M, Nawaz S, Sahi ST, Pervaiz S (2008) Growth and ionic composition of salt-stressed *Eucalyptus camaldulensis* and *Eucalyptus tereticornis*. Pak J Bot 40(2): 799–805.

Nisa M, Khan MA, Weber DJ (2007) Dormancy, germination and viability of *Salsola imbricata* seeds in relation to light, temperature and salinity. Seed Sci Technol 35: 595–606.

Noreen Z, Ashraf M, Hassan MU (2007) Inter-accessional variation for salt tolerance in pea (*Pisum sativum* L.) at germination and seedling stage. Pak J Bot 39(6): 275–285.

Özçoban M, Demir İ (2002) Longevity of pepper (*Capsicum annuum*) and watermelon (*Citrullus lanatus*) seeds in relation to seed moisture and storage temperature. Indian J Agr Sci 72: 589–593.

Öztürk M (1996) Lectures in Economic Botany. Ege University, Science Faculty, Botany Department, Bornova, Izmir, Turkiye.

Öztürk M, Gökçeoğlu M, Pirdal M (1983) Bitki-toprak ilişkileri Uygulama Kilavuzu. Ege University, Science Faculty, Publ 32, Bornova, Izmir, Turkiye

Öztürk O, Gemici M, Yilmazer Ç, Özdemir F (1993) Alleviation of salinity stress by GA_3, KIN and IAA on seed germination of *Brassica campestris* L. Doga Turk J Bot 17: 47–52.

Öztürk M, Eşiyok D, Özdemir F, Olcay G, Öner M (1994) Studies on the effects of growth substances on the germination and seedling growth of *Brassica oleracea* L.var. *acephala* (Kara lahana). Ege University Science Faculty Journal 16: 63–70.

Öztürk M, Özdemir F, Eser Adiyahşi OI, Ilbi H (1995) Studies on the salt-hormone interactions in the germination and seedling growth of some vegetable species. In: Khan MA, Ungar IA (eds) Biology of Salt Tolerant Plants. Book Crafters, Chelsea, MI, pp. 59–64.

Öztürk M, Başlar S, Doğan Y, Mert HH (1997) Alleviation of salinity stress in the germination of *Eruca sativa* mill. Cruciferae Newslett 19: 69–70.

Öztürk M,Yücel E, Yarci C, Çelik A, Aksoy A (1998) Plant diversity in the Turkish black sea region and strategies for its conservation. In: Kotlyakov V, Uppenbrink M, Metreveli V (eds) NATO-ARW. Kluwer, Batumi, GA, pp. 155–173.

Öztürk M, Güvensen A, Aksoy A, Beyazgül M (2006a) An overview of the soils and sustainable land use in Turkiye. In: Proceedings of the Fifth International GAP Engineering Congress, Şanliurfa Turkey, pp. 1548–1555.

Öztürk M, Güvensen A, Sakçali S, Bahadir H (2006b) An overview of the land degradation problems in East Anatolia. In: Proceedings of the Fifth International GAP Engineering Congress, Şanliurfa, Turkey, pp. 1556–1561.

Öztürk M, Waisel Y, Khan MA, Görk G (2006) (eds) Biosaline Agriculture and Salinity Tolerance in Plants. Birkhauser Verlag-AG (Springer Science), Basel, pp. 205.

Pezeshki SR (1994) Plant responses to flooding. In: Wilkinson RE (ed) Plant-Environment Interactions. Marcel Dekker, New York, pp. 289–321.

Prior LD, Grieve AM, Bevington KB, Slavich PG (2007) Long-term effects of saline irrigation water on 'Valencia' orange trees: relationships between growth and yield, and salt levels in soil and leaves. Aust J Agr Res 58: 349–358.

Sabir P, Ashraf M (2008) Inter-cultivar variation for salt tolerance in proso millet (*Panicum miliaceum* L.) at the germination stage. Pak J Bot 40(2): 677–682.

Shannon MC (1997) Adaptation of plants to salinity. Adv Argon 60: 75–120.

Shirai T, Hagimori M (2004) A multiplication method of sweet pepper (*Capsicum annuum* L.) by vegetative propagation. J Jpn Soc Hort Sci 73: 259–265.

Smith BD (2005) Reassessing Coxcatlan Cave and the early history of domesticated plants in Mesoamerica. Proc Natl Acad Sci USA 102: 9438–9445.

Suh HD, Cho KY, Park SK, Lee KH (1987) Effect of flooding on the growth and yield of hot pepper (*Capsicum annuum* L.). Research Report of RDA (Horticulture) 29: 1–9.

Szabolcs I (1994) Soils and salinisation. In: Pessarakali M (ed), Handbook of Plant and Crop Stress. Marcel Dekker, New York, 311 pp.

Szafirowska A (1984) Effect of seed osmoconditioning on germination, regulation of emergence and yield of carrot roots. Biuletyn Instytutu Hodowli Aklimatyzacji Roslin 153: 251–257.

Szafirowska A, Khan AA, Peck NH (1981) Osmoconditioning of carrot seeds to improve seedling establishment and yield in cold soil. Agron J 73: 845–848.

Ulfat M, Athar HR, Ashraf M, Arkam NA, Jamil A (2007) Appraisal of physiological and biochemical selection criteria for evaluation of salt tolerance in canola (*Brassica napus* L.). Pak J Bot 39(5): 1593–1608.

USDA-ARS (2008) Research Databases. Bibliography on Salt Tolerance. US Department of Agriculture, Agriculture Reserve Service, Riverside CA. www.ars.usda.gov/Services/docs.htm?docid=8908

Yanmaz R, Özdil AH (1992) Domates ve Havuç Tohumlarinda Ekim Öncesi PEG (Polyethylenglyçol) Uygulamalarinin Çimlenme ve Çikiş Orani İle Çikiş Süresi Üzerine Etkileri. Türkiye I. Ulusal Bahçe Bitkileri Kongresi II. 25–27, İzmir, Turkiye

Votava EJ, Nabhan GP, Bosland PW (2002) Genetic diversity and similarity revealed via molecular analysis among and within an in situ population and ex situ accessions of chiltepín (*Capsicum annuum* var. *glabriusculum*). Conser Genet 3: 123–129.

Waisel Y (1991) Adaptation to salinity. In: Raghavendra AS (ed), Physiology of Trees. Wiley, New York, pp. 359–383.

Watkins JT, Cantliffe DJ (1983a) Hormonal control of pepper seed germination. Hort Sci 18: 342–343.

Watkins JT, Cantliffe DJ (1983b) Mechanical resistance of the seed coat and endosperm during germination of *Capsicum annuum* at low temperature. Plant Physiol 72: 146–150

Watkins JT, Cantliffe DJ, Huber HB, Nell TA (1985) Gibberellic acid stimulated degradation of endosperm in pepper. J Am Soc Hort Sci 110: 61–65.

Whitmore TM, Turner BL (2002) Cultivated Landscapes of Middle America on the Eve of Conquest-II. Oxford University Press, New York, 338 pp.

Yamamoto S, Nawata E (2005) *Capsicum frutescens* L. in Southeast and East Asia, and its dispersal routes into Japan. Econ Bot 59: 18–28.

Yoon JB, Do JW, Yang DC, Park HG (2004) Interspecific cross compatibility among five domesticated species of *Capsicum* genus. J Kor Soc Hort Sci 45: 324–329.

Chapter 7
Triticeae: The Ultimate Source of Abiotic Stress Tolerance Improvement in Wheat

S. Farooq

Abstract Salinity of arable land is one of the major abiotic stresses, which along with the world population is increasing simultaneously at a very rapid pace. In some of the developing countries especially those located in the arid regions, more than 50% of their arable land is affected while it is anticipated that about 6.8 billion of anticipated 8 billion people would be living in these countries. Wheat and wheat based products are their major staple food which needs to be increased by 40% if food security is to be ensured to this much population. This is possible only through cultivation of saline lands provided; salt tolerant wheat varieties are available. Efforts made so for in this direction have not produced results of any practical significance despite the fact that tribe *Triticeae* to which bread wheat belongs; possess tremendous potential for salt tolerance that has been extensively and practically identified, tested and transferred to wheat cultivars with proven expression of tolerance. In this paper we are discussing (i) the potential of salt tolerance in wild wheat grasses and genome contributing species of *Triticeae*, (ii) success related with practical utilization of this potential and (iii) future prospects of using *Triticeae* as potential source of salt tolerance improvement in wheat.

Keywords Annual species • aridity • salinity • wheat • wild wheat grasses

S. Farooq
Nuclear Institute for Agriculture and Biology (NIAB), P. O. Box 128, Jhang road, Faisalabad, Pakistan
e-mail: shafqat_niab@hotmail.com
Present address: Director (Technical-V) PAEC Head Office, Opposite K Block Secretariat, Islamabad

1 Introduction

Salinity of arable land is a global problem that has restricted productivity on 955 million hectares of land (Farooq and Azam 2005). It is developed due to accumulation of water soluble salts in the soil and its interactions with groundwater (Rengasamy 2006) to a level that is significantly affecting agricultural production, environmental health, and economic welfare of the countries especially those located in the arid regions. Water availability is another problem and Pakistan is one of the most water-stressed countries in the world (Anonymous 2005). Also in West, Central and South Asia, and in Middle East, water will drop to <1,000 m^3 per person per annum by 2025 (Anonymous 2000). Due to such stresses, inhabitants of these areas are migrating to other places in search of lively hood. This is one of the most important reasons for creating poverty in the developing countries of the arid regions (FAO 2005). The yield and growth rates in these countries have also fallen to the level below those needed to both substantially alleviate the serious malnutrition and poverty. The new challenge of inevitable but poorly predictable global climate change that will begin to have profound effect on food production around the globe during the 21st century (Guy et al. 2006) has further aggravated the situation. The uncertainty surrounded by GM crops in the form of safety, control and access to the technology, and by the perceived inequity in access to the essential Intellectual Property Right (IPR) for crop research (Swaminathan 2004) is also haunting the farmer's communities in the developing countries. All this and the anticipated addition of 30% more salt affected land in the current estimates (Table 7.1) within next 25 years and 50% by the year 2050 (Wang et al. 2003a) is posing a serious threat to the future of agriculture.

M. Ashraf et al. (eds.), *Salinity and Water Stress*,

Table 7.1 Estimates of salt affected soils (Mha) around the globe and of irrigated lands affected by salinization in selected countries

Continent wise estimates		Countries wise estimates			
Regions	Salt affected area	Country	Salt affected %	Country	Salt affected %
Africa	80.5	Algeria	10–15	India	27
Australia	357.3	Egypt	30–40	Iran	<30
Europe	50.8	Senegal	10–15	Iraq	50
Mexico and Central America	2.0	Sudan	>20	Israel	13
North America	15.7	USA	20–25	Jordan	16
North and Central Asia	211.7	Colombia	20	Pakistan	>40
South America	129.2	Peru	12	Sri Lanka	13
South Asia	87.6	China	15	Arab Republic	30–35
South East Asia	20.0				
Total	954.8				

Source: Data table 19.3 of World Resources 1987, a report by International Institute of Environment Development and the World Resources Institute, published by Basic Book Inc, New York

This is simultaneously happening at a time when population is expected to be increased from 6 to 8 billion in 2025 of which 6.8 billion would be living in developing countries (Anonymous 1998). Wheat and wheat based products as their major stable cereal especially those living in South and Southeast Asia. Assuring food security to this much population would require 1 billion metric tons of wheat compared to the current production of 600 million metric tons (Rajaram 2001), which means an increase in productivity of about 40–50% during next 30 years. This is an uphill task and can be achieved either through increasing yield or cultivating saline lands with suitable salt tolerant wheat germplasm, which can be selected or evolved through various approaches.

One of such approaches is improving salt tolerance of commercial wheat cultivars through screening and breeding programmes, which is feasible because tremendous variability exists for salt tolerance in wheat (Mass and Poss 1989; Munns et al. 2000; Noori and McNeilly 2000; Singh and Singh 2000; Wilson et al. 2002; Flowers 2004; Sairam and Tyagi 2004; El-Hendawy et al. 2005; Munns 2005; Munns et al. 2006). Member of the tribe *Triticeae* which harbor hexaploid wheat (*Triticum aestivum* L.), its wild relatives and genome contributing species also possess tremendous potential for salt tolerance. In this paper we are discussing (i) the potential of salt tolerance in wild wheat grasses and genome contributing species of *Triticeae*, (ii) success related with practical utilization of this potential and (ii) future prospects of using *Triticeae* as potential source of salt tolerance improvement in wheat.

2 Salt Tolerance Potential in *Triticeae*

2.1 Perennial Triticeae

The tribe *Triticeae* with almost 350 species is an excellent source of gene pool for certain environmental stresses like salinity, alkalinity and disease resistance. These species are mostly perennial grasses possessing excellent forage quality and are distributed into various genera including *Agropyron*, *Pseudoroegnaria*, *Psathyrostachys*, *Thinopyrum*, *Elytrigia*, *Elymus*, *Leymus and Pascopyrum* (Table 7.2).

Studies to explore salt tolerance potential of these grasses began in 1960 when Dewey tested 25 strains of tall wheat grasses and found *Agropyron elongatum* host. (*Elytrigia pontica* Pod. Holub.) as most salt tolerant species with significant inter-specific variations. Such variations were also observed when *A. intermedium* and *A. cristatum* were tested (Dewey 1962; Hunt 1965). Later studies repeatedly confirmed high salt tolerance potential of *E. pontica* (Elzam and Epstein 1969; Moxley et al. 1978; Shannon 1978). McGuire and Dvorak (1981) tested different accessions belonging to *A. elongatum* (*E. elongate*), *A. intermedium* (*E. intermedia*) and *A junceum* (*E. junciformis*) and found species of the *elongatum* (*E. pontica* and *E. scirpea*) and *junceum* complexes (A. *junceum*) the most salt tolerant. Gorham et al. (1985) tested and found another salt tolerant diploid species: *Thinopyrum bessarabicum*, which can withstand prolonged exposure to 350 mol m^3 NaCl. The most salt tolerant genera (*A. junceum* and *A. elongatum*)

Table 7.2 The habitat and significance of some important genera of the tribe *Triticeae*

Genus	Habitat	Significance
Agropyron	Saline and arid rangeland	Tolerant to drought and cold
Pseudoroegnaria	Rangeland and rocky hills	Extremely tolerant to alkalinity
Psathyrostrachys	Arid rangeland and rocky hills	Tolerant to alkalinity and drought
Thinopyrum	Coastal inlands	Tolerant to inland salinity
Elytrigia	Coastal saline regions	Tolerant to moderate salinity
Elymus	Moderately saline and alkaline soils	Moderately tolerant to salinity and alkalinity
Leymus	Saline and alkaline soils	Extremely tolerant to salinity and alkalinity
Pascopyrum	Heavily saline and alkaline soils	Extremely tolerant to salinity and alkalinity

Source: Farooq 1990

have now been combined into one genus i.e. *Thinopyrum* (Dewey 1984). Farooq et al. (1988) tested about 100 different accessions of various perennial genera and found species of the genera *Leymus* (*L kerelenii*), and *Thinopyrum* (*Th. scripeum* and *Th. junceum*) as most salt tolerant. These species showed 100% and 83% survival at EC 54 dS m^{-1}.

The mechanisms imparting salt tolerance to perennial *Triticeae* have also been thoroughly investigated with major emphasis on ion accumulation in leaf and high K^+/Na^+ ratio but little emphasis on tissue tolerance of accumulated Na^+ and Cl^+ (Colmer et al. 2006). Studies by Gorham et al. (1984, 1985) have indicated that salt tolerance in *Thinopyrum* and *Leymus* is achieved by (i) strictly controlling the influx of Na^+ and Cl^- to shoot, (ii) attaining high glycine betaine concentration (Hitz and Hanson 1980), (iii) reduced transpiration rate coupled with constant water use efficiency under salt stress and (iv) maintenance of high K^+/Na^+ ratio in leaves. The genetics of some of these mechanisms is only partly known. For example, the genes controlling K^+/Na^+ ratio are located on long arm of chromosome 4D of *Ae. tauschii* (Shah et al. 1987) and of *Triticum aestivum* (Gorham et al. 1990). In perennial species, these genes are located on several chromosomes (Zhong and Dvorak 1995). The genes for Na^+ and Cl^- exclusion are located on chromosomes 5 J of *Th. junceum* (Forster et al. 1988) and can be transferred to wheat for improvement its salt tolerance.

2.2 *Practical Utilization of Salt Tolerance Available in Perennial* Triticeae

The first successful attempt to transfer salt tolerance from decaploid *E. pontica* was made by Jan Dvorak and his group (1985) who selected hybrid plant derivatives showing superior salt tolerance compared to wheat parents in hydrponics. Diploid *E. elongata* has also been hybridized with wheat cultivar Chinese spring and the hybrid tested for salt tolerance (Storey et al. 1985). At higher external NaCl concentration (120 mM), it behaves like *Elytrigia* which is known to restrict salt accumulation in shoots. The amphiploid was grown in saline conditions for only three weeks. Hence its performance at maturity and yield data was not reported. However, study did indicate that salt tolerance of *E. elongata* can be transferred to wheat. This was re-confirmed by Dvorak and Ross (1986) after testing amphiploid of *E. elongata* and Chinese spring in a solution of NaCl, KCl, $MgSO_4$, K_2SO_4, and in sea water. Addition lines of *E. elongata* (*Th. elongatum*) were also tested confirming that several chromosomes contribute towards salt tolerance (Dvorak et al. 1988). None of these studies showed field performance of the material developed by transferring salt tolerance form *Th. elongatum* into wheat.

Attempts have also been made to transfer salt tolerance from diploid *A. junceum* (*Th. bassarabicum*) to wheat (Alonso and Kimber 1980; Forster and Miller 1985; Mujeeb-Kazi et al. 1987, 1989). The amphiploid survived but produced nonviable seeds at salt concentration (250 mM NaCl solution) lethal to wheat (Forster et al. 1987). Later studies (Dvorak et al. 1988) indicated that gene(s) responsible for ion regulation in *Thinopyrum* species have been transferred to wheat. These genes were found on chromosomes 5 J and 2 J of *Thinopyrum*. Line 5 J survived at 200 mM NaCl and produced shriveled grains which indicated that chromosomes 5 J carries major gene(s) for ion regulation, which can be transfer to wheat (Gorham et al. 1986).

Genes were also transferred from *Th. bassarabicum* to tetraploid wheat (King et al. 1997). The fertile amphiploid (*Tritipyrum*) survived at 150 mM NaCl and performed better than any of the salt tolerant wheats. *Lophopyron elongatum* was used to transfer salt tolerance to hexapoloid wheat (Omielan et al. 1991). Its amphiploid with Chinese spring and disomic substitution line (3E) when tested under saline field appeared more tolerant than Chinese spring.

Farooq et al. (1993) transferred salt tolerance from *Th. scirpeum* to various hexaploid wheat varieties including *ph1b* mutant of Chinese spring. F1 hybrids regenerated through embryo culture were back crossed and BC1 seeds with various chromosomes number were allowed to self twice. One of the lines with 44 chromosomes exhibited more vigorous growth and grain yield at EC 15 dS m^{-1} compared to 42 chromosomes lines and indicated the potential of *Th. scripeum* for improvement of salt tolerance of wheat cultivars.

Salt tolerance has also been transferred from hexaploid *Th. junceum* (Charpentier 1992). Upon testing for salt tolerance, one of the addition lines (AJDAj5) survived at EC 42 dS m^{-1} (Wang et al. 2003b) and showed salt tolerance comparable to that of amphiploid. The addition line AJDAj5 is reported to have a pair of chromosomes (E^bE^b) from *Th. junceum* (2n = 42: $E^bE^bE^e$). In order to introduce salt tolerance from this addition line it was crossed with hexaploid wheat carrying the Ph^1 gene. Three F5 families were selected and tested for salt tolerance of which two lines (4909 and 4910) showed salt tolerance greater than AJDAj5 and can be used as gene source for breeding salt tolerant wheat cultivars (Wang et al. 2003b).

2.3 *Annual* Triticeae

Salt tolerance in annual *Triticeae* has been investigated very extensively. In *Aegilops* species, it is known when Farooq et al. (1989) tested more than 100 different accessions of various species and found *Ae. squarrosa* (*Ae. tauschii*) the most salt tolerant. *Ae. ovata, Ae. cylindrica, Ae. tricuncialis and Ae. bicornis* were also found salt tolerant for the first time. *Ae. squarrosa* that contributes DD genome to hexaploid wheat (Kimber and Zhao 1983) is more tolerant than *Ae. speltoide.*

The latter is one of the probable BB genome donors to hexaploid and tetraploid wheats (Alonso and Kimber 1983). Shah et al. (1987) have indicated that hexaploid wheat is more salt tolerance than diploid wheats (*T. monococcum*) which is the AA genome contributor to hexaploid and tetraploid wheats. Durum wheat that lacks the DD genome tends to accumulate more Na^+ and less K^+ than bread wheat under salinity stress. The high tolerance of bread wheat compared to durum wheat was found to be related with K^+/Na^+ discrimination (Gorham 1990a) which is genetically controlled and the gene(s) are located on long arm of chromosome 4D of *Ae. squarrosa* (Gorham 1990b).

No significant progress has so far been made to transfer salt tolerance from *Aegilops* species to bread or durum wheat cultivars. However, synthetic hexaploids have been produced by crossing various accessions of *Ae. tauschii* (*Ae. squarrosa*) with durum wheat via bridge crosses technology (Schachtman et al. 1991). In this method, *T. turgidum* (2n = 4× = 28: AABB) was crossed with *Ae. tauschii* (2n = 2× = 14: DD) to produce F1 hybrid having 21 chromosomes (ABD) which are doubled using colchicine to produce hexaploid wheat with 42 chromosomes. Approximately 800 such synthetic hexaploids have been produced at CIMMYT of which about 95 have been studied for various characteristics including tolerance to abiotic stresses (Mujeeb-Kazi et al. 1996). Most of these synthetics possess K^+/Na^+ ratios in the range of 1–5. Those with K^+/Na^+ ratios above 4 could be particularly useful for transferring salt tolerance to sensitive *T. aestivum* cultivars. This material has been tested under saline field in various countries including Pakistan. However, till to-date, neither any of these synthetics or its derivatives, have officially or unofficially been released for commercial cultivation nor any one of them has reached to the farmer's field despite the fact that the material has proven drought tolerance as well (Trethowan et al. 2000).

2.4 *Practical Utilization of Salt Tolerance Available in Annual* Triticeae

Unlike perennial *Triticeae* examples of successful transfer of salt tolerance from annual *Triticeae* or *Aegilops* species are significantly low. The reason could be that salt tolerance in *Aegilops* species was reported for the first time in 1989 (Farooq et al. 1989). Hence the first successful attempt was also made by Farooq et al. (1990a, b, c) by transferring genes from *Ae. cylindrica*. Genes for salt tolerance were also transferred (Schachtum et al. 1991) from *Ae. tauschii*. The fate of this material is not known however, the material produced by Farooq et al. (1992) has reached the farmers field (Farooq and Azam 2001; Farooq 2004). This germplasm has its stress tolerance increased by many fold. It survived up to maturity at EC 25 dS m^{-1} under gravel culture, and between EC 15–20 dS m^{-1} under saline fields (Farooq et al. 1995). Wheat lines WL-1076 and WL-41 out yielded LU-26: the salt tolerant local check and one of the parents of these lines (Farooq et al. 1992). These lines require only three irrigations instead of six given to the commercial cultivars and half the recommended dose of both urea and phosphate fertilizers. This material is being used in national and international field trials (Hollington 1998) and has shown much better performance than most of the locally recommended and other known salt tolerant genotypes. It is being used by the farmers especially those residing in the areas beset with water shortage and by the resource poor farmers who cannot afford to purchase expensive fertilizers. It is anticipated that cultivation of such genotypes will reduce the import especially of phosphate fertilizer thus relieving the burden on the economy while less use of nitrogenous fertilizer will improve the environment (Farooq 2004). The lesser number of irrigations will help saving the precious commodity like water that can be used for some other purposes.

The material is also being cultivated in the southern Punjab, which is a cotton belt and farmers cannot vacate the fields before January, which is not a normal wheat sowing time. After having the stress tolerant material produced at NIAB, they are cultivating wheat inside standing cotton, which again is not a normal practice. During recent field trials, farmers in the southern Punjab have produce grain yield equal 3,300–4,000 kg ha^{-1} and demonstrated that (a) diversity in agriculture does play dividends and that (b) for sustainable agriculture; diversity must be created, collected, characterized, and utilized continuously. This is essential in order to meet the ever-changing demand of present day agriculture and unforeseen requirements of future. NIAB is the only institution in the country working on its indigenous program of creation of diversity for stress tolerant wheat germplasm through transferring gene(s) from the annual *Triticeae*.

3 Conclusion

In the *Triticeae*, there exists a huge genetic variability for salinity tolerance yet, a real salt tolerant wheat genotype has not been produced that can go to the farmer's field. There could be many factors responsible for this bottleneck. It could be the differences in type of salinity, climatic conditions and source of irrigation water, agricultural practices, and disease/pest incidence in different regions of the world, where each and every gene involved in controlling salt tolerance interacts with these factors differently. It is understood that interaction of all these factors with various salinity controlling genes would probably require different variety for different saline area. Therefore, the farmers will have to rely on extensive breeding efforts made over the years to produce many different varieties showing desired traits in regionally adapted crops. These varieties are in harmony with local environment and its ecology. Salt tolerant varieties have been, are being and will perhaps be produced in similar way till the time when GM varieties will come to the field and compete with these varieties.

References

Alonso LC, Kimber G (1980) A hybrid between diploid *Agropyron junceum* and *Triticum aestivum*. Cereal Res Commun 8: 355–358.

Alonso LC, Kimber G (1983) A study for genomic relationships in wheat based on telocentric chromosome pairing. Z. Pflanzensuchtung 90: 23–31.

Anonymous (1998) World agriculture towards 2010. FAO, Rome, 83.

Anonymous (2000) Projected water scarcity in 2025. International Water Management Institute (IWMI). www.iwmi.cgiar.org/home/wsmap.htm, Colombo, Sri Lanka

Anonymous (2005) Pakistan: country water resources assistance strategy. Water Economy Running Dry. Report No. 34081-PK. South Asia Region. Agriculture and Rural Development Unit of the World Bank.

Charpentier A (1992) Production of disomic addition lines and partial amphiploids of *Thinopyrum junceum* on wheat. C R Acad Sci Paris 315: 551–557.

Colmer TD, Flower TJ, Munns R (2006) Use of wild relatives to improve salt tolerance in wheat. J Exp Bot 57: 1059–1078.

Dewey DR (1960) Salt tolerance of twenty five strains of *Agropyron*. Agron J 52: 631–635.

Dewey DR (1962) Breeding crested wheat grasses for salt tolerance. Crop Sci 2: 403–407.

Dewey DR (1984) The genomic system of classification as a guide to inter-generic hybridization with the perennial *Triticeae*. In: Gustafson JP (ed) Gene manipulation in plant improvement. Plenum, New York.

Dvorak J, Ross K (1986) Expression of tolerance of Na^+, K^+, Mg^+ Cl^-, and SO_4 ions and sea water in the amphiploid of *Triticum aestivum* x *Elytrigia elongata*. Crop Sci 26: 658–660.

Dvorak J, Katheleen R, Mendlinger S (1985) Transfer of salt tolerance from *Elytrigia pontica* to wheat from amphiploid of an incomplete *Elytirigia* genome. Crop Sci 25: 306–309.

Dvorak J, Edge M, Ross K (1988) On the evolution of the adaptation of *Lophopyrum elongatum* to growth in saline environment. Proc Natl Acad Sci USA 85: 3805–3809.

Elzam OE, Epstein E (1969) Salt tolerance of two grass species differing in salt tolerance. I. Growth and salt contents at different salt concentrations. Agrochimica 13: 187–195.

FAO (2005) Agriculture 21: water use in agriculture. Food and Agriculture Organization of the United Nations, Rome.

Farooq S (1990) Salt tolerance potential of wild resources for crop improvement. Ph.D. thesis. University of the Punjab, Lahore, Pakistan.

Farooq S (2004) Salt tolerance in *Aegilops* species: a success story from research and production to large scale utilization of salt tolerant wheat. In: Taha FK, Ismail S, Jaradat A (eds) Prospect of saline agriculture in Arabian Peninsula. Amherst Scientific Publisher, Amherst, MA.

Farooq S, Azam F (2001) Production of low input and stress tolerant wheat germplasm through the use of biodiversity residing in the wild relatives. Hereditas 135: 211–215.

Farooq S, Azam F (2005) Salt tolerance in *Triticeae*. Czech J Genet Plant Breed 41: 252–262.

Farooq S, Aslam Z, Niazi MLK, Shah TM (1988) Salt tolerance potential of wild resources of tribe *Triticeae*-I. Screening of perennial genera. Pak J Sci Ind Res 31: 506–511.

Farooq S, Niazi MLK, Iqbal N, Shah TM (1989) Salt tolerance potential of wild resources of tribe *Triticeae*-II. Screening of species of the genus *Aegilops*. Plant Soil 119: 255–260.

Farooq S, Shah TM, Iqbal N (1990a) Variation in cross-ability among inter-generic hybrids of wheat and salt tolerant accessions of 3 *Aegilops* species. Cereal Res Commun 18: 335–338.

Farooq S, Iqba N, Shah TM (1990b) Inter-generic hybridization for wheat improvement-II. Utilization of *Ph1b* mutant for direct alien introgression into cultivated wheat and production of backcross seeds. Cereal Res Commun 18: 21–26.

Farooq S, Iqbal N, Shah TM (1990c) Inter-generic hybridization for wheat improvement-III. Genetic variation in *Triticum* species affecting homoeologous chromosomes pairing. Cereal Res Commun 18: 233–237.

Farooq S, Iqbal N, Asghar M, Shah TM (1992) Intergeneric hybridization for wheat improvement-VI. Production of salt tolerant wheat germplasm through crossing wheat (*Triticum aestivum* L.) with *Aegilops cylindrica* and its significance in practical agriculture. J Genet Breed 46: 125–132.

Farooq S, Shah TM, Asghar M (1993) Inter-generic hybridization for wheat improvement-VII. Transfer in hexaploid wheat of salt tolerance gene(s) from *Thinopyrum scirpeum*. J Genet Breed 47: 191–198.

Farooq S, Asghar M, Iqbal N, Askari E, Arif M, Shah TM (1995) Production and evaluation of salt tolerant wheat germplasm produced through crossing wheat (*Triticum aestivum* L.) with *Aegilops cylindrica*-II. Field evaluation of salt tolerant germplasm. Cereal Res Commun 23: 275–282.

Flowers TJ (2004) Improving crop salt tolerance. J Exp Bot 55: 307–319.

Forster BP, Miller TE (1985) A hybrid between diploid *Agropyron junceum* and *Triticum aestivum*. Cereal Res Commun 8: 355–358.

Forster BP, Gorham J, Miller TE (1987) Salt tolerance of an amphiploid between *Triticum aestivum* and *Agropyron junceum*. Plant Breed 98: 1–8.

Forster BP, Miller TE, Law CN (1988) Salt tolerance of two wheat-*Agropyron junceum* disomic additions lines. Genome 30: 559–564.

Gorham J (1990a) Salt tolerance in the *Triticeae*: K^+/Na^+ discrimination in *Aegilops* species. J Exp Bot 41: 615–621.

Gorham J (1990b) Salt tolerance in the Triticeae: K^+/Na^+ discrimination in synthetic hexaploid wheats. J Exp Bot 41: 623–627.

Gorham J, McDonnell E, Wyn Jones RG (1984) Salt tolerance in the *Triticeae: Lymus sabulosus*. J Exp Bot 35: 1200–1209.

Gorham J, Wyn Jones RG, McDonnel E, Wyn Jones RG (1985) Salt tolerance in the *Triticeae*: Growth and solute accumulation in leaves of *Thinopyrum bessarabicum*. J Exp Bot 36: 1021–1031.

Gorham J, Forster BP, Budrewicz E, Wyn Jones RG, Miller TE, Law CN (1986) Salt tolerance in the *Triticeae*: solute accumulation and distribution in an amphiploid derived from *Triticum aestivum* cv. Chinese Spring and *Thinopyrum bessarabicum*. J Exp Bot 37: 1435–1449.

Gorham J, Wyn Jones RG, Bristol M (1990) Partial characterization of the trait for enhanced K^+-Na^+ discrimination in the D genome of wheat. Planta 180: 249–268.

Guy C, Porat R, Hurry V (2006) Plant cold and abiotic stress gets hot. Physiol Plant 126: 1–4.

El-Hendawy SE, Yuncai Hu, Gamal M, Yakout, Ahmed M, Awad, Salah Hafiz E Schmidhalter U (2005) Evaluating salt tolerance of wheat genotypes using multiple parameters. Eur J Agron 22: 243–253.

Hitz WD, Hanson AD (1980) Determination of glycine betaine by pyrolysis-gas chromatography in cereals and grasses. Phytochemistry 19: 2371–2374.

Hollington PA (1998) Technological breakthrough in screening/breeding wheat varieties for salt tolerance. National conference on salinity management in agriculture. CSSI, Karnal, India.

Hunt OJ (1965) Salt tolerance in intermediate wheat grasses. Crop Sci 5: 407–409.

Kimber G, Zhao YH (1983) The D genome of the *Triticeae*. Can J Genet Cytol 25: 589–581.

King IP, Law CN, Cant KA, Orford SE, Reader SM, Miller TE (1997) *Tritipyrum*: a potential new salt tolerant cereal. Plant Breed 116: 127–132.

Mass EV, Poss JA (1989) Salt sensitivity of wheat at various growth stages. Irrigation Sci 10: 29–40.

McGuire PE, Dvorak J (1981) High salt tolerance potential in wheat grasses. Crop Sci 21: 702–705.

Moxley MG, Berg WA, Barrau EM (1978) Salt tolerance of five varieties of wheat grasses during seedling growth. J Range Manage 31: 54–55.

Mujeeb-Kazi A, Roldan S, Suh DY, Stich LA, Farooq S (1987) Production and cytogenetic analysis of hybrids of *Triticum aestivum* and some caespetose *Agropyron* species. Genome 29: 537–553.

Mujeeb-Kazi A, Rolden S, Suh DY, Kulie NT, Farooq S (1989) Production and cytogenetics of *Tricticum aestivum* L. hybrids with some rhizomatous *Agropyron* species. Theor Appl Genet 77: 162–168.

Mujeeb-Kazi A, Rosas V, Roadan S (1996) Conservation of the genetic variation of *Triticum tauschii* (Coss.) Schmalh (*Aegilops squarrosa* auct. Non L) in synthetic hexaploid wheats (*T. turgidum* L. s. lat. × *T. tauschii*; 2n=6x=42, AABBDD) and its potential utilization for wheat improvement. Genet Res Crop Evol 43: 129–134.

Munns R (2005) Genes and salt tolerance: bringing them together. New Phytol 167: 645–663.

Munns R, Hare RA, James RA, Rebetzke GJ (2000) Genetic variation for improving the salt tolerance of durum wheat. Aust J Agric Res 51: 69–74.

Munns R, James RA, Lauchli A (2006) Approaches to increasing salt tolerance of wheat and other cereals. J Exp Bot 57: 1025–1043.

Noori SAS, McNeilly T (2000) Assessment of variability in salt tolerance based on seedling growth in *Triticum durum* Desf. Genet Res Crop Evol 47: 285–291.

Omielan JA, Epstein E, Dvorak J (1991) Salt tolerance and ionic relationship of wheat as affected by individual chromosomes of salt-tolerant *Lophopyrum elongatum*. Genome 34: 961–974.

Rajaram S (2001) Prospects and promise of wheat breeding in 21st century. Euphytica 119: 3–15.

Rengasamy P (2006) World salinization with emphasis on Australia. J Exp Bot 57: 1017–1023.

Sairam RK, Tyagi A (2004) Physiology and molecular biology of salinity stress tolerance in plants. Curr Sci 86: 407–421.

Schachtman DP, Munns R, Whitecross MI (1991) Variation in sodium exclusion and salt tolerance in *Triticum tauschii*. Crop Sci 31: 992–997.

Shah SH, Gorham J, Forster BF, Wyb Jones GR (1987) Salt tolerance in Triticeae - the contribution of D genome to cation selectivity in hexaploid wheat. J Exp Bot 38: 254–269.

Shannon MC (1978) Testing salt tolerance variability among tall wheat grass lines. Agron J 70: 719–722.

Singh S, Singh M (2000) Genotypic basis response to salinity stress in some crosses of spring wheat *Triticum aestivum* L. Euphytica 115: 209–214.

Storey R, Gorham RD, Shepherd KW (1985) Modification of the salinity response of wheat by the genome of *Elytrigia elongate*. Plant Soil 83: 327–330.

Swaminathan MS (2004) Stocktake on cropping and crop science for a diverse planet. 4th International Crop Science Conference, September 26 to October 1 2004. Brisbane, Australia.

Trethowan R, Van Ginkle M, Mujeeb-Kazi A (2000) Performance of advanced bread wheat × synthetic hexaploid derivatives under reduced irrigation. Ann Wheat Newslett 46: 87–88.

Wang W, Vinocur B, Altma A (2003a) Plant responses to drought, salinity and extreme temperatures: towards genetic engineering for stress tolerance. Planta 218: 1–14.

Wang RC, Li XM, Hu ZM, Zhang JY, Larson SR, Zhang SY, Grieve CM, Shannon. MC (2003b) Development of salinity tolerant what recombinant lines from a wheat disomic addition line carrying a *Thinopyrum junceum* chromosome Int J Plant Sci 164: 25–33.

Wilson C, Read JJ, Abo-Kassem E (2002) Effect of mixed salt salinity of growth and ion relations of a Quinoa and a wheat variety. J Plant Nutr 25: 2689–2704.

Zhong GY, Dvorák J (1995) Chromosomal control of the tolerance of gradually and suddenly imposed salt stress in the *Lophopyrum elongatum* and wheat, *Triticum aestivum* L., genomes. Theor Appl Genet 90: 229–236.

Chapter 8
Water Loss and Gene Expression of Rice (*Oryza sativa* L.) Plants Under Dehydration

T.-R. Kwon, J.-O. Lee, S.-K. Lee, and S.-C. Park

Abstract This study aims to determine physiological and molecular alterations as exposed to dehydration stress in rice plant. Rice seedlings were grown in a nutrient solution within a managed environment chamber prior to the imposition of the dehydration stress. Dehydration was imposed through uprooting and exposing to controlled environment condition (25°C, RH 50%, and 290 PAR). Water loss of intact plant was determined by continuous weightings with every minute interval till 300 minutes after starting the imposition of dehydration. The imposition of dehydration caused significant loss of internal water, resulting in 44% out of initial water content at 300 minutes-long dehydration. The dehydration imposition also reduced the rate of water loss per minute per gram dry weight from 14.2 to 2.1 mg min^{-1} g dry $weight^{-1}$. These results indicate that the dehydration imposition could causes osmotic stress due to water loss in tissue. The dehydration stress also reduced significantly relative water content and osmotic potentials over the time. The dehydration stress induced the mRNA expression of drought-induced protein (Dip1), drought-induced hydrophobic protein (DRR2) and mitogen-activated protein kinase (MAPK). MAPKs were mostly expressed before 20% water loss out of the initial water content. However, Dip1 and DRR2 were strongly expressed after 20–40 water loss out of the initial water content. These results indicate that the physiological parameters such as water loss rate and water status can be used a physiological scale, especially to explore the genes related to simple response and/or tolerance against internal water deficit.

Keywords Dehydration • geneexpression • osmotic stress • rice • water loss

1 Introduction

Water scarcity is a major constrain in rice production in the world. One-half of the rice production area is in the rainfed environment causing the severe reduction of productivity in rice plants (Garrity and O'Toole 1994; Ouk et al. 2006). So far, great deals of molecular works have been done to understand the rice plant's tolerance and/or responses under water deficit environments. Expression profiling of genes is an essential step to identify candidate genes responding to or tolerating against water deficit condition.

For the exploration of the genes, plants are usually exposed to dehydration condition resulting from limitation of water uptake and increment of water loss. There are a few ways to impose dehydration stress to explore gene expression of plants. First, whole plants or detached organs are transferred directly to an air space in a certain temperature, humidity and light (Chen et al. 2005; Huang et al. 2008; Campbell et al. 2001). Second, plant's roots are exposed to hypertonic solution such as polyethylene glycole (PEG) 6000 (Zhao et al. 2007; Zhang et al. 2007). Third, watering is withheld in the soil (Rodriguez et al. 2006). First two stress treatments may cause acute dehydration in plants while third one is a way to induce slower water loss. There were certain differences among these stress-imposing methods but all of them induce water loss from plants together with morphological, physiological and/or biochemical alterations caused by the water-deficit status. This study focuses on the physiological and molecular changes of rice seedling plants in an acute dehydration

T.-R. Kwon (✉), J.-O. Lee, S.-K. Lee, and S.-C. Park
The National Institute of Agricultural Biotechnology/RDA, 225 Seodun-dong, Suwon 441–707, Korea
e-mail: trkwon@rda.go.kr

M. Ashraf et al. (eds.), *Salinity and Water Stress*,

imposition via uprooting the plants from nutrient culture solution and leaving them in a controlled temperature, humidity and light.

The dehydration imposition through uprooting of plants from soil or water culture system has been used to explore responsive/tolerant genes to the stress imposition (Campbell et al. 2001; Chen et al. 2005; Huang et al. 2008) because of fast and easiness to achieve the target stress. Verslues et al. (2006) postulated an experimental technique to determine leaf water loss via changes of fresh weight in detached leaf over time. Plants can response to the acute dehydration stress presumably with complicated alterations in internal physiological and molecular processes primarily due to remarkable disturbances of water relation. Dehydration-induced response of plants depend on the discharging amount of internal water, water loss rate per unit time, and exposing time period to the stress condition (Bray 1997). Field drought environment causes multiple physiological changes in rice plants including loss of internal water, closure of stomata (Dingkuhn et al. 1999; Cabuslay et al. 2002), reduction of tissue water potential (Jongdee et al. 2002), induction of osmotic adjustment and changes of leaf morphology (Turner et al. 1986), leaf rolling (Dingkuhn et al. 1999), accumulation of abscisic acid (ABA) (Dingkuhn et al. 1999), and reduction of photosynthetic ability (Turner et al. 1986). These multiple physiological alterations must be used in dehydration-induced expression profiling of relevant genes.

Plant responses to dehydration should be understood at the molecular levels such as the process of stress recognition, the transduction of signals, and the regulation of gene expression (Bray 1997). A plant cell recognizes water deficit condition first and then switches the recognized physical stress into biochemical responses. These biochemical responses works as cellular signal transduction to trigger the expression of specific genes. Dehydration induces the complex cascades of gene expression, working for stress tolerance as well as only for stress response (Shinozaki et al. 2003). Abscisic acid (ABA) is a water-deficit induced signal, which induces MAP kinase (mitogen-activated protein kinase) under dehydration condition (Knetsch et al. 1996; Mizoguchi et al. 1996). An integrated approach from cellular to whole plant should be applied to identify target genes working for the dehydration (Verslues et al. 2006; Bray 1997; Verslues et al. 2006).

It is well known that plants close their stomata upon the imposition of dehydration stress to prevent water loss. However, one of the most significant physiological phenomenons is water loss especially as plants exposed directly to dehydration condition. So, precise determinations of water loss and water status are important to understand direct mechanical or physical components of target stress. Liang and Sun (2002) reported that degree of dehydration is associated with desiccation tolerance of isolated cocoa (*Theobroma cacao*) and ginkgo (*Ginkgo biloba*) embryonic tissues. Rice plants have a well-developed vascular system for a dynamic water relation as grown in an optimal condition. Under dehydration stress, whole rice plants can lose significantly internal water upon the stress imposition. Rapid dehydration-induced water loss may trigger physiological changes and expression of responsive genes over time in rice plants. So, this study aims to link between water loss and gene expression of rice seedling plants when they are under dehydration condition.

2 Materials and Methods

2.1 Plant Growth

Eight cultivars of rice were used for this study. However, a Korean popular cultivar, 'Dongjin', used to determine parameters of water relation and gene expression under dehydration. Dr. E.L. Javier, International Rice Research Institute, Lagua, Philippines, kindly provided seven rice cultivars, 'IR55419-04', 'IR 55423-01 (APO)', 'Vandana', 'IR 74371-54-1-1', 'IR 64', 'IR 72', and 'N22', differing in drought tolerance in the field (Atlin et al. 2004). They were used to find relationship between initial water content just prior to dehydration imposition and dehydration-induced alteration of water loss. Seeds were germinated in distilled water at 30°C. At 10 days after seed germination, infant young seedlings with two young leaves were transplanted in a water culture system containing Yoshida's nutrient solution. young seedlings were kept in a plant growth chamber for 10 days setting at 28°C ± 1°C and 290 PAR during 13 hours day time and 22°C ± 1°C during night time. Relative humidity was 60% ± 10% all the day.

2.2 Dehydration Treatment

At the growth stage of five or six foliar leaves, seedlings were uprooted from the water culture solution. Wet roots were blot with paper to remove surface water on root tissue. Then, in order to impose dehydration stress, uprooted and blot-dried whole plants were placed in a controlled environment chamber, such as 25°C ± 1°C, RH 50% ± 5% and 290 PAR. After starting the imposition of dehydration stress, a series of samplings were taken for concurrent physiological and molecular studies over time.

2.3 Determination of Water Loss, Relative Water Content and Osmotic Potentials

Intact whole plants were continuously weighted every minute from 0 to 300 minutes after starting the dehydration stress on a chemical balance in a controlled environment chamber. Changes of plant's fresh weight were logged on a spreadsheet using a communication software and interface (RS-232). Downloaded data on the spreadsheet were used to determine the water loss parameters of the whole plants such as (1) changes of fresh weight, (2) cumulative amounts of water loss, (3) percentage of water loss and (4) rates of water loss.

1. Changes of fresh weigh (ΔFW)

$$\Delta FW = FW_{t0} - FW_{t1}$$

where FW_{t0} is the fresh weight of whole plants at a time and FW_{t1} is the fresh weight of the same whole plants at a minute later than the FW_{t0}.

2. Cumulative amounts of water loss ($\Sigma\,\Delta FW$)

$$\Sigma\Delta FW = \Delta FW_{0-1} + \Delta FW_{1-2} + \Delta FW_{2-3} + \ldots\ldots + \Delta FW_{(n-1)-n}$$

where ΔFW_{0-1} is the reduction of fresh weight between FW_{t0} and FW_{t1}; $\Delta FW_{(n-1)-n}$ is the reduction of fresh weight between $FW_{t(n-1)}$ and FW_{tn}

3. Percentage of water loss (%WL)

$$\%WL = \{\Sigma\, DFW/(FW-DW)\}*100$$

where FW is the fresh weight of plants at the starting the imposition of dehydration and DW is the dry weight of the same plants dried at 85°C for 3 days.

4. Rates of water loss (g g DW^{-1} $minute^{-1}$)

$$\text{Rates of water loss} = \Delta FW/DW/t0{-}t1$$

While doing the determination of water loss, another set of plants were sampled to measure relative water content and osmotic potentials of expressed cell sap. Relative water content (RWC, %) was determined using a formula, $RWC = \{(FW-DW)/(FW-TW)\} \times 100$, where FW, fresh weight of dehydrated plants; TW, weight of plants in full turgid achieved by putting dehydrated plants in distilled water at 100% humidity for 8 hours, and DW, weights of plants after drying at 85°C for 3 days. Upper parts of plants were separated from roots. Immediately after then, the upper parts were wrapped with a peace of parafilm and then aluminum foil. The wrapped tissues were put into liquid nitrogen to freeze. The freeze tissues were thawed at room temperature prior to the expression of tissue sap. The tissue saps of dehydrated and full-turgid plants were expressed at 1,200 g relative centrifugal forces for 20 minutes at 4°C. Osmotic potentials of the expressed tissue sap were determined using a vapor pressure osmometer (5520, Wescor Inc., Utah, USA).

2.4 Expression of Dehydration-Responsive Genes

Preparation of total RNA. For Northern blot analysis, total RNA samples were isolated from upper part (leaf and stem) and lower part (root) of seedling plants (cvs. 'Dongjin', 'IR64' and 'Vandana') using the method described by Sambrook et al. (2001).

Northern blotting. For Northern blot analysis, total RNA samples (15 µg each) were separated in 1.2% formaldehyde agarose gel, and transferred to Hybond-N^{+} membranes (Amersham Biosciences). RNA blot was hybridized by the methods described by Sambrook et al. (2001).

Probes for Northern blot hybridization. Several double-stranded 32P-labeled DNA probes were used to detect each specific mRNA. The 547 bp DRR2 (Accession NO. AY554051) probe used in this study was amplified by polymerase chain reaction with an appropriated set of primer (DRR2-F: 5′-tactttactttgcagctattt-3′,

DRR2-R: 5′-attcacacgaaagcaacaga-3′). The others used about 700 bp Dip1 (Accession NO. AY587109) digested *Pst* I, 760 bp MAPK s20051 digested *Sac* I, and 445 bp MAPK s20028 digested *Xba* I and *EcoR* V, cDNA fragments. Each ^{32}P-labeled probe was prepared by the Ladderman™ Labeling Kit (TAKARA BIO INC, Japan).

3 Results

3.1 Changes of Fresh Weight and Water Loss Induced by Dehydration

The dehydration imposition for 300 minutes caused significant reduction of fresh weight with remarked water loss in the intact rice plants tested (Fig. 8.1). The dehydration-induced reduction of fresh weight was 40% as compared with the fresh weight of plants before starting the stress imposition. Water loss was a primary cause the dehydration-induced reduction of fresh weight. The amount of water loss by the dehydration was 45% out of the initial water content in the rice plants before starting dehydration. The rate of water loss, as expressed the amount of water loss per unit gram dry weight per unit time, made an inverse polynomial curve. The plants lost internal water with two distinguished phases, first plunged loss and later stable low loss. The plunged water loss occurred during the first 100 minutes after starting dehydration. After then, the water loss rates remain stably low.

Upon starting dehydration imposition, the intact rice plants altered physiologically as well with the reduction of relative water content and osmotic potentials

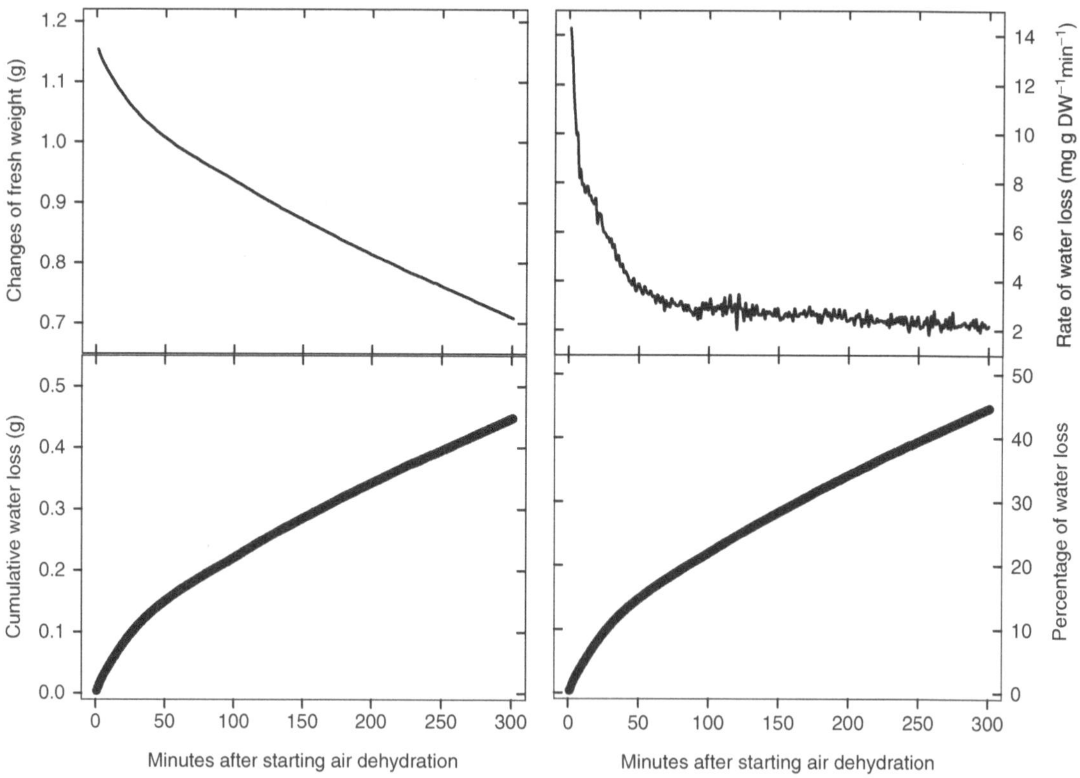

Fig. 8.1 Dehydration-induced alterations of the water relation in the seedling plants of cv. 'Dongjin' with 4 to 5 foliar leaves including the changes of fresh weight (A), rates of water loss as expressed in the amount of water loss per gram dry weight per minute (B), the amount of cumulative water loss (C) and the percentage of water loss (D), as the plants being exposed to controlled dehydration over 300 minutes (25°C, 50% humidity, 290 PAR). These curves on the above graph were obtained from the mean of 10 replicates. Each replicate had six individual plants (total n = 60)

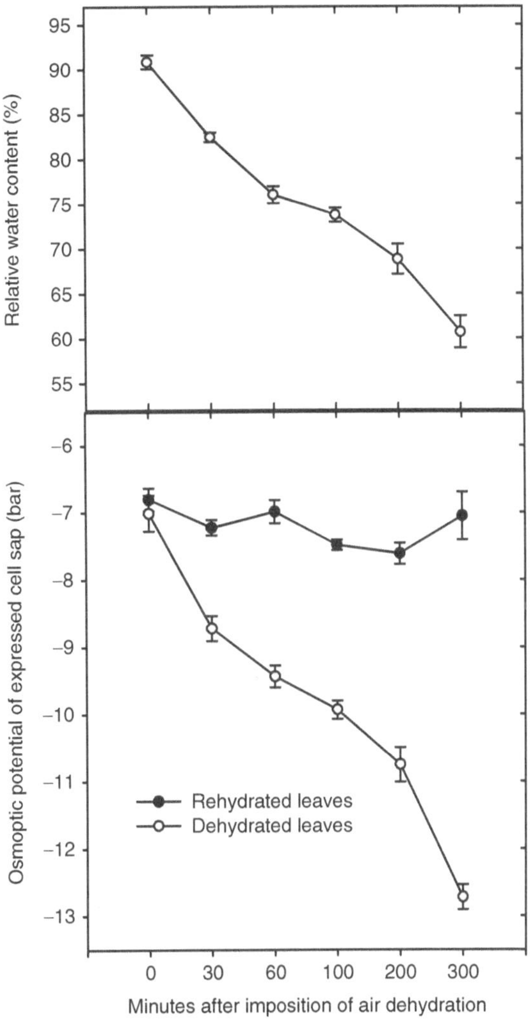

Fig. 8.2 Concurrent determinations of the relative water content of whole plants (A) and osmotic potentials of expressed sap from dehydrated or rehydrated tissue (B) when the seedling plants of cv. 'Dongjin' with 4 to 5 foliar leaves were exposed to controlled dehydration condition (25°C, 50% humidity, 290 PAR) at 0, 30, 60, 100, 200 and 300 minutes

(Fig. 8.2). The plants under dehydration reduced from 92% relative water content at the initial time of the treatment 60% at the 300 minutes after starting the treatment. Dehydration over 300 minutes reduced the osmotic potentials of cell sap expressed from leaves from −7.15 bar to −12.80 bar. However, rehydration of dehydrated leaves recovered the dehydration-induced decrease of osmotic potential to the level of osmotic potential at initial time of dehydration.

3.2 Influence of Initial Water Content on Water Relation under Dehydration

Supplement of 200 mM mannitol to growth medium caused significant changes of water loss rate, relative water content and osmotic potentials of rice plants (Fig. 8.3). Plants were treated with or without 200 mM

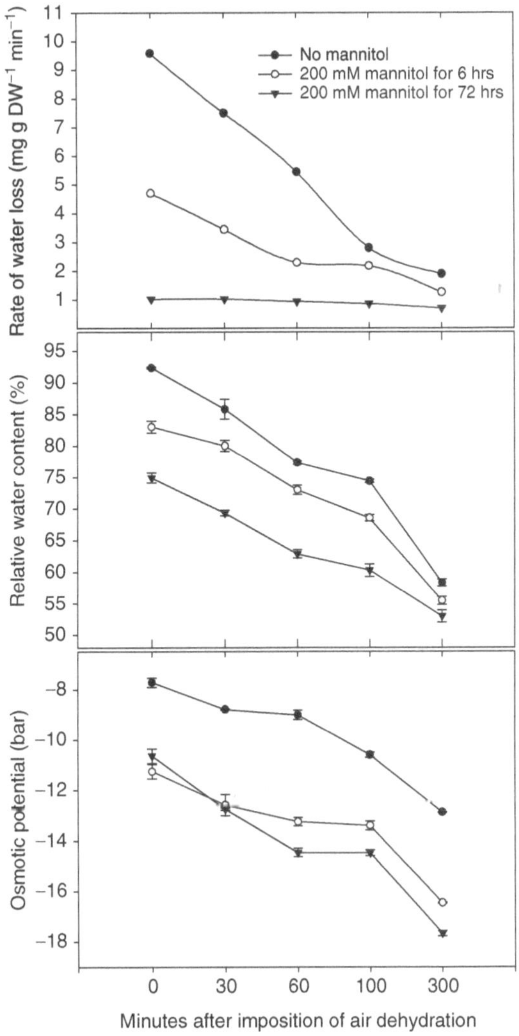

Fig. 8.3 Rate of water loss (A), relative water content (B) and osmotic potential of expressed tissue's sap (C) of the seedling plants under controlled dehydration condition (25°C, 50% humidity, 290 PAR) at 0, 30, 60, 100 and 300 minutes. Cv. 'Dongjin' with 4 to 5 foliar leaves were pretreated with 200 mm mannitol for 6 and 72 hours prior to determine these parameters. Each mark represents the mean of three replicate (n = 18). Vertical bar is standard error

mannitol prior to dehydration imposition to determine water loss rate, relative water content and osmotic potential. Over 300 minute-long dehydration, there were also significant difference between plants without previous mannitol treatment and plants with the precious mannitol treatment for 6 and 72 hours. The mannitol-treated plants for 72 hours showed lowest rate of water loss over the 300 minutes without remarkable changes. Six hours treatment of 200 mM mannitol also reduced water loss rate but the late dehydration treatment caused severe reduction of water loss rate over 300 minutes. The late dehydration imposition caused further reduction of relative water content over 300 with maintaining parallel difference caused by prior reduction by the mannitol treatment. Previous treatment of 200 mM mannitol for 6 and 72 hours caused significant reduction of osmotic potential. This reduction was maintained even during late dehydration imposition over 300 minutes.

Initial water content of eight rice genotypes was highly correlated with total water loss (r^2 = 0.94), water loss rate (r^2 = 0.81) and remained water content during 300 minute dehydration (Fig. 8.4). Initial water contents were obtained from the difference between fresh weight before dehydration treatment and dry weight after drying at 85°C for 3 days. Remained water content was determined by the deduction of the cumulative amount of water loss from initial water content. Genotypes with greater initial water content showed greater water loss rate and amount but maintained more water even after 300 minute dehydration.

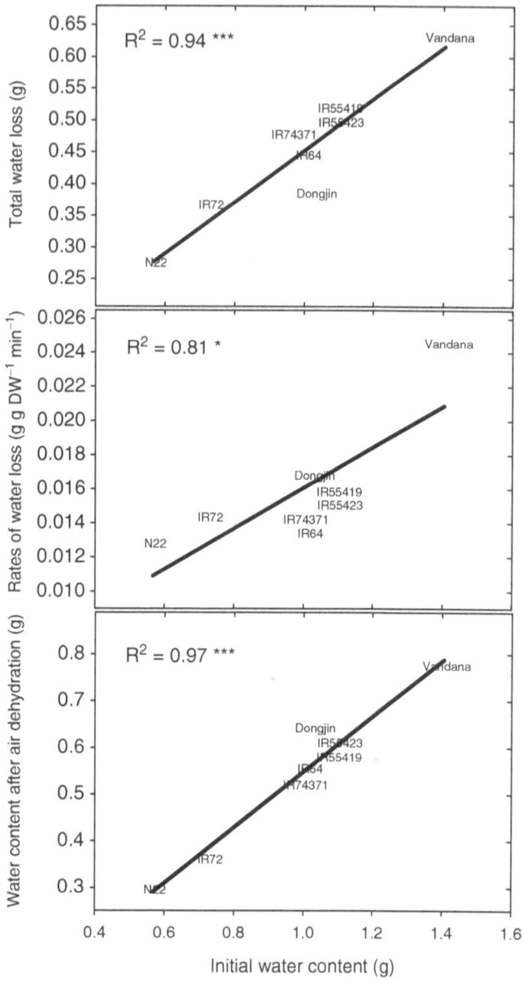

Fig. 8.4 Relationship of initial water content with either total amount of water loss (A), rate of water loss (B) or remained water content at the end of the stress (C) in eight rice genotypes, differing in drought tolerance in the field, under controlled dehydration condition (25°C, 50% humidity, 290 PAR) over 300 minutes

3.3 Time-Course Expression of Genes

The expression level of each gene in the rice seedlings was examined. With RNA samples from rice young seedlings, northern blot hybridization analyses were carried out with appropriate probes for dehydration. We investigated time-course mRNA expression level of the drought-induced genes under dehydration and 200 mM mannitol. After imposing dehydration, mRNA levels of the genes such as Dip and DRR2 increased proportionally with increment of dehydration time (Fig. 8.5). It showed that the Dip1 and DRR2 were rapidly increased its transcription level after 200 minutes and reached their maximum expression level till 300 minutes. These dehydration-induced expressions of those genes were maintained till 500 minutes after starting the dehydration. In contrast, MAPK s20028 were expressed slightly at early stage of dehydration imposition. The MAPK s20028 completed it expression till 100 minutes-long dehydration. The dehydration stress did not trigger the expression of MAPK s20051. Also, addition of 200 mM mannitol to nutrient solution also induced clear expression of the Dip1 and DRR2 gene (Fig. 8.6). Significant expressions of their transcription levels were found within 1 hour and

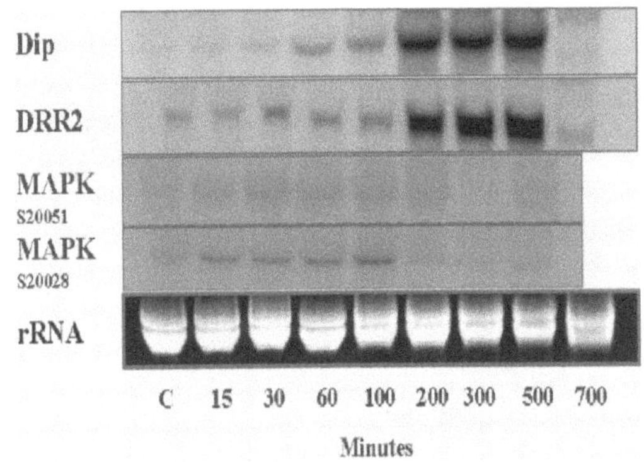

Fig. 8.5 Northern blot analysis for the mRNA expression of the genes (Dip, DRR2, MAPK s20051 and MAPK s20028) induced by controlled dehydration condition (25°C, 50% humidity, 290 PAR) over 700 minutes in cv. 'Dongjin'. RNA of 15 µg per sample was denaturated in formamide-dormaldehyde and separated on 1.2% formaldehyde agarose gel. RNA gel was stained with EtBr

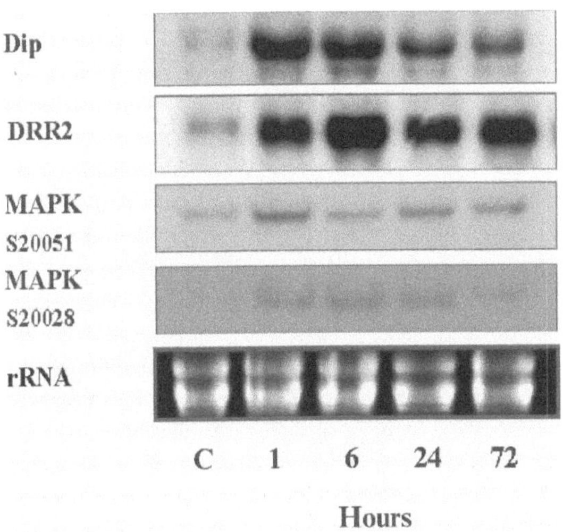

Fig. 8.6 Northern blot analysis for the mRNA expression of the genes (Dip, DRR2, MAPK s20051 and MAPK s20028) induced by 200 mm mannitol addition to Yoshida's nutrient solution for 1, 6, 24 and 72 hours in cv. 'Dongjin'. RNA of 15 µg per sample was denaturated in formamide-dormaldehyde and separated on 1.2% formaldehyde agarose gel. RNA gel was stained with EtBr

reached their maximum level within 6 hour. The high osmolarity treatment with 200 mM mannitol caused slight expression of MAPK s20051 but less response of MAPK s20028.

3.4 Water Relation and Gene Expression Related to Drought Tolerance

Two genotypes, cvs. 'Vandana' and 'IR64', compared each other in changes of water loss rate and osmotic potential under dehydration (Fig. 8.7). Cv. 'Vandana' is known as strong drought tolerant genotype in the field while cv. 'IR64' is known as moderately tolerant one. Cv. 'Vandana' showed greater water loss rate than cv. 'IR 64' did, especially, during the first 60 minutes of dehydration. However, cv. 'Vandana' maintained much greater internal osmotic potential than cv. 'IR64' did over the dehydration treatment time.

Two genotypes, cv. 'Vandana' (strong drought tolerant) and cv. 'IR64' (moderate drought tolerant), showed clear differences in the dehydration-induced expression of MAPK s20028 and MAPK s20051 (Fig. 8.8). The expression of two MAPK genes were observed in upper parts (shoots) as well as lower part

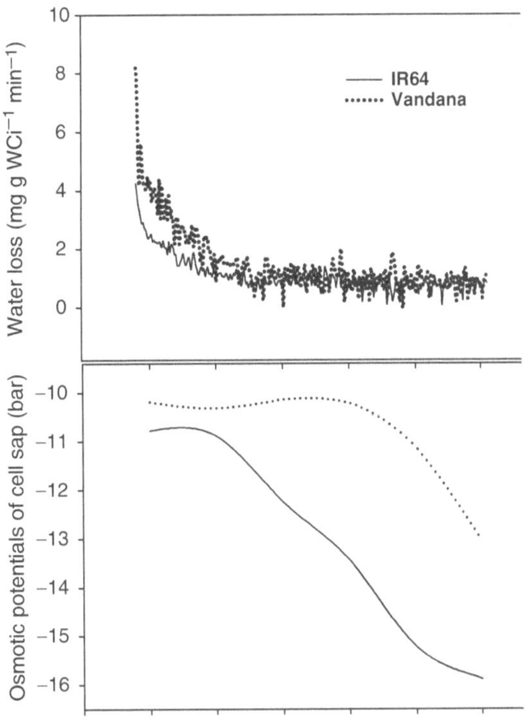

Fig. 8.7 Rate of water loss (A) and osmotic potential of expressed tissue's sap (C) of cvs. 'Vandana' (greater drought-tolerant) and 'IR64' (less drought tolerant) under controlled dehydration condition (25°C, 50% humidity, 290 PAR) over 300 minutes

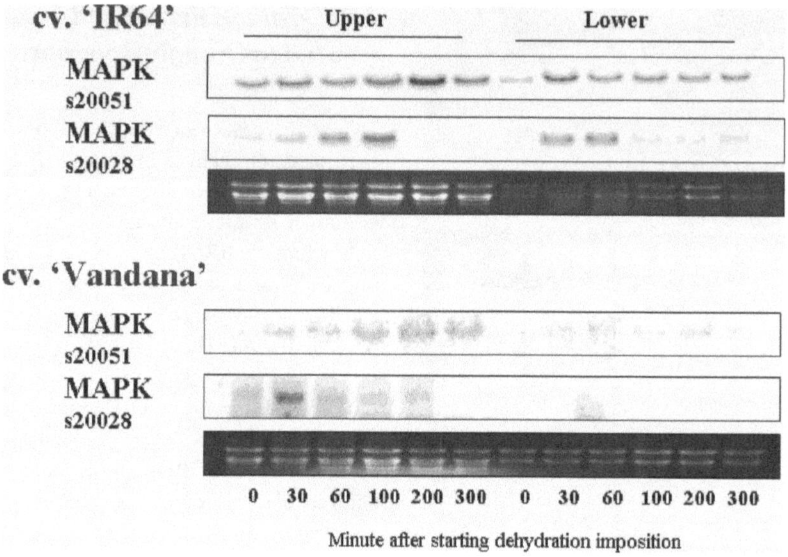

Fig. 8.8 Northern blot analysis for the mRNA expression of the genes (MAPK s20051 and MAPK s20028) induced by controlled dehydration condition (25°C, 50% humidity, 290 PAR) over 300 minutes in the upper and lower parts of cv. 'Dongjin' RNA of 15 μg per sample was denaturated in formamide-dormaldehyde and separated on 1.2% formaldehyde agarose gel. RNA gel was stained with EtBr

(roots) of rice plants with the time course of dehydration (0, 30, 60, 100, 200, 300 minutes). In cv. 'IR64', MAPK s20028 worked at early (30 minute after dehydration) in lower part and increased its transcription level within 60 minute. Its maximum expression level was reached within 100 minute. The MAPK s20051 was expressed continuously in lower and upper parts by the dehydration over time. In contrast, cv. 'Vandana' did not express both MAPKs in the dehydration treatment, except a slight up of transcriptional level of MAPK s20028 at 30 minute-long dehydration.

4 Discussion

Dehydration forces rice plants to loose internal water. As mentioned by Turner et al. (1986) and Bray (1997), the dehydration can cause multiple responses depending on the amount and rate of water loss and exposing time to the water deficit condition. When intact whole plants uprooted from water culture medium, they firstly confront to direct-physical water deficit condition due to transpiration via stomata. This direct dehydration causes acute stress to plants including immediate water loss. This dehydration-induced water loss accompanies with quantitative and qualitative changes of other physiological responses. Water deficit in plant tissues leads physical disintegration and instability of cell membrane, a concentration of solute, reduction of water potential due to increased negative osmotic potential, and denaturation of protein (Bray 1997). In this study, the results showed a sharp reduction of fresh weight of the intact plants upon uprooting from the water culture system. A continuous weighting every 1 minute for 300 minutes can make a curve expressing the changes of fresh weight. The low data for the 300 minute-long changes of fresh weight is set of element values to obtain cumulative water loss as well as water loss rate, which is defined as water loss per unit time interval (1 minute). Verslues et al. (2006) also displayed an experimental technique to determine leaf water loss via continuous weighing of detached leaves rather than intact plant over time. This technique may have a flaw in the profiling of gene expression. Cutting leaf for the gene expression can also trigger genes working for wound rather than solely dehydration.

After starting dehydration, rice plants plunged their water loss rate till the first 100 minute and then maintained low profile till 300 minute. So, dehydration-induced water loss made an identical inverse polynomial curve over time. The first sharp reduction may occur by two reasons. First, plant transpire internal water especially in apoplastic vascular spaces. Imposition of dehydration causes a first loss of free water in the intracellular apoplastic spaces in tissues (Bray 1997). Turner et al. (1986) estimated that rice plants have about 18% apoplastic portion out of total volume at full tugor status. After 100 minute after starting dehydration in this study, the rice plants lost more or less 20% water, agreeing with the apoplastic free water content. Hoekstra et al. (2001) postulated a drought tolerance mechanism working at 20% water loss. Cells have no free bulky water. This apoplastic water can emit to air via open stomata. Verslues et al. (2006) also support the usefulness of determination of water loss rate to understand stomatal responses as well as ABA-governed mechanism under dehydration condition. Dehydration-induced water loss is largely regulated by stomatal closure. Rice plants start to close partially their stomata below a soil matrix potential of –5 bar and then close them completely at –20 bar soil matrix potential or leaf water potential in the early morning (Dingkuhn et al. 1999). Second, plants close stomata in response to the dehydration. Dehydration-induced close of stomata decrease water loss rate per unit time probably in order to prevent further severe water loss. Plants can maintain tissue water content via limiting water loss, mainly by stomata closure in acute dehydration (Sharp and Le Noble 2002; Verslues et al. 2006). After closing most stomata, plants can maintained low level water loss rate but still loose certain amount of internal water because of partially small space opening of stomata and direct emit from epidermal cell contacting to air. In the first reduction, rice plants may express genes related to stomatal regulation and a signal transduction. However, after stomatal closure by dehydration, plants may change membrane integration due to the further water loss. So, this second phase of water loss may trigger genes related to membrane integration.

Dehydration-induced water loss results in concurrent reduction of internal water content resulting in decrease of osmotic potentials of expressed tissue sap. The 300 minute imposition of dehydration reduced the relative water content of rice plants under the controlled environment described in material and method section from 92% to 60%. Hoekstra et al. (2001) reported that critical level of water status causing water deficit differ among plant species. Generally, plants have a critical water status causing no bulk cytoplasmic water present. This critical point can be reached more or less 23% water content on a fresh weight basis. Rice plants reached to minimal stomatal conductance at the 20 bar soil matrix potential (Dingkuhn et al. 1999) and to a permanent wilting point at the volumetric soil water content of 15 bar (Kato et al. 2007). The further water loss can cause irreversible membrane damage unless proper protection mechanism by compatible solute. In case of rice plants, the critical point reached at 100 minute after starting dehydration stress. Rice plants usually keep as much as high osmotic potential of more or less –7 to –8 bar. The direct dehydration for 300 minute reduced up to –12.8 bar. However, the dehydration-induced reductions of osmotic potential were reversed to the level of that of dehydration-free plants, indicating that the reduction of osmotic potential by dehydration was caused by 'concentration effect' rather than osmotic adjustment. These results indicate this acute direct dehydration may not associate with any active solute accumulation and did not force the treated plants to reach a permanent wilting. This direct dehydration imposition is not proper approach to explore the genes and traits related to osmotic adjustment.

Dehydration also triggers expression of genes related to signal transduction such as MAPK and downstream genes such as DRR2 and Dip. MAPK gene is known to respond drought stress. Water deficit condition induced the expression of MAPK in *Arabidopsis thaliana* (Mizoguchi et al. 1996). It was revealed that Abscisic acid (ABA) activates the MAPK in water deficit condition (Knetsch et al. 1996). DRR2 and Dip are known in the drought responses in plant species. Rate and amount of water loss could be a physiological scale in line with the expression profiling of drought responsive genes. Bray (1997) proposed that plants may response to water loss within a few seconds with an alteration of the phosphorylation status of protein.

Initial water content is a critical factor in the alteration of water status caused by dehydration. Water status before starting dehydration imposition was highly associated with rate and amount f water loss as well as water content during and after the 300 minute-long dehydration. Plants with greater water content prior to dehydration treatment can loss more internal water with higher rate of water loss but maintained greater water content after. In this study, eight genotypes, differing in drought tolerance, revealed the existence of certain diversity among them in initial water content. When plants confronted to 200 mM mannitol in growth medium, they had adverse water status even prior to the dehydration treatment by uprooting, resulting in the reduction of water loss rate, relative water content and osmotic potential. This mannitol-induced adverse water relation affected significantly the further change of water loss, relative water content and osmotic potential over 300 minute-long dehydration. So, these results indicate that initial water status must be counted in the studies on dehydration-induced physiological and molecular changes.

Drought-tolerant cultivar has not only higher initial water status but also greater water loss rate and osmotic potential under the dehydration imposition. Cv. 'Vandana', which was selected as a drought tolerant breeding genotype in the field, showed greater absolute water contents prior to and during dehydration treatment as compared with cv. 'IR64', which is known as moderate drought tolerant genotype. This result agrees well with the findings that drought tolerant rice genotype had great transpiration ability under water deficit condition (Cabuslay et al. 2002). However, cv. 'Vandana' had higher water loss rate and osmotic potential than cv. 'IR64' did during the dehydration condition. The dehydration condition triggers the expression of MAPKs in the upper and lower parts of cv. 'IR64' but not in cv. 'Vandana' except in the upper part at 30 minute dehydration. These results indicate that the low absolute water content of cv. 'IR64' can be a critical element in the reduction of osmotic potential and expression of MAPKs under dehydration condition even though lower water loss rate. These findings lead a suggestion that the expression of drought-responsive genes could be associated with the absolute status of internal water.

In conclusion, dehydration significantly altered internal water relation of rice plants primarily through acute water loss and solute concentration due to the water loss. Dehydration imposition also triggers the expression of drought-responsive genes. In this study, the precise status of water loss and absolute internal can be determined in line with a few drought-responsive genes. These results indicate that the physiological parameters such as water loss rate and water status can be used a physiological scale, especially, to explore the genes related to simple response and/or tolerance against internal water deficit.

Acknowledgement This work is partial results of a research project (RIMS code: 200803101010096) financially supported by the National Institute of Agricultural Biotechnology/RDA, Suwon, Korea. Authors greatly appreciate Prof. M. Ashraf, Faisalabad Agricultural University, Pakistan for his insightful comments.

References

Atlin GN, Lafitte R, Venuprasad R, Kumar R, Jongdee B (2004) Heritability of rice yield under reproductive-stage drought stress, correlations across stress levels and effects of selection: Implications for drought tolerance breeding. CIMMYT/Drought/Rockefeller Foundation Workshop 2004, pp 85–87.

Bray EA (1997) Plant responses to water deficit. Trends Plant Sci 2: 48–54.

Cabuslay GS, Ito O, Alejar AA (2002) Physiological evaluation of responses of rice (*Oryza sativa* L.) to water deficit. Plant Sci 163: 815–827.

Campbell JL, Klueva NY, Zheng HG, Nieto-Sotelo J, Ho T-HD, Nguyen HT (2001) Cloning of new members of heat shock protein HSP101 gene family in wheat (*Triticum aestivum* (L.) Moench) inducible by heat, dehydration and ABA. Biochimic Biophys Acta 1517: 270–277.

Chen BJ, Wang Y, Hu YL, Wu Q, Lin JP (2005) Cloning and characterization of a drought-inducible MYB gene from *Boea crassifolia*. Plant Sci 168: 493–500

Dingkuhn M, Audebert AY, Jones MP, Etienne K, Sow A (1999) Control of stomatal conductance and leaf rolling in *O. sativa* and *O. glaberrima* upland rice. Field Crop Res 61: 223–236.

Garrity DP, O'Toole JC (1994) Screening rice for drought resistance at the reproductive phase. Field Crop Res 39: 99–110.

Hoekstra FA, Golovinia EA, Buitink J (2001) Mechanisms of plant desiccation tolerance. Trends Plant Sci 6: 1360–1385.

Huang, B, Jin L, Liu JY (2008) Identification and characterization of the novel gene GhDBP2 encoding a DRE-binding protein from cotton (*Gossypium hirstum*). J Plant Physiol 165:214–223.

Jongdee B, Fukai S, Cooper M (2002) Leaf water potential and osmotic adjustment as physiological traits to improve drought tolerance in rice. Field Crops Res 76: 153–163.

Kato Y, Kamoshita A, Abe J, Yamagishi J (2007) Improvement of rice (*Oryza sativa* L.) growth in upland conditions with deep tillage and mulch. Soil Till Res 92: 30–44.

Knetsch MLW, Wang M, Snaar-Jagalska BE, Heimovaara-Dijkstra S (1996) Abscisic acid induces mitogen-activated protein kinase activation in barley aleurone protoplasts. Plant Cell 8: 1061–1067.

Liang YH, Sun WQ (2002) Rate of dehydration and cumulative desiccation stress interacted to modulate desiccation tolerance of recalcitrant cocoa and ginkgo embryonic tissues. Plant Physiol 128: 1323–1331.

Mizoguchi T, Irie K, Hirayama T, Hayashida N, Yamaguchi-Shinozaki K, Matsumoto K, Shinozaki K (1996) A gene encoding a mitogen-activated protein kinase kinase kinase is induced simultaneously with genes for a mitogen-activated protein kinase and an S6 ribosomal protein kinase by touch, cold, and water stress in *Arabidopsis thaliana*. PNAS 93: 765–769.

Ouk M, Basnayake J, Tsubo M, Fukai S, Fisher KS, Cooper M, Nesbitt H (2006) Use of drought response index for identification of drought tolerant genotypes in rainfed lowland rice. Field Crop Res 99: 48–58.

Rodriguez M, Canales E, Borroto CJ, Carmona E, Lopez J, Pujol M, Borras-Hidalgo O (2006) Identification of genes induced upon water-deficit stress in a drought-tolerant rice cultivar. J Plant Physiol 163: 577–584.

Sambrook J, Fritsh EF, Maniatis T (2001) Molecular Cloning: A Laboratory Manual, 3rd ed. Cold Spring Harbor Laboratory Press, Cold Spring Harbor, New York.

Sharp RE and Le Noble ME (2002) ABA, ethylene and the control of shoot and root growth under water stress. J Exp Bot 53: 33–37.

Shinozaki K, Yamagichi-Shinozaki K, Seki M (2003) Regulatory network of gene expression in the drought and cold stress responses. Curr Opin Plant Biol 6: 410–417.

Turner NC, O'Toole JC, Cruz RZ, Yambao EB, Ahmad S, Namuco OS, Dingkuhn M (1986) Responses of seven diverse rice cultivars to water deficits. II. Osmotic adjustment, leaf elasticity, leaf extension, leaf death, stomatal conductance and photosynthesis. Field Crop Res 13: 273–286.

Verslues PE, Agarwal M, Katiyar-Agarwal S, Zhu JH, Zhu JK (2006) Methods and concepts in quantifying resistance to drought, salt and freezing, abiotic stresses that affect plant water status. Plant J 45: 523–539.

Chapter 9
Effect of Different Water Table Treatments on Cabbage in Saline Saemangeum Soil

M. Jamil and E.S. Rha

Abstract In the Saemangeum tide embankment that will connect the cities of Gunsan and Buan a large area is specified for grain and horticultural crops. However, salinity and waterlogging are two main problems of this area. In order to assess up to what extent, this area can be efficiently utilized by growing cabbage, in the present study cabbage (*Brassica oleracea var capitata* L.) was subjected to various water table treatments (20, 30, 50 and 70 cm) in Saemangeum soil (marginally saline soil, ECe 3.8 dS m^{-1}) area. It was observed that with increasing water table treatments, relative growth rate (RGR), leaf area ratio (LAR), number of leaf, leaf area and net assimilation rate (NAR) increased significantly. Significant increased in the maximal quantum yield of PSII (Fv/Fm), electron transport rate (ETR) was observed while there was no change in non-photochemical quenching coefficient (NPQ). Chlorophyll content (SPAD value) increased significantly with the increase in water table treatments. Correlation shows that growth attributes had a significant positive relationship with Fv/Fm and ETR while non significant relationship was found between growth attributes and NPQ.

Keywords *Brassica oleracea capitata* L. • water depth • salinity • relative growth • photochemistry

M. Jamil
Department of Biotechnology and Genetic Engineering, Kohat University of Science and Technology (KUST), Kohat 26000, Pakistan

E.S. Rha (✉)
College of Agriculture & Life Sciences, Sunchon National University, Suncheon 540-742, Republic of Korea
e-mail: euishik@sunchon.ac.kr

1 Introduction

The Saemangeum project is the construction of a 33 km tide embankment that will connect the cities of Gunsan and Buan and create 28,300 ha of land and 112,800 ha freshwater lake. Out of total area 11,800 ha is for grain crops and 2,500 for horticultural crops. Two main problems of this reclaimed area are salinity and water logging. Cultivation of horticultural crops that can tolerate salinity and or waterlogging is one of the most promising strategies to economically and efficiently utilize waterlogged and/or salt affected soil (Ashraf 1994, 2004; Ashraf and Foolad 2007). Of various horticultural crops, cabbage (*Brassica oleracea*) has been ranked as moderately sensitive to salinity (Bernstein and Ayers 1949; Osawa 1961). There has been a renewed interest in cultivating *Brassica* species as a consequence of the high concentration of isothiocyanates and similar compounds found in cabbage, which have been proved to possess anticancer properties (Hecht et al. 1996; Wargovich 2000).

It is well established that salt stress reduced the crop growth and productivity by reducing photosynthetic capacity (Ashraf 2004; Munns 2005). However, the RGR is a function of net assimilation rate (NAR), which is an index of photosynthetic capacity of the plant per unite area, and leaf area ration (LAR), which is an index of the leafiness of the plant (Hunt 1990). These growth attributes make it possible to clarify whether genotypic variation under saline condition can be attributable to morphological changes or photosynthetic response (Ishikawa et al. 1991). Therefore, relative growth rate (RGR) has been considered a key parameter under saline condition to allow more appropriate comparison of growth (Cramer et al. 1994).

Likewise, change in water table height can cause changes in biomass allocation to above- and below-ground

M. Ashraf et al. (eds.), *Salinity and Water Stress*,

tissues, thus affecting the amount of above ground photosynthetic tissue and ecosystem productivity (Mann and Wetzel 1999). A change in water table height may also affect the growth and changes in the rates of evapotranspiration (Klimesova 1994, 1995). Kalita and Kanwar (1992) reported that water-table depths from 0.6 to 1 m increased yield, while water-table depths of 0.2–0.3 m reduced grain yields due to waterlogging. In view of this information, water depth is an important parameter for the prediction of plant growth under saline condition.

Thus, the present study was aimed to determine the interactive effects of water depth and salinity on growth of cabbage. Moreover, to draw the relationships between growth and PSII photochemistry to identify the critical growth component(s) attributed to the PSII photochemistry of cabbage plants under saline condition.

2 Materials and Methods

2.1 Plant Material and Water Table Treatments

Seeds of the cabbage (*Brassica oleracea capitata* L. cv Gaeul baechu) were obtained from Jeollabuk-do Agricultural and Extension Services, Iksan, Korea. Seeds were grown in plastic trays. After 3 weeks, seedlings were transferred into plastic pots of different heights (25, 35, 55 and 75 cm) containing saline soil. The soil for this study was obtained from Saemangeum saline area. Electrical conductivity of the soil solution paste extract (EC_e) was used to describe soil salinity levels. The samples were analyzed for EC_e using established procedures (Rhoades 1982). Electrical conductivity of the soil samples was 3.8 dS m^{-1}. The pots had drainage holes in the bottom. These holes were covered with nets to prevent the soil seepage and allowed water to enter into the pot easily. All the pots were placed in blocks containing 5 cm of water levels. All measurements on the youngest and expanded leaves were made after 6 weeks. The average temperature for day/night was 25°C/15°C and photoperiod for the day/night cycle was 16/8 h.

2.2 Growth Measurements

Leaf area of individual plant leaves was measured by using Area meter (AM-200, ADC Bio Scientific Ltd., England). The relative growth rate (RGR), net assimilation rate (NAR) and Leaf area ratio (LAR) were calculated by using the following equations (Hunt 1990).

1. $RGR = 1/W * \Delta W/\Delta T$
2. $NAR = 1/L_A * \Delta W/\Delta T$
3. $LAR = L_A/W$

Where W, T and L_A are plant dry weight (g), time (day) and leaf area (cm^2) respectively.

2.3 Measurements of Chlorophyll Fluorescence

Measurements were made with a portable Mini PAM fluorometer (PAM-2000, Walz, Germany) on the upper surface of leaves, which had been predarkened for at least 30 min. The data acquisition software (DA-2000, Walz) was used to connect the fluorometer with computer. The experimental protocol of Genty et al. (1989) was basically followed. The minimal fluorescence level (Fo) was measured by the measuring modulated light, which was sufficiently low (<0.1 µmol m^{-2} s^{-1}) not to induce any significant variable fluorescence. The maximal fluorescence level (Fm) was measured by a 0.8 s saturating pulse at 8,000 µmol m^{-2} s^{-1}. The measurements of Fo were performed with the measuring beam set to a frequency of 0.6 kHz, whereas measurements of Fm were performed with the measuring beam automatically switching to 20 kHz during the saturating flash. The leaves were continuously illuminated with white actinic light at the intensity of 300 ∝mol m^{-2} s^{-1}. By using fluorescence parameters determined in leaves, following calculations were made of: the maximal quantum yield of PSII photochemistry (Fv/Fm), non-photochemical quenching (NPQ) was calculated as $NPQ = Fm - Fm'/Fm'$ and electron transport rate (ETR) ($ETR = yield \times PFD \times 0.5 \times 0.84$) (the standard factor 0.84 corresponds to the fraction of incident light absorbed by leaf).

2.4 Measurements of Leaf Area and Leaf Chlorophyll Content

Leaf chlorophyll content was measured using a handheld chlorophyll content meter (CCM-200, Opti-Science, USA). The chlorophyll content was measured 6 times from leaf tip to the base and then average for analysis.

2.5 Statistical Analysis

Analysis of variance was performed by using MS-Excel Statistical software. Mean values for growth and photosynthesis parameters were compared by LSD using Tukey's *t* (Li 1964). Correlation between growth and chlorophyll fluorescence was obtained by using Minitab version 14.0 statistical software package.

3 Results

3.1 Water Table Effect on Growth

Growth contribution was the highest under the deepest water table (70 cm) conditions, which gradually reduced with decreasing water-table depth (Fig. 9.1). Increase in water table depth significantly increased the relative growth rate (Table 9.1). Relative growth rate was maximum (18.22 g/g/day) at 70 cm depth. With the water tables, below this level, relative growth rate was reduced. This reduction however, was more pronounced at 20 cm depth (Table 9.1). The other growth parameter followed the same trend of growth reduction when water tables were less than 70 cm (Table 9.1). Maximum leaf area ratio (185 cm^2/day) was observed at 70 cm while minimum leaf area ratio (67 cm^2/day) was found at 20 cm water table depth (Table 9.1). Maximum net assimilation rate (15.07 g/m^2/day) was obtained with water table at 70 cm or below with gradual reduction in net assimilation rate when water table was lowered from 70 cm (Table 9.1).

Leaf number and area were the highest under the deepest water table conditions, which gradually reduced with decreasing water-table depth (Table 9.2). The number of leaves was the highest at 70 cm water table depth for cabbage. It decreased with decrease in water-table depth and attained a minimum value at 20 cm depth. Similarly leaf width and length increased with increasing water table depth (Table 9.2). Leaf area was maximum (213.70 cm^2) at 70 cm water table depth. Below this level, leaf area decreased with decreasing water table depth. However, reduction was more pronounced at 20 cm (Table 9.2).

3.2 Water Table Effect on PSII Photochemistry

Increase in water table depth caused a significant increase in maximal quantum yield of PSII (Fv/Fm) in cabbage. At the lowest water table depth, cabbage had lower Fv/Fm than that of the other treatments (Fig. 9.2A). It was observed that electron transport rate (ETR) increased with increasing water table treatment. Maximum ETR was obtained with water table at 70 cm or below with drastic reduction in ETR when water table was lowered from 70 cm (Fig. 9.2B). However, salinity had non-significant effect on non-photochemical quenching coefficient (NPQ) (Fig. 9.2C).

Table 9.1 Effect of various water table treatment on relative growth rate, net assimilation rate and leaf area rate in cabbage under 3.8 ds m^{-1} saline condition after 5 weeks

Water table	Growth parameters		
(cm)	RGR (g/g/day)	LAR (cm^2/g)	NAR (g/m^2/day)
20	5.58 ± 0.06	67 ± 2	11.45 ± 0.05
30	9.75 ± 0.05	98 ± 2	13.76 ± 0.06
50	14.50 ± 0.05	154 ± 4	14.11 ± 0.05
70	18.22 ± 0.07	185 ± 5	15.07 ± 0.07

Fig. 9.1 Effect of various water table treatments on cabbage growth under saline condition (3.8 ds m^{-1}) after 5 weeks

Table 9.2 Effect of various water table treatments on number of leaf, leaf length, leaf width and leaf area in cabbage under 3.8 ds m^{-1} saline condition after 5 weeks

	Growth parameters			
Water table	Number of leaf	Leaf length (cm)	Leaf width (cm)	Leaf area (cm)
20	8.9 ± 0.05	6.03 ± 0.03	4.59 ± 0.04	28.68 ± 0.07
30	9.2 ± 0.02	11.44 ± 0.04	6.70 ± 0.05	77.65 ± 0.05
50	11.4 ± 0.05	16.21 ± 0.04	9.84 ± 0.04	160.61 ± 0.10
70	13.6 ± 0.03	19.02 ± 0.02	11.13 ± 0.06	213.70 ± 0.10

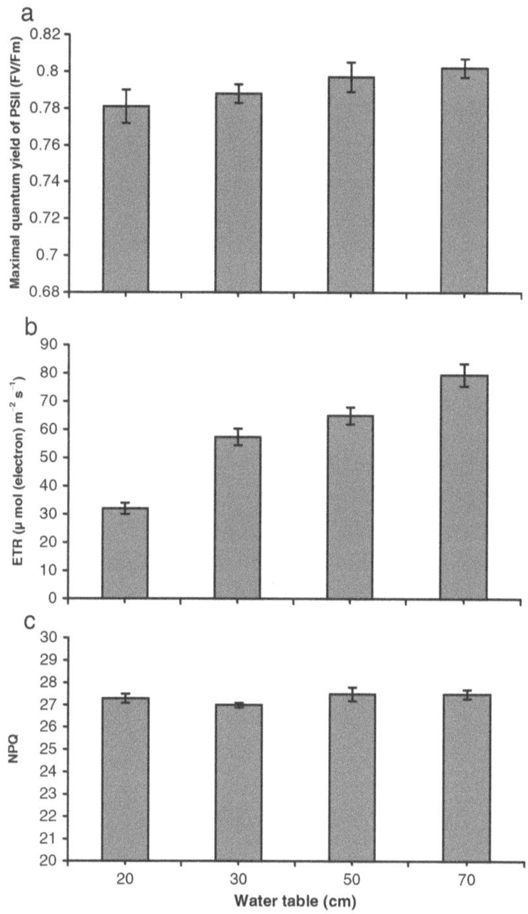

Fig. 9.2 Effect of various water table treatments on Fv/Fm (A) ETR (B) and NPQ (C) in cabbage under saline condition (3.8 ds m^{-1}) after 5 weeks

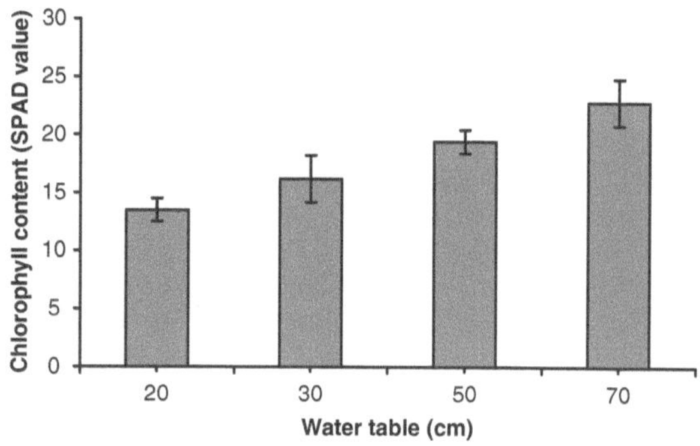

Fig. 9.3 Effect of various water table treatments on chlorophyll contents in cabbage under saline condition (3.8 ds m^{-1}) after 5 weeks

3.3 Water Table Effect on Chlorophyll Content

The leaf chlorophyll content increased with increase in water table depth but magnitude of increase was more at 70 cm (Fig. 9.3). Chlorophyll content had significant positive relationship with growth and PSII photochemistry but it had non significant relation with NPQ (Table 9.3).

3.4 Relationship Between Growth Attributes and PSII Photochemistry

Correlation revealed a significant positive relationship between growth attributes, PSII photochemistry and chlorophyll content (Table 9.1). There was a positive relationship between RGR, NAR, LAR, leaf area and Fv/Fm, ETR, NPQ (Table 9.1). Correlation also revealed a strong (R^2 = 0.998, P = 0.002) significant positive relationship between RGR and Fv/Fm. Table 9.1 also showed a weak (R^2 = 0.96, P = 0.04) significant positive relationship between LAR and ETR. There was also a non-significant relation between growth parameters and NPQ (Table 9.1).

4 Discussion

An understanding of plant response to the changes of water depths under salt stress is important strategy in dealing with salinity problem and to improve the agronomic

Table 9.3 Relationships among relative growth attributes and PSII photochemistry in cabbage under various water table treatments in saline soil

	Growth attribute			PSII photochemistry			
	RGR	LAR	NAR	Leaf area	Fv/Fm	ETR	NPQ
LAR	0.996*						
NAR	0.940*	0.910*					
Leaf area	0.997*	1.000**	0.913ns				
Fv/Fm	0.998*	0.997*	0.935ns	0.997*			
ETR	0.982*	0.962*	0.987*	0.965*	0.977*		
NPQ	0.619ns	0.683ns	0.320ns	0.674ns	0.634ns	0.458ns	
CC	0.997**	0.993**	0.927ns	0.998**	0.991**	0.975*	0.625ns

**, * Significant at $P = 0.01$ and $P = 0.05$, respectively; ns = non significant
RGR; Relative growth rate, NAR; net assimilation rate, LAR; leaf area ratio, Fv/Fm; maximal quantum yield of PSII, ETR; electron transport rate, NPQ; non-photochemical quenching coefficient, CC; chlorophyll content

performance of crops in saline water irrigation (Zeng et al. 2003). The crop growth under saline condition basically depends on the environment in the root zone, rooting depth, sensitivity of crop for water and salt. In the present study, decreasing water depth under saline condition did not affect plant survival but strongly inhibited plant growth. Total leaf area, number of leaf, RGR, NAR and LAR were maximum at 70cm depth, below this level growth was reduced (Table 9.1). This reduction however, was more pronounced at 20cm depth probably due to reduced aeration in the root zone and water logging. The increases observed in RGR with increasing water depth could be attributed to NAR and LAR depending on genotype (Hunt 1990). These findings are consistent with those of Zeng et al. (2003) who found water depth reduces plant growth under saline condition. Kahlown et al. (1998) found a linear relationship between the wheat yield and water table depth. They obtained optimum yield of wheat, when the water-table depth was less than 1 m whereas 30.5cm of irrigation water was required when the water-table depths ranges from 1 to 2m. However, if the groundwater is saline generally more than 4.0dS m^{-1}, the water table of 1–2m or less results in decreased wheat and sugarcane yields (Kahlown and Azam 2002). Kalita and Kanwar (1992) reported that water-table depths from 0.6 to 1m increased yield, while water-table depths of 0.2–0.3m reduced grain yields due to waterlogging. At clay soils and water tables shallower than 90cm, the lack of air limited the plant development and crop yield significantly (Tille and Mueller 1992).

Photosynthesis is a major factor in the determination of growth under saline condition (Heuer and Plaut 1989). Maximal quantum yield of PSII (Fv/Fm) and electron transport rate (ETR) increased with increasing depth of water level but water table had no significant relation with non-photochemical quenching coefficient (NPQ) (Fig. 9.2). The results indicate that the increased in growth under saline condition was primarily a result of increase in the photosynthesis. There was also a positive relation between growth and PSII photochemistry (Table 3). These results are in agreement with the reports by Cramer et al. (1994). The productivity observed for many plant species subjected to saline condition is often associated with the photosynthesis (Long and Baker 1986). Change in water table height can cause changes in biomass allocation to above and below ground tissues, thus affecting the amount of above ground photosynthetic tissue and ecosystem productivity (Mann and Wetzel 1999).

The increase in photosynthesis under saline condition can also be attributed to an increase in chlorophyll content. The results indicate that the leaf chlorophyll content increased with increase in water depth under saline condition (Fig. 9.3) may be due to increase in chlorophyll content may be due to the accumulation of NaCl in the chloroplast. Chlorophyll content had significant positive relationship with growth and PSII photochemistry (Table 3). It has been reported that chlorophyll content has been increased in salt tolerant plants such as pearl millet (Reddy and Vora 1986) and mustard (Singh et al. 1990) but chlorophyll content decreases in salt susceptible plants such as pea (Hamada and El-Enany 1994) and soybean (Seemann and Critchley 1985) under saline condition.

Acknowledgment This present research was made possible by the research fund of Rural Development Administration, Korea in 2005.

References

Ashraf M (2004) Some important physiological selection criteria for salt tolerance in plant. Flora 199: 361–376.

Ashraf M, Foolad MR (2007) Roles of Glycine betaine and proline in improving plant abiotic stress resistance. Environ. Exp. Bot. 59: 206–216.

Bernstein L, Ayers A D (1949) Salt tolerance of cabbage and broccoli. United States Salinity Laboratory Report to Collaborators. Riverside, CA, pp 39.

Cramer G R, Alberico G J, Schmidt C (1994) Salt tolerance is not associated with the sodium accumulation of 2 maize hybrids. Aust J Plant Physiol 21: 675–692.

Genty B, Briantais J M, Baker N R (1989) The Relationship between the Quantum Yield of Photosynthetic Electron Transport and Quenching of Chlorophyll Fluorescence. Biochim. Biophys. Acta 99: 87–92

Hamada A M, El-Enany A E (1994) Effect of NaCl salinity on growth, pigment and mineral element contents, and gas exchange of broad bean and pea plants. Biol Plant 36: 75–81.

Hecht S S, Trushin N, Rigotty J, Carmella S G, Borukhova A, Akerkar S, Desai D, Amin S, Rivenson A (1996) Inhibitory effects of 6-phenylhexyl isothiocyanate on 4-(methylnitrosamino)-1-(3-pyridyl)-1-butanone metabolic activation and lung tumorigenesis in rats. Carcinogenesis 17: 2061–2067.

Heuer B, Plaut Z (1989) Photosynthesis and osmotic adjustment of two sugar beet cultivars grown under saline condition. J Exp Bot 40: 437–440.

Hunt R (1990) Basic growth analysis: plant growth analysis for beginners. Academic, London.

Ishikawa S, Oikawa T, Furukawa A (1991) Responses of photosynthesis, leaf conductance and growth to different salinities in 3 coastal dune plants. Ecol Res 6: 217–226.

Kahlown M A, Azam M (2002) Individual and combined effect of waterlogging and salinity on crop yield in the Indus basin. Irrig Drain 51: 329–338.

Kahlown M A, Iqbal M, Skogerboe G V, Rehman S U (1998) Water logging, salinity and crop yield relationships. Mona Reclamation Experimental Project, WAPDA Report No. 233.

Kalita P K, Kanwar R S (1992) Shallow water table effects on photosynthesis and corn yield. Trans ASAE 35: 97–104.

Klimesova A J (1994) The effects of timing and duration of floods on growth of young plants of Phalaris arundinacea L. and Urtica dioica L.: an experimental study. Aquat Bot 48: 21–29.

Klimesova A J (1995) Population dynamics of Phalaris arundinacea L. and Urtica dioica L. in a flood plain during a dry period. Wetlands Ecol Manage 3: 79–85.

Long S P, Baker N R (1986) Saline terrestrial environments. In: Baker N R, Long S P (eds) Photosynthesis in contrasting environments. Elsevier, New York, pp. 63–102.

Li C C (1964) Introduction to Experimental Statistics. McGraw Hill Book Company. New York.

Mann C J, Wetzel R G (1999) Photosynthesis and stomatal conductance of Juncus effusus in a temperate wetland ecosystem. Aquat Bot 63: 127–144.

Munns R (2005) Genes and salt tolerance: bringing them together. New Phytol 167: 645–663.

Osawa T (1961) Studies on the salt tolerance of vegetable crops in sand cultures. II. Leafy vegetables. J. Jpn. Soc. Hort. Sci. 30: 48–56.

Reddy M P, Vora A B (1986) Changes in pigment composition. Hill reaction activity and saccharides metabolism in bajra (*Penisetum typhoides* S & H) leaves under NaCl salinity. Photosynthetica 20: 50–55.

Rhoades J D (1982) Soluble salts. In: Page A L (ed) Methods of soil analysis. Part 2. Chemical and microbiological properties. Vol 9, 2nd edn. SSSA Monograph, ASA/CSSA/SSSA, Madison, WI, pp. 167–179.

Seemann J R, Critchley C (1985) Effects of salt stress on the growth, ion content, stomatal behavior and photosynthetic capacity of a salt-sensitive species, *Phaseolus vulgaris* L. Planta 164: 151–162.

Singh M P, Pandey S K, Singh M, Ram P C, Singh B B (1990) Photosynthesis, transpiration, stomatal conductance and leaf chlorophyll content in mustard genotypes grown under sodic conditions. Photosynthetica 24: 623–627.

Tille P, Mueller L (1992) Einfluss des Grundwassers auf Bodendurchlueftung und Ertrag auf einem Auentonstandort. Arch. Acker-Pflanzenbau Bodenkd 36: 391–401.

Wargovich M J (2000) Anticancer properties of fruits and vegetables. Hortic Sci 35: 573–575.

Zeng L, Lesch S M, Grieve C M (2003) Rice growth and yield respond to changes in water depth and salinity stress. Agric Water Manage 59: 67–75.

Chapter 10
How Does Ammonium Nutrition Influence Salt Tolerance in *Spartina alterniflora* Loisel?

K. Hessini, M. Gandour, W. Megdich, A. Soltani, and C. Abdely

Abstract *Spartina alterniflora* (Poaceae), commonly known as smooth cordgrass is originated from hydromorph and salted marginal lands. In these regions, hypoxic and salt stress conditions strongly inhibit nitrification and promote denitrification. Distribution and yield of this species depend on its capacity to tolerate salinity and to use ammonium as the predominant source of nitrogen. *S. alterniflora* grown on medium containing varying ammonium:nitrate ratio were treated with 500 mol m^{-3} NaCl. In absence of NaCl, the highest growth rate was observed in the mixed medium and the lowest one in medium containing nitrate alone. However, growth was higher in ammonium than in nitrate medium, indicating that *S. alterniflora* was one of the scarce species which prefered ammonium as nitrogen source. The addition of 500 mol m^{-3} NaCl affected plant growth in both nitrate and mixed-media, but did not affect when ammonium was supplied. In absence or in presence of 500 mM NaCl, differences in growth between the treatments were due to difference in photosynthesis activity. In the same way the mean secretion of ions, especially Na^+ by shoots of *S. alterniflora* was increased by ammonium treatment. Ammonium treatment probably improved salt tolerance of *S. alterniflora* by increasing secretion activity of salt glands leading to a decrease in salt content of leaves mesophyll and consequently to avoiding toxic buildup of Na^+ in the apoplastic tissues of the leaves.

Keywords Ammonium • nitrate • salinity • *Spartina alterniflora* Loisel

K. Hessini, M. Gandour, W. Megdich, A. Soltani, and C. Abdely
Laboratoire d'Adaptation des Plantes aux Stress Abiotiques, Centre de Biotechnologie à la Technopole Borj Cédria BP 901, Hammam-Lif 2050, Tunisia
e-mail: hessini_kamel@yahoo.fr

1 Introduction

Under natural conditions of growth and development, plants are inevitably exposed to different types of stresses, such as drought, salinity, low and high temperatures, flooding and high radiation. In salted marginal zones, salinity reduces agricultural yield throughout the world (Yokoi et al. 2002; Cant et al. 2007). Salt stress inhibits plants growth by disrupting ion and water homeostasis as well as causing oxidative stress (Cant et al. 2007). The extent of growth and yield reduction due to salt stress depends on the type of species, duration and severity of salt stress. Application of nitrogen fertilizers to plants growing under saline conditions might have some beneficial effects on plant growth and yield since nitrogen is involved in biochemical and physiological processed related to osmotic adjustment (Sagi et al. 1997). Some studies showed that salinity affects nitrogen mineralization, and this effect is more marked on nitrification than on ammonification (Laura 1977; Sahrawat 1982; McClung and Frankenberger 1987). Under this condition, other stable nitrogen forms in soil, such as ammonium, are also available to plants (Cantera et al. 1999). The interaction between nitrogen and salinity has been studied in several plant species, such as peanut, wheat, maize and barley (Lewis et al. 1989; Soltani et al. 1992; Cant et al. 2007). The response of plant growth to nitrogen fertilization under saline and non-saline conditions varies according to whether the nitrogen is supplied as nitrate or ammonium and also depends on plant species. Ammonium was a suitable nitrogen source under non-saline conditions for some species, whereas for other species (e.g. cotton), NH_4^+ caused much less growth than did NO_3^- (Yokoi et al. 2002). However, NO_3^- alone may not be beneficial, especially under salt stress, where its rate of uptake is reduced in many plant

M. Ashraf et al. (eds.), *Salinity and Water Stress*,

species due to high Cl^- content of saline soil. Several plant species show enhanced growth when provided with both nitrate and ammonium, as compared to growth on a sole nitrate or ammonium source and in at least one plant species, tomato, a combined nitrate/ammonium regime can partially restore salinity-mediated decreases in plant growth (Flores et al. 2001; Cant et al. 2007). However, there are substantial differences among plant species in the extent to which NH_4^+ levels in the NO_3^-/NH_4^+ mixture is beneficial (Sandoval-Villa et al. 1999; Flores et al. 2001).

The purpose of our research was to examine whether a form of nitrogen is able to reduce the adverse affects of salinity on the growth of the salt-tolerant Poaceae, *Spartina alterniflora* and to determine the mechanisms by which ammonium might improved salt tolerance of this species.

2 Material and Methods

2.1 Plant Material, Growth Conditions, Stress Imposition and Elemental Analysis

Cuttings of *Spartina alterniflora* Loisel from its native range (USA) were cultivated in five outdoor containers filled with a mixture of sandy soil and organic matter at our Experimental Station near the Mediterranean sea shore, 35 km north-east of Tunis. Uniform cuttings (25 cm height) were taken from these mother plants and were washed before transplanting them individually into 4 L blow-molded (one plant plug per pot) filled with sandy sol and irrigated with nutritive solution, under non-saline condition for the first month after transplanting. During this time, some plants performing poorly were replaced. The experiment was conducted in a greenhouse, with average day temperatures of 25°C/18°C day/night and a relative humidity of 65%/90%. Midday photon flux density was 900–1,000 µmol m^{-2} S^{-1}. Nitrogen was applied as either NO_3^- or NH_4^+ or a mixture of NO_3^- and NH_4^+ in a ration of 50: 50 (of the total 7.5 mM nitrogen) with or without 500 mM NaCl. Salt was added in weekly steps of 100 mmol L^{-1} day^{-1} in order to avoid an osmotic shock. The concentration of micro-elements and macro-elements used in the nutrient solution is given by Hewitt (1966). Iron (50 µM Fe-EDTA) was added separately. The irrigation with nutritive solution (pH 6.0 ± 0.1) was renewed every 3 days. The medium containing NH_4^+ as the only N source was buffered by $CaCO_3$ (Cantera et al. 1999). Plants were harvested between 10:00 and 14:00 h, 60 days after the start of the salinity treatment. Plants were separated into above- and below-ground components. Fresh and dry masses were determined, after counting the leaf number and determining their surface area with portable area meter (LI-3000A). The dry matter weights were determined after drying shoots and roots for 72 h in a thermo-ventilated oven at 80°C.

Salt secretion and accumulation rates (mmol g^{-1} DW) of leaves exposed to tow levels of salinities (0 and 500 mM) under various nitrogen sources (NO_3^-, NH_4^+, NO_3/NH_4) were determined in the same sample at the end of the experiment (after 90 days). The secreted salt fraction was determined as the amounts of salt washed off the leaves into 100 ml of deionized water during 5 min rinsing (Waisel et al. 1986; Youssef and Ghanem 2002), while the accumulated fraction was determined as the salt kept in washed leaves. Sodium and potassium were determined by a flame spectrophotometer (Corning) and Ca^{2+} and Mg^{2+} by an atomic absorption spectrophotometer (Varian 06).

2.2 Statistical Analysis

All data presented are the mean values. The measurements were done with five replicates on shoot ionic content and ten replicates on biomass determination and water contents. Statistical analyses were carried out with ANOVA test, and means were compared by the Duncan's multiple range test. Comparisons with *P*-values of <0.05 were considered significantly different.

3 Results and Discussion

3.1 Plant Growth and Gas Exchange Measurements

In order to study the effect of nitrogen ionic form on the growth of *S. alterniflora* treated or not with 500 mol m^{-3} NaCl, plant growth was assessed by measuring leaf surface and number as well as whole plant dry weight (Fig. 10.1). Significant differences in absolute growth rate between treatments were observed (Fig. 10.1). Under non saline conditions, the highest growth rate was observed in the medium contained both nitrate and ammonium and the lowest in medium contained nitrate

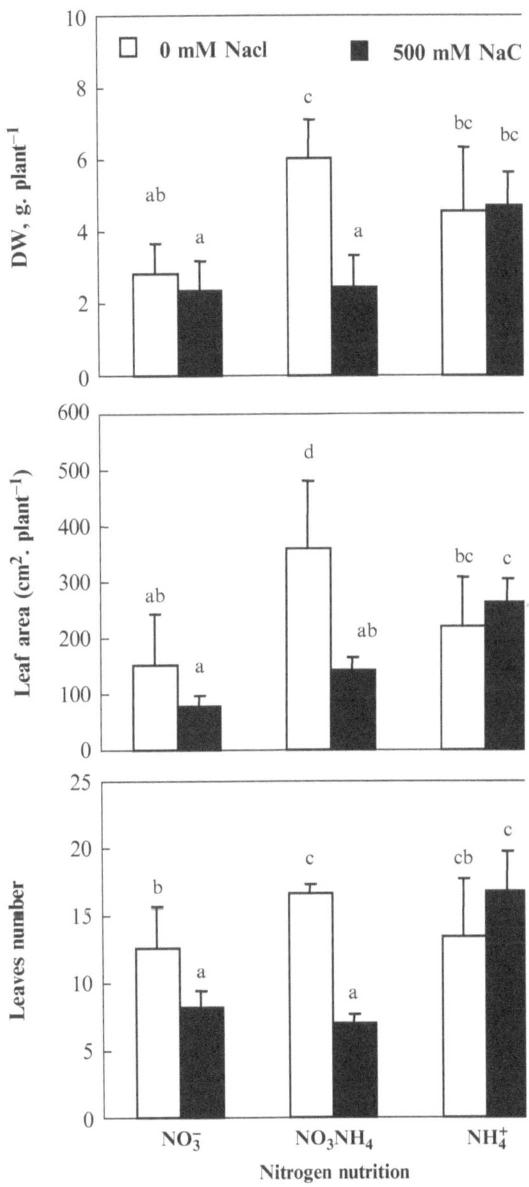

Fig. 10.1 Effects of salinity (NaCl = 0 or = 500 mol m^{-3}) and nitrogen regimes on plant growth. Upper panel: biomass (dry weight); Median panel: leaf area and lower panel: number of living leaves. Data are means ± S.E. (n = 10)

alone, but, growth was better with ammonium than with nitrate nutrition indicating that *S. alterniflora* was one of the scarce species witch prefered ammonium as nitrogen source. The benifical effect of NH_4^+ over NO_3^- was observed in some wetland and marine species (Munzarova et al. 2006; Thursby and Harlin 1984) as well as in plants colonising terrestrial habitats with the prevailing NH_4^+-N form (Garnett et al. 2001; Munzarova et al. 2006). The addition of 500 mol m^{-3} NaCl to the different media caused a reduction in all growth parameters in plants grown under both nitrate and mixed-media. However, this addition did not affect, growth of ammonium fed-plants; the application of an ammonical regime not only reduced the deleterious effects of the salt treatment, there is even a slight increase in dry matter accumulation (Fig. 10.1). This slight improvement was due to an increase in both leaf area and number of living leaves (Fig. 10b, c). This results is in agreement with some studies (Flores et al. 2001; Cant et al. 2007) but not with others (Speer et al. 1994; Speer and Kaiser 1994, I; II) and this is for great economical and ecological importance for at least two reasons: (1) when ammonium is added or replaced nitrate, it can induce a decrease in the nitrate concentration of plant tissues. Considering the high capability for accumulation of nitrate in leaves (Santamaria et al. 1999) and the high toxicity of this anion to human (Gangolli et al. 1994, I; II) and animal health (Bruning-Fann and Kaneene 1993), ammonium fertilization may be a desirable source of nitrogen nutrition under certain conditions, (2): the detrimental effects of salinity on plant growth and productivity can be reduced or scant by partial or totally substitution of NO_3^- with NH_4^+ in the nutrient solution.

In order to test whether the differences in growth between the three nitrogen regimes could be correlated with differences in carbon assimilations, photosynthesis rates were determined (Table 10.1). Except, in ammonium fed-plant, when the rate of photosynthesis and transpiration were not changed, salinity significantly reduce the two later parameters in both nitrate and mixed-media. However, irrespective to nitrogen form, the stomatal conductance was reduced by an average of 50%. This suggests that in spite of the large decrease of stomatal aperture, salt-treated plants in ammonium media were able to maintain a high level of photosynthetic and transpiration rate. This result is in agreement with previous findings on barley (Cant et al. 2007) and tomato (Flores et al. 2001). On the other hand, the growth

Table 10.1 Effect of nitrogen ionic form and salinity on photosynthesis (Pn) (μmol CO_2 m^{-2} s^{-1}), transpiration rate (E) (mmol H_2O m^{-2} s^{-1}) and stomatal conductance (gs) (mmol H_2O m^{-2} s^{-1}) in *Spartina alterniflora*

NO_3^-/NH_4^+	PN	E	gs
No salinity			
100/0	12.48 ± 0.66 b	2.05 ± 0.26 b	153.00 ± 9.06 c
50/50	18.12 ± 0.79 c	2.21 ± 0.36 cb	213.13 ± 14.22 d
0/100	17.76 ± 0.94 c	2.55 ± 0.2 c	234.53 ± 3.58 e
500 mM NaCl			
100/0	8.02 ± 1.19 a	1.02 ± 0.21 a	75.21 ± 5.24 a
50/50	7.31 ± 0.61 a	1.26 ± 0.38 a	93.62 ± 6.47 b
0/100	16.92 ± 0.45 c	2.21 ± 0.06 cb	104.09 ± 4.94 b

Lower case letters in the sum column indicates differences ($P \leq 0.05$, Duncan test) between treatments

increase observed in the present study with the ammonium regime could be correlated with an increase in both photosynthesis per unit area and total leaf area available for photosynthesis (Table 10.1 and Fig. 10.1). The benefit effect of ammonium on total leaf area confirm our previous results (unpublished observations) who showed that ammonium nutrition affected only slightly individual leaves area surface of *Beta macrocarpa* Guss, but had not significant effect on the initiation rates of leaves. The mechanism by witch ammonium increase total leaf area (initiation of new organs or extension of individual leaves) needs to be more elucidated.

3.2 Mineral Composition

Previous studies have centred mainly on the comparison of nitrate and ammonium form on the diffusible ionic balance in leaves (Sagi et al. 1997). Our results showed that when *S. alterniflora* plants were grown on ammonium medium contained 500 mol m^3 NaCl, ammonium compared to nitrate, decreased the leaf concentrations of Na^+ (Fig. 10.2a). Ammonium decreased K^+ concentration either in absence or in presence of salt in the medium (Fig. 10.2c), but it increased Ca^{2+} concentration, this later effect was more pronounced in salt added medium (Fig. 10.3a). Mg^{2+} content was independent of nitrogen source, but it was significantly decreased by NaCl (Fig. 10.3b).

It has been suggested that in ryegrass and in barely, competition between NH_4^+ and Na^+ for root uptake sites reduces Na^+ uptake and transport from roots to shoots under NH_4^+ nutrition, thereby minimizing the Na^+ in leaves (Sagi et al. 1997; Cant et al. 2007). Our results suggest that this is also the case with *S. alterniflora* since Na^+ content in leaves was reduced by 15% under the ammonium nitrogen medium compared to NO_3^- alone. This competition between NH_4^+ and Na^+ for root uptake sites was observed also, for K^+, but not for Ca^{2+}; there is even a significant increase of Ca^{2+} content (Figs. 10.2a, c, and 10.3a). When plants are exposed to stress, an essential function of Ca^{2+} is that of a second messenger in stress signalling (Knight and Knight 2001). Its crucial role in triggering a signalling cascade to active Na^+/H^+ antiporters has been demonstrated (Liu and Zhu 1998). Exposure to salt initially increased Ca^{2+} uptake in some species like in *P. euphratica* (Chen et al. 2001).

3.3 Salt Recretion

Salt recretion represents in *S. alterniflora*, like several plant genera and families, an avoidance strategy that permits control and regulation of salt content in plant organs, and especially in photosynthetic ones (Atkinson et al. 1967). Salt recretion has been shown to be mediated by specific glands scattered on the leaf surfaces. Specificity and the effect of nitrogen form on salt recretion in halophytes species, especially; *S. alterniflora* is not documented. In this study mean recretion of ions, especially Na^+ by shoots of *S. alterniflora* was increased by ammonium treatment (Fig. 10.3b). However, these of potassium was not or less affected leading to an increase of the ratio Na_{exo}/K_{exo} compared to control one. Thus, the recretion mechanism is characterized by high selectivity in favor of sodium opposing

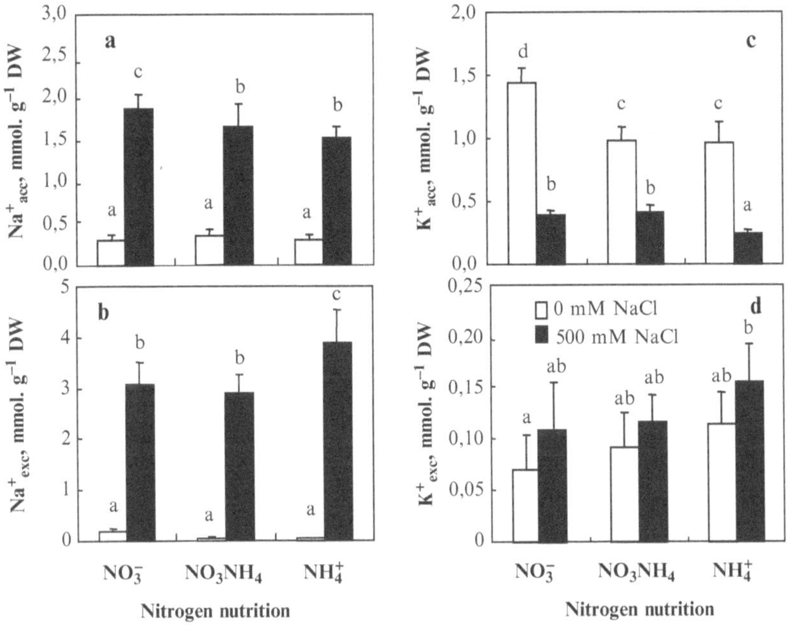

Fig. 10.2 Effects of salinity (NaCl = 0 or = 500 mol m^{-3}) and nitrogen regimes on inorganic content concentrations (mmol. g^{-1} DW): Na^{+}_{acc} (accumulated) (a), Na^{+}_{exc} (excreted) (b), K^{+}_{acc} (c) and K^{+}_{exc} (d). Data are means ± S.E. (n = 10)

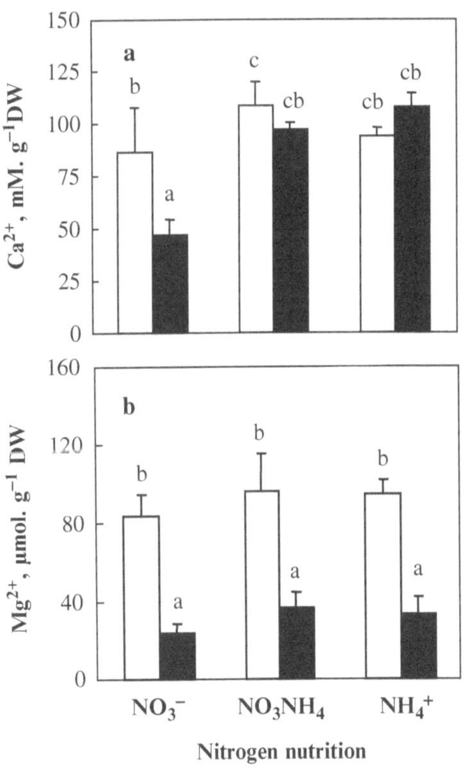

Fig. 10.3 Effects of salinity (NaCl = 0 or = 500 mol m^{-3}) and nitrogen regimes on inorganic content concentrations Ca^{2+} and Mg^{2+}. Data are means ± S.E. (n = 10)

potassium and calcium (Pollak and Waisel 1979; Barhoumi et al. 2007). It summed that ammonium treatment improved the salt tolerance of *S. alterniflora* by increasing recretion activity of salt glands in favor of sodium leading to a decrease in salt content of leaves mesophyll (Fig. 10.2). The higher growth of ammonium fed-plants (compared to nitrate fed-plants) was probably due in part to lower Na^+ level in mesophyll.

In summary, Ours results clearly showed that growth of *S. alterniflora* was significantly enhanced by ammonium tratment. This was achieved probably by the increase of the recretion activity of salt glands leading to a decrease in salt content of leaves mesophyll and consequently to a better photosynthetic activity.

References

Atkinson MR, Findlay GP, Hope AB, Pitman MG, Saddler HDW, West K (1967) Salt regulation in the mangroves *Rhizophora mucronata* Lam. and *Aegialitis annulata* R. Aust J Biol Sci 20: 589–599.

Barhoumi Z, Djebali W, Smaoui A, Chaïbi W, Abdelly W (2007) Contribution of NaCl excretion to salt resistance of *Aeluropus littoralis* (Willd) Parl. J Plant Physiol 164: 842–850.

Bruning-Fann C, Kaneene JB (1993) The effects of nitrate, nitrite, and N-nitroso compounds on animal health. Vet Hum Toxicol 35: 237–253.

Cant S, Kant P, Lips H, Barak S (2007) Partial substitution of NO_3^- by NH_4^+ fertilization increases ammonium assimilating

enzyme activities and reduces the deleterious effects of salinity on the growth of barely. J Plant Physiol 164: 303–311.
Cantera R, Urdaniz A, Lerga J, Garcia-Mina JM (1999) Acidifying product as a source of stabilized ammonia nitrogen under greenhouse conditions. In: Bech J (Ed.) Soils with Mediterranean type of climate. UB Publicaciones: Barcelona, pp. 233–235.
Chen S, Li J, Wang S, Hüttermann A, Altman A (2001) Salt, nutrient uptake and transport, and ABA of *Populus euphratica*: a hybrid in response to increasing salt. Trees (Berl) 15: 186–194.
Flores P, Carvajal M, Cerda A, Martines V (2001) Salinity and ammonium/nitrate interactions on tomato plant development, nutrition, and metabolites. J Plant Nutr 24: 1561–1573.
Gangolli SD, Van Den Brandt PA, Feron VJ, Jan-Zowsky C, Koeman JH, Speijers GJA, Spiegelhalder B, Walker R, Winshnok JS (1994) Nitrate, nitrite and N-nitroso compounds. Eur J Pharm Environ Toxicol Pharm Sect 292: 1–38.
Garnett TP, Shabala SN, Smethurst PJ, Newman IA (2001) Simultaneous measurement of ammonium, nitrate and proton fluxes along the length of eucalypt roots. Plant Soil 236: 55–62.
Hewitt EJ (1966) Sand and water culture methods used in the study of plant nutrition. Commonw Bur Hortic Tech Com 22: 431–446.
Knight H, Knight MR (2001) Abiotic stress signalling pathways: specificity and cross-talk. Trends Plant Sci 6: 262–267.
Laura RD (1977) Salinity and nitrogen mineralization in soil. Soil Biol Biochem 9: 333–336.
Lewis OAM, Leidi EO, Lips SH (1989) Effect of nitrogen source on growth response to salinity stress in maize and wheat. New Phytol 111: 155–160.
Liu JP, Zhu JK (1998) A calcium sensor homolog required for plant salt tolerance. Science 280: 1943–1945.
McClung G, Frankenberger WTJ (1987) Nitrogen mineralization rates in saline vs. salted amended soil. Plant Soil 104: 13–21.
Munzarova M, Lorenzen B, Brix H, Vojtiskova L, Votrubova O (2006) Effect of NH_4^+ /NO_3^- availability on nitrate reductase activity and nitrogen accumulation in wetland helophytes *Phragmites australis* and *Glyceria maxima*. Environ Exp Bot 55: 49–60.
Pollak G, Waisel Y (1979) Ecophysiology of salt excretion in *Aeluropus littoralis* (Graminae). Physiol Plant 47: 177–184.
Sagi M, Dovrat A, Kipnis T, Lips H (1997) Ionic balance, biomass production, and organic nitrogen as affected by salinity and nitrogen source in annual ryegrass. J Plant Nutr 20(10): 1291–1316.
Sahrawat KL (1982) Nitrification in some tropical soils. Plant Soil 65: 281–286.
Sandoval-Villa M, Wood CW, Guertal EA (1999) Effects of nitrogen form, nighttime nutrient solution strength, and cultivar on greenhouse tomato production. J Plant Nutr 22: 1931–1945.
Santamaria P, Elia A, Serio F, Gonella M, Parente A (1999) Comparison between nitrate and ammonium nutrition in fennel (fenouil), celery (céleri) and swiss chard (blette). J Plant Nutr 22(7): 1091–1106.
Soltani A, Hajji M, Grignon C (1992) Bilan des échanges ioniques en milieu NO_3/NH_4 et coûts énergétiques de la croissance chez l'orge (*Hordeum vulgare* L.). Agronomie 12: 723–732.
Speer M, Kaiser WM (1994) Remplacement of nitrate by ammonium as the nitrogen source increases the salt sensitivity of pea plant. II. Inter- and intracellular solute compartimentation in leaflets. Plant Cell Environ 17: 1223–1231.
Speer M, Brune A, Kaiser WM (1994) Remplacement of nitrate by ammonium as the nitrogen source increases the salt sensitivity of pea plant. I. Ion concentrations in roots and leaves. Plant Cell Environ 17: 1215–1221.
Thursby GB, Harlin MM (1984) Interaction of leaves and roots of Ruppia maritima in the uptake of phosphate, ammonia and nitrate. Marine Biol 83: 61–67.
Waisel Y, Eshel A, Agami M (1986) Salt balance of leaves of the mangrove *Avicennia marina*. Physiol Plant 67: 67–72.
Yokoi S, Bressan RA, Hasegwa (2002) Salt stress tolerance of plants. JIRCAS Working report, pp. 25–33.
Youssef T, Ghanem A (2002) Salt secretion and stomatal behaviour in *Avicennia marina* seedlings fumigated with the volatile fraction of light Arabian crude oil. Environ Pollut 116: 215–223.

Part II
Improving Crop Efficiency

Chapter 11
Strategies for Crop Improvement in Saline Soils

R. Munns

Abstract With increasing salinization and desertification of previously productive land, new sources of salt tolerance are needed for crops grown in areas with saline sub-soils, or with rising water tables that bring salts to the surface. Salt tolerance is needed in perennial species that might be used to lower the water tables and so control salinization, and also for annual crops providing food and forage.

Salts in the soil water inhibit plant growth for two reasons. First, the presence of salt in the soil solution reduces the ability of the plant to take up water, and this leads to reductions in the growth rate. This is the osmotic or water-deficit effect of salinity. Second, if excessive amounts of salt enter the plant in the transpiration stream there will be injury to cells in the transpiring leaves and this may cause further reductions in growth. This is the salt-specific or toxic effect of salinity. As salinity is often caused by rising water tables, it can be accompanied by waterlogging. Waterlogging itself inhibits plant growth and also reduces the ability of the roots to exclude salt, thus increasing the uptake rate of salt and its accumulation in shoots.

Vast natural variation exists within crop species and their close relatives which is largely unexplored. This biodiversity could provide improved germplasm for salt-affected land. However, screening large germplasm collections is difficult and more targeted and feasible selection techniques are required. Knowledge of the target environment and understanding of the genetic basis for improvement will help to choose the most appropriate screening method. Towards this end, the different types of salinity and the physiological and molecular mechanisms for salt tolerance are briefly summarised in this review. There is great scope for improving the tolerance of important food and feed crops.

Keywords Sodium • chloride • wheat • NaCl • Na^+ • Cl^-

1 Introduction

Salinity affects over 6% of the world's land. Of the current 230 million hectares of irrigated land, 45 million hectares are salt-affected (19.5%) and of the 1,500 million hectares under dryland agriculture, 32 million are salt-affected to varying degrees (2.1%). These data come from the FAO Land and Plant Nutrition Management Service (http://www.fao.org/ag/agl/agll/spush).

2 The Nature of Salinity

Salinity occurs through natural or human-induced processes that result in the accumulation of dissolved salts in the soil water to an extent that inhibits plant growth. Sodicity is a secondary result of salinity in clay soils, where leaching due to rainfall (or in some cases, irrigation water) has washed soluble salts into the subsoil, and left sodium bound to the negative charges of the clay.

2.1 Types and Causes of Salinity

A *saline soil* is defined as having a high concentration of soluble salts, high enough to affect plant growth.

R. Munns
CSIRO Plant Industry, G.P.O. Box 1600,
Canberra ACT 2601, Australia

M. Ashraf et al. (eds.), *Salinity and Water Stress*,

Salt concentration in a soil is measured in terms of its electrical conductivity. The USDA Salinity Laboratory defines a saline soil as having an EC_e of 4 dS/m or more. EC_e is the electrical conductivity of the 'saturated paste extract'. However, many crops are affected by soil with an EC_e less than 4 dS/m, particularly in regions of high evapotranspiration, leading to low soil moistures. The actual salinity of a field whose soil has an EC_e of 4 dS/m could be over 8 dS/m most of the time. As described below, this would severely limit yield of most crops.

2.1.1 Natural or Primary Salinity

Primary salinity results from the accumulation of salts over long periods of time, in the soil or groundwater. It is caused by two natural processes. The first is the weathering of parent materials containing soluble salts. Weathering processes break down rocks and release soluble salts of various types, mainly chlorides of sodium, calcium and magnesium, and to a lesser extent, sulphates and carbonates. Sodium chloride is the most soluble salt. The second is the deposition of oceanic salt carried in wind and rain. 'Cyclic salts' are ocean salts carried inland by wind and deposited by rainfall, and are mainly sodium chloride. Rainwater contains from 6 to 50 mg/kg of salt, the concentration of salts decreasing with distance from the coast. If the concentration is 10 mg/kg, this would add 10 kg/ha of salt for each 100 mm of rainfall per year. Accumulation of this salt in the soil is considerable over millennia.

The term '*transient salinity*' denotes the seasonal and spatial variation of salt accumulation in the root zone; salinity that is not influenced by groundwater processes and rising watertable (Rengasamy 2002). This transient salinity fluctuates in depth, due mainly to seasonal rainfall patterns. Transient salinity is extensive in many landscapes dominated by subsoil sodicity (Rengasamy 2002).

2.1.2 Secondary or Human-Induced Salinity

Secondary salinisation results from human activities that change the hydrologic balance in the soil between precipitation (irrigation or rainfall) and water used by crops (transpiration). The most common causes are (i) land clearing and the replacement of perennial vegetation with annual crops, and (ii) irrigation schemes using irrigation water containing salts, or having insufficient drainage. Prior to human activities, in arid or semi-arid climates, the water used by natural vegetation was in balance with the rainfall, with the deep roots of native vegetation ensuring that the watertables were well below the surface. Clearing and irrigation have changed this balance, so that rainfall on the one hand, and irrigation water on the other, provides more water than the crops could use. The excess water raises water tables and mobilises salts previously stored in the subsoil and brings them up to the root zone. Plants use the water and leave the salt behind until the soil water becomes too salty for further water uptake by roots. The watertable continues to rise, and when it comes close to the surface, water evaporates and leaves salts behind on the surface, forming a salt scald. The mobilised salt can also move beneath the soil, and into water courses.

In many irrigated areas, the water table has risen due to excessive amounts of applied water coupled with poor drainage. Irrigation water adds appreciable amounts of salt, even with good quality irrigation water containing only 200–500 mg/kg of soluble salt. Irrigation water with a salt content of 500 mg/kg (i.e. 500 mg/l) contains 0.5 t of salt per 1,000 m^3. Since crops require 6,000–10,000 m^3 of water per hectare each year, one hectare of land will receive 3–5 t of salt. Because the amount of salt removed by crops is negligible, salt will accumulate in the root zone, and must be leached by supplying more water than is required by the crops. If drainage is not adequate, the excess water causes the water table to rise, mobilising salts which accumulate in the root zone. When the crop is unable to use all the applied water, waterlogging occurs.

Land clearing also changes the hydrological balance. In its natural state, native deep-rooted and perennial vegetation use almost all the rainwater that falls on the land. In arid or semi-arid climates the growth rate of the natural vegetation is limited by the availability of fresh rainwater. Salts naturally present in the soil will be flushed down by rain, and accumulate at the bottom of the root zone. Clearing the deep-rooted native vegetation, and replacing it with shallow-rooted annual species that do not use all the rainfall, allows rainwater to escape below the roots, and 'recharge' the groundwater. Clearing of native vegetation for dryland agriculture can increase the rate of drainage by 100 times. In the Mallee Region of southern Australia, measurements show that drainage under native vegetation is only 0.1 mm per year, but under annual crops it is 10 mm per year. This additional rainwater enters the groundwater or aquifers, causing the watertable to rise. In Australia, 2 million hectares of land have been damaged by rising watertables due to land clearing, and another 15 million hectares are at risk of salinisation by rising watertables over the next 50 years (National Land and Water Resources Audit, http://audit.ea.gov.au).

2.2 Soil Sodicity

Sodic soils have a low concentration of soluble salts, but a high percent of exchangeable Na^+; that is, Na^+ forms a high percent of all cations bound to the negative charges on the clay particles that make up the soil complex. Sodicity is defined in terms of the threshold ESP (exchangeable sodium percentage) that causes degradation of soil structure. The negatively charged clay particles are held together by divalent cations. When monovalent cations displace the divalent cations on the soil complex, and the concentration of free soluble salts is low, the complex swells and the clay particles separate (*'disperse'*). The USDA Salinity Laboratory defines a sodic soil as having an ESP greater than 15 (www.ussl.ars.usda.gov), but in Australia it is considered sodic when the ESP is greater than 6. This lower threshold is due to Australian soils having a low content of other soluble cations, particularly Ca^{2+}, which help to stabilise clay colloids during leaching.

If the concentration of soluble salts is sufficiently low, hydrolysis of the sodic clay will occur, creating a highly alkaline soil. *Alkaline soils* are a type of sodic soil with a high pH due to carbonate salts, and are defined as having an ESP of 15 or more with a pH of 8.5–10.

The process of sodicity is complex and occurs over a long period of time. Initially, salts that have accumulated within the soil profile, from either airborne deposition or mineral weathering, cause the clay fraction of the soil to become saturated with sodium. Subsequently, leaching of the profile, either by rainwater over prolonged periods, or by irrigation with fresh water, lowers the electrolyte concentration and the clay particles disperse. Further leaching washes the dispersed clay particles deeper into the profile where they block pores and hinder infiltration of water. The soil then is very slow to drain, and is readily waterlogged.

In semi-arid environments, soil profiles are commonly *saline/sodic*, where the salt has accumulated due to the low permeability of the sodic subsoil. In theory, if sufficient salts accumulate, the threshold electrolyte concentration for flocculation will be exceeded and the clay will flocculate and take on pseudo-structure. However, given that permeability and leaching will then increase, the subsequent dilution of salts will cause colloids to disperse. Consequently, a quasi-steady state between flocculation and dispersion processes is maintained.

Saline/sodic soils are widespread in arid and semi-arid lands of the world. Water infiltration is slow, and salts derived from rainfall or weathering reactions accumulate in saturated zones in the subsoil.

3 Effects of Salinity on Crops

3.1 The Effect of Salinity on Plant Growth

Salts in the soil water inhibit plant growth for two reasons. First, the presence of salt in the soil solution reduces the ability of the plant to take up water, and this leads to reductions in the growth rate. This is referred to as the osmotic or water-deficit effect of salinity. Second, if excessive amounts of salt enter the plant in the transpiration stream there will be injury to cells in the transpiring leaves and this may cause further reductions in growth. This is called the salt-specific or ion-excess effect of salinity (Greenway and Munns 1980). The definition of salt tolerance is usually the percent biomass production in saline soil relative to plants in non-saline soil, after growth for an extended period of time. For slow-growing, long-lived, or uncultivated species it is often difficult to assess the reduction in biomass production, so percent survival is often used.

As salinity is often caused by rising water tables, it can be accompanied by waterlogging. Oxygen deficiency inhibits plant growth and also reduces the ability of the roots to exclude salt, thus increasing the uptake rate of salt and its accumulation in shoots (Barrett-Lennard 2003).

3.2 Causes of the Growth Reduction Under Saline Conditions

The salt in the soil solution (the "osmotic stress") reduces leaf growth and to a lesser extent root growth, and decreases stomatal conductance and thereby photosynthesis (Munns 1993). The cellular and metabolic processes involved are in common to drought-affected plants (Munns 2002). The rate at which new leaves are produced depends largely on the water potential of the soil solution, in the same way as for a drought-stressed plant. Salts themselves do not build up in the growing tissues at concentrations that inhibit growth: meristematic tissues are fed largely in the phloem from which salt is effectively excluded, and rapidly elongating cells can accommodate the salt that arrives in the xylem within their expanding vacuoles. So, the salt taken up by the plant does not directly inhibit the growth of new leaves.

The salt within the plant enhances the senescence of old leaves. Continued transport of salt into transpiring leaves over a long period of time eventually results in very high Na^+ and Cl^- concentrations, and they die. The rate of leaf death is crucial for the survival of the plant. If new leaves are continually produced at a rate greater than that at which old leaves die, then there are enough photosynthesising leaves for the plant to produce flowers and seeds, although in reduced numbers. However, if the rate of leaf death exceeds the rate at which new leaves are produced, then the plant may not survive to produce seed. For an annual plant there is a race against time to initiate flowers and form seeds, while the leaf area is still adequate to supply the necessary photosynthate. For perennial species, there is an opportunity to enter a state of dormancy, and thus survive the stress.

The two responses give rise to a two-phase growth response to salinity over time. The first phase of growth reduction is quickly apparent, and is due to the salt outside the roots. It is essentially a water stress or osmotic phase, for which there is surprisingly little genotypic difference. Then there is a second phase of growth reduction, which takes time to develop, and results from internal injury. The two-phase growth response is illustrated in Fig. 11.1. The experiment was conducted with two genotypes with contrasting rates of Na^+ uptake, and known differences in salt tolerance; previous experiments had shown that the genotype with the low Na^+ uptake rate had a higher survival of high salinity. Figure 11.1 shows that during the first 3–4 weeks after the soil was salinised, there was a large growth reduction in both genotypes. This is called the 'Phase 1' response, and is due to the osmotic effect of the salt. Then after 4 weeks, the genotypes separated; the one with the low Na^+ uptake rate continued to grow, although still at a reduced rate compared to the controls in non-saline solution, but the one with the high Na^+ uptake rate produced little biomass and many individuals died. This is the 'Phase 2' response, and is due to genotypic differences in coping with the Na^+ or Cl^- ions in the soil, as distinct from the osmotic stress.

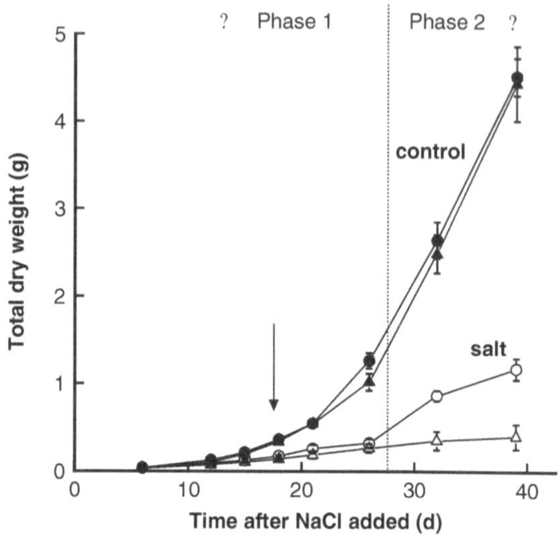

Fig. 11.1 Two accessions of the diploid wheat progenitor *Ae. Tauschii* grown in supported hydroponics in control solution (closed symbols) and in 150 mm NaCl (open symbols). Circles denote the tolerant accession, triangles the sensitive one. The arrow marks the time at which symptoms of salt injury could be seen on the sensitive accession; at that time the proportion of dead leaves was 10% for the sensitive and 1% for the tolerant accession (Adapted from Munns et al. 1995)

These results illustrate the principle that the initial growth reduction is due to the osmotic effect of the salt outside the roots, and that what distinguishes a salt-sensitive plant from a more tolerant one is the inability to prevent salt from reaching toxic levels in the transpiring leaves, which takes some time.

3.3 Variation in Salt Tolerance Between Species

Species vary in their capacity to tolerate salinity. Amongst the major food crops, barley, cotton, sugar beet and canola

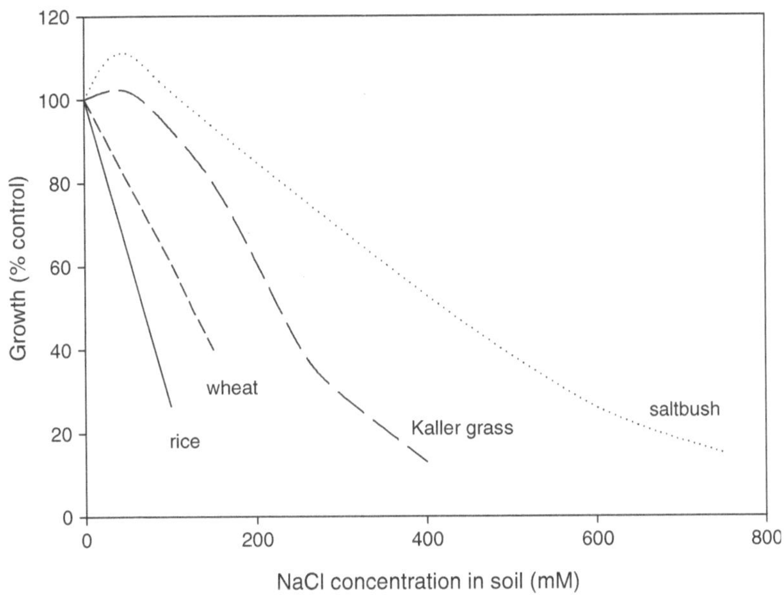

Fig. 11.2 Biomass production of four diverse and important plant species in a range of salinities. Wheat is one of the more salt-tolerant crops, and rice is one of the more salt-sensitive crops. Two halophytes: a saltbush species *Atriplex amnicola* and a grass *Diplachne* (syn. *Leptochloa*) *fusca* or Kallar grass. Both halophytes show outstanding salt tolerance with high growth rates and are being used in Australia and Asia for grazing on saline land

are the most tolerant; bread wheat is moderately tolerant, while rice and most legume species are sensitive. For a more complete list of crops see Maas and Hoffman (1977) and Kotuby-Amacher et al. (2000).

The three most widely grown crops in the world are wheat, rice and maize. Figure 11.2 shows the growth of rice in saline soil in comparison to that of more tolerant species. Wheat can be grown in salinities up to 150 mM NaCl (15 dS/m) as long as rainfall or irrigation can rescue the crop at critical stages such as pollen meiosis and fertilisation. Rice is more salt-sensitive. Maize falls in between these two species in terms of salt sensitivity.

Another criterion of salt tolerance of crops is their yield reduction in increasingly saline soils. A survey of salt tolerance of crops, vegetables and fruit trees was made by the USDA Salinity Laboratory. This shows for each species a threshold salinity below which there is no reduction in yield, and then a regression for the reduction in yield with increasing salinity (Fig. 11.3). Full details are available at http://www.ussl.ars.usda.gov. The data in some cases are for a single cultivar of the species, or a limited number of cultivars at a single site, so they are not necessarily representative of the species. Further, the data are related to an EC_e value, which is not an appropriate reference point for a sandy soil. However, the data are useful in that they show the wide range of tolerance across species, and also show that yield has a different pattern of response than does vegetative biomass. Yield always shows a threshold in response to a range of salinities, but with young plants a threshold is rarely seen. With plants exposed to salinity at an early stage of seedling deve-lopment there are linear reductions in both leaf area expansion and total plant biomass with increasing salinity, as shown in Fig. 11.2.

3.4 Mechanisms of Control of Salt Transport

To grow in saline conditions, plants must maintain a high water status in the face of soil water deficits and potential ion toxicity. A plant can only grow or survive in a saline soil if it can both continue to take up water *and* exclude a large proportion of the salt in the soil solution.

Roots must exclude most of the Na^+ and Cl^- in the soil solution or the salt will gradually build up with time in the shoot and become so high that it kills the leaves. To prevent salt building up with time in the shoot, roots

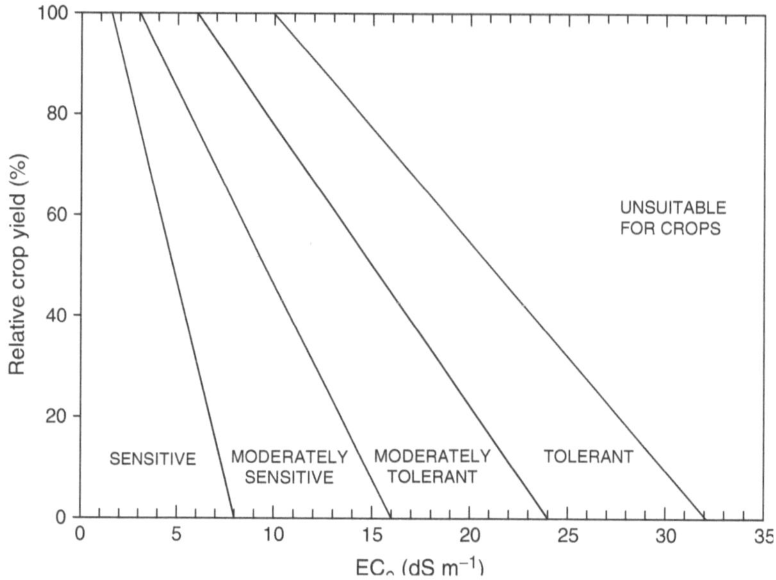

Fig. 11.3 Categories for classifying crop tolerance to salinity (http://www.ussl.ars.usda.gov). Note that the EC_e is more applicable to an irrigated than a rainfed field as, in the latter the soil moisture content might be two to four time less than in a saturated soil

should exclude 98% of the salt in the soil solution, allowing only 2% to be transported in the xylem to the shoots. As described in Munns et al. (2006), plants retain only about 2% of the water they transpire, i.e. they take up about 50 times more water from the soil than they retain in their shoot tissues. In order to prevent the salt concentration in the shoot increasing above that in the soil, then only 2% of the salt should be allowed into the shoot, i.e. 98% should be excluded.

Most plants in fact do exclude about 98% of the salt in the soil solution, allowing only 2% to be transported in the xylem to the shoots. Differences between cereal genotypes with contrasting rates of Na^+ uptake, when grown in 50 mM NaCl, range from 99% for Janz to 98% for other bread wheats (Munns 2005). Durum wheat, rice and barley are not such good excluders, yet they still exclude at least 94% of the soil Na^+ from the transpiration stream (Munns 2005).

3.5 *Other Mechanisms of Salt Tolerance*

There are other mechanisms of salt tolerance, such as partitioning the salt arriving in the shoot, either by retaining it in the leaf base or stem, or directing it salt away from younger leaves and towards older leaves. This is particularly important for species that cannot exclude more the 98% of the salt from the transpiration stream. Other traits are needed to improve tolerance to the osmotic effect of the salt outside the roots. Such traits include water use efficiency, osmotic adjustment, and morphological or developmental patterns that conserve water and advance the flowering date (Colmer et al. 2005).

If a plant cannot exclude 98% of the salt from the transpiration stream, it must be able to compartmentalise the salt in vacuoles, thereby protecting the cytoplasm from ion toxicity, and avoiding build-up in the cell wall which would cause dehydration (Flowers and Yeo 1986). Otherwise, in older leaves, the salt concentration would eventually become high enough to kill the cells. If Na^+ and Cl^- are sequestered in the vacuole of the cell, K^+ and organic solutes should accumulate in the cytoplasm and organelles to balance the osmotic pressure of the Na^+ and Cl^- in the vacuole. This mechanism has been shown to be enhanced in barley versus wheat (James et al. 2006).

4 Mitigation of Salinity Stress by Management Practices

The solution is different in the three types of salinity.

4.1 Irrigated Agriculture

Irrigated agriculture can be sustained by better irrigation practices. For example with fruit trees or grapevines, adoption of different methods such as regulated deficit irrigation, partial root zone drying methodology, or drip or micro-jet irrigation, can optimise use of water (Stirzaker 2003). With field crops, appropriate use of ridges or beds for planting can prevent waterlogging due to irrigation.

Salt build-up in the soil can be controlled by leaching fractions, where fresh irrigation water is available, and by drains. The disposal of saline drainage water from salt-affected irrigated land has been a controversial issue, and recycling of such waters has been considered for further crop irrigation. Feasibility studies indicate that re-use of drainage water is suitable for irrigation of moderately salt-tolerant crops. Up to 4 dS/m can be used for irrigation of moderately tolerant crops provided that the ground is leached with fresh water before sowing (Rhoades et al. 1999; Goyal et al. 1999). With the more salt tolerant crops like sugar beet and cotton, the use of water up to 9 dS/m can be sustained for 3 years, but for a longer period the salinity must be reduced to 5 dS/m (Goyal et al. 1999).

Mulching reduces evaporation from the soil surface which in turn reduces the upward movement of salts. Reduced evaporation also reduces the need to irrigate. Incorporating crop residues or green-manure crops improves soil structure and improves water infiltration which provides safeguard against adverse effects of salinity.

4.2 Dryland Agriculture

The cause of 'dryland salinity' is the clearing land for dryland agriculture, and replacement of native perennial vegetation by annual crops. The replacement of perennial deep-rooted native vegetation by shallow-rooted annual crops or pastures results in wetter subsoils and accompanying larger deep drainage beyond the reach of shallow roots, leading eventually to rising water tables. If the ground water is saline, which it commonly is in semiarid environments, salt scalds appear when the water tables reach the soil surface. Even if the water tables are only slightly rather than saline the surface can become saline owing to the salt concentrating as water evaporates.

Mitigation of dryland salinity in cropping lands requires control of drainage beyond the reach of the crop roots. There is no single solution. There is, however, a range of options that farmers can select from, including growing longer-season crops, which tend to have deeper roots, and various techniques for incorporating some deep-rooted perennial species into cropping systems to tap the water in the deep subsoil that may have accumulated during a wet season (Black et al. 1981). Re-introduction of perennials to use more water and reduce recharge is recommended (http://crcsalinity.com).

'Phase farming' is one effective way of incorporating perennials into a cropping system. Phase farming involves the tactical rotation of herbaceous perennial pasture, such as alfalfa (lucerne) which can be grazed or harvested for hay, with a series of annual crops. The perennial pasture dries the subsoil below the roots of annual crops, thereby creating a buffer zone in which water and nutrients that leak below the crops can be held for a few seasons.

Strips of woody perennials can also help, though their effectiveness is limited by the maximum lateral movement of water through unsaturated soil to their roots, which is usually no more than about 1 m. Even though surface roots may spread from the tree for a distance several times the spread of the canopy (and thereby reduce yields of adjacent crop), the deeper roots do not spread so far and the strips may do little more than control deep drainage only in a strip of soil little larger than the width of the canopy (Stirzaker et al. 2002).

Another possibility is to identify areas giving consistently poor yield. Such areas, which are especially prone to deep drainage, can be excluded from cropping, and put under either permanent perennial pasture or trees.

If these various options for reducing deep drainage are effective in lowering water tables so that any salt scalds dry out, there is still the problem that the salt remains in the root zone. Further rehabilitation requires a succession of plants, starting with halophytes, with can take up the water and thereby create space for rain to wash the salt deeper into the soil profile. Salt tolerant crops may then be able to grow there, and enable further leaching of the salt. If the soil has become sodic, chemical amelioration (e.g. with gypsum) may also be necessary.

The question of whether plants can remove the salt from the soil should be considered.

An estimation was made by Barrett-Lennard (2002) of the efficacy of saltbush in removing salt from soils. His calculation was for a more saline soil that is typical of Western Australia (86 t ha^{-1} of salt in the upper 2 m of the soil profile), two rates of growth (2 and 10 t ha^{-1} per annum) and a salt content of 25% salt. The calculation showed that even after 20 years the salt concentration in the soil would be 89% and 45% of the starting levels (at 2 and 10 t ha^{-1} production respectively). This is an optimistic outcome as saltbush, even though the most productive of all halophytes, rarely produces more than 2 t/ha per annum without irrigation (Barrett-Lennard 2002). Furthermore, the salt that accumulates in the leaves will probably return to the soil through leaf fall or animal urine.

With grain crops, the grain does not contain any salt, so the normal harvesting methods do not remove any salt.

4.3 *Natural Salinity*

Natural salinity occurs in the low rainfall zones, of 300 mm and below, where even an annual crop can use all the rain, so water tables are not disturbed. For natural salinity, as well as for irrigated and dryland, improving the salt tolerance of crops is a useful approach.

5 Increasing the Salt Tolerance of Crops

5.1 *Conventional Breeding for Salt Tolerance in Crops and Pastures*

Not only is there much variation in salinity tolerance between species, there is also variation in salinity tolerance within species, especially in out-crossing species like lucerne. Approaches to plant improvement in the past have been to (1) screen collected germplasm for salinity tolerance, (2) cross the identified tolerant types with adapted cultivars and (3) select the desired plant types having salinity tolerance as well as other agronomic traits from the advancing and segregating generations. By exploiting the naturally occurring genetic variability that exists within a species, some relatively tolerant cultivars have been developed for crops including rice, wheat, lucerne, white clover and citrus.

There is probably a great diversity in salinity tolerance within species that has not been fully explored and exploited. One reason for this is the difficulty of screening large numbers of individuals for small, repeatable and quantifiable differences in biomass production, let alone yield. Obtaining a wide range of germplasm with potential genetic differences in salinity tolerance is not difficult, because international collections are usually available to any scientist. However, the difficulty lies in how to measure salinity tolerance.

Salinity tolerance is difficult to measure because of its complexity. Not only are there a number of genes controlling salinity tolerance whose effect interacts strongly with environmental conditions, but there are two major and distinct components of salinity tolerance which can often be difficult to distinguish (Munns 2002).

Screening based on growth needs to allow for the two distinct mechanisms for salinity tolerance: tolerance to the osmotic effect of the saline soil solution, and tolerance to the salt-specific nature of the saline solution.

The osmotic effect occurs instantly the soil water potential decreases, and recovers instantly it increases (Passioura and Munns 2000). The genetic variation in the growth response to the osmotic effect of salinity is likely to be small, both within a species, and across similar species. For example, we found little difference between the effect of salinity on leaf elongation in 15 cultivars of bread wheat, durum wheat, barley and triticale, cultivars with established reputations in salinity tolerance. All 15 cultivars had the same decrease when the salt was increased from 0 to 250 mM NaCl over 10 days (Munns et al. 1995). Even the most salinity-sensitive genotype, a durum wheat cultivar, had the same percentage reduction as the most salinity-tolerant genotype, a barley cultivar. In fact, there is a remarkable similarly between different species in the osmotic response in saline solution. For instance, 100 mM NaCl causes approximately 50% decrease in leaf elongation rate of the salinity-sensitive species maize and rice, and nearly as much in the salinity-tolerant species bread wheat and barley. Longer term growth responses clearly differ.

Growth can be measured as leaf elongation, root elongation, leaf area expansion, or shoot biomass. Of these indices of growth, leaf area expansion or shoot biomasses are the most sensitive and comprehensive, as this includes production of tillers or lateral shoots. The number of lateral shoots is more sensitive to water stress than the elongation rate of any given leaf. Whole plant survival at high salinity may reflect the salt-specific effect rather than the osmotic effect, as the plant's ability to control salt uptake may break down at high salinity. Germination is easy to measure, but little or no relation between salinity tolerance at germination and that of the seedling or adult plant has been found in any species examined. These methods are summarised in Munns and James (2003)

The salt-specific effect shows up in old leaves that have accumulated excessive amounts of Na^+ and Cl^-. While it may be visible, in that old leaves die earlier, it may not affect the rate of new leaf production for some time. The length of time required before the growth differences between genotypes due to the salt-specific effect can be seen depends on the salinity and the degree of tolerance of the species. This represents a 'second phase' of the growth reduction. The second phase will start earlier in plants that are poor excluders of Na^+, and will start earlier when root temperatures are higher. For plants such as rice that are grown at high temperatures, 2 weeks days in salinity is sufficient to generate differences in biomass between genotypes that correlate well with differences in yield. However for temperate crops, as at least a month is needed for genotypic differences in the response of biomass production to take effect.

The labour and space demands of these long experiments makes them impractical for screening large numbers of genotypes, or selecting salinity-tolerant progeny resulting from crosses with cultivars. Not only do plants need to be grown for lengthy periods of time, but controls need to be included. Control plants consume a large area of glasshouse space if they are to be grown at their optimum rate, with sufficient space so that radiation is not limiting. In the field, the major drawback is the heterogeneous nature of soil salinity and other soil constraints.

Screening for a trait associated with a specific mechanism is preferable to screening for salinity tolerance itself. Screening for specific traits can reduce the time needed to grow plants in salinity, and can eliminate the need to grow plants under control conditions, thus making savings on glasshouse space and labour.

5.2 Traits for Salt-Specific Effect

Because of the complex nature of salinity tolerance, as well as the difficulties in maintaining long-term growth experiments, trait-based selection criteria have been recommended for screening techniques. Specific traits are less subject to environmental influence than growth rates.

Further, this allows for different traits to be pyramided, especially when molecular markers for specific traits have been identified.

5.2.1 Salt Exclusion

The most successful trait for plant breeding is the rate of Na^+ or Cl^- accumulation in leaves, measured as the increase in salt in a given leaf over a fixed period of time. This trait has a high heritability and has been used to develop cultivars of rice, white clover, and lucerne with increased tolerance to saline soil. Sometimes K^+/Na^+ discrimination instead of Na^+ exclusion is used for screening, however the uptake of K^+ and the resultant K^+/Na^+ discrimination may be the result of genetic differences in the regulation of Na^+ uptake, and not independent of it. If so, there is nothing to be gained by measuring K^+ as well as Na^+.

A correlation between Na^+ or Cl^- accumulation and salinity tolerance is found in most species, however not all species contain significant genetic variation in Na^+ or Cl^- accumulation (Munns et al. 2006). Durum wheat is one species in which there is significant genetic variation in Na^+ but not Cl^- uptake. Genotypes with low Na^+ uptake are more salinity tolerant, as indicate by greater biomass production over 1–2 months in saline solution, in comparison to non-saline solution. In some species, Na^+ is retained in roots and stems in exchange for K^+, and only Cl^- progresses through to the leaves, balanced by K^+. In those cases, Cl^- exclusion correlates with salinity tolerance.

Proving that Na^+ (or Cl^-) exclusion confers salinity tolerance in terms of yield is not so easy. A comparison between durum landraces with very low and very high rates of Na^+ accumulation showed that, at moderate salinity, the yield of genotypes with low Na^+ was greater than those with high Na^+, but at high salinity there was no yield advantage. The osmotic effect of the salinity then dominated (Husain et al. 2003).

5.2.2 Tissue Tolerance

Tissue tolerance, i.e. tolerance of high internal Na^+ concentrations, cannot be measured directly, and is difficult to quantify. Tolerance of high internal Na^+ levels is evidenced by an absence of leaf injury despite high leaf concentrations of Na^+. Concentrations of Na^+ above 100 mM will start to inhibit most enzymes, so when tissue concentrations are over 100 mM, which corresponds to about 0.5 mmol g^{-1} DW (assuming a leaf water content of 5 g H_2O g^{-1} DW), the Na^+ must be compartmentalised in vacuoles, and be a higher concentration there than in the cytoplasm. Glycophytes are able to compartmentation Na^+ in vacuoles to some extent, as levels of Na^+ up to 1 mmol g^{-1} DW (200 mM) are quite common in photosynthetically active leaves of many species. In a study in wheat genotypes, Na^+ became potentially toxic only when leaf concentrations exceeded 1.25 mmol g^{-1} DW (250 mM), as judged by the onset of non-stomatal reductions in photosynthesis in durum wheat at this concentration (James et al. 2002).

There may be genetic variation for tolerance of high internal Na^+ concentrations in many species, as indicated by 'out-liers' in correlations between leaf Na^+ concentrations and salinity tolerance. This has been found in rice and wheat and probably many other species. However, quantitative assess of the genetic variation is difficult. The trait is characterised by leaf longevity, lack of necrosis, and prolonged growth despite very high accumulation of Na^+. Leaf injury, however could arise from a number of reasons. First there would be the osmotic effects of salt in the soil solutions, causing accelerated senescence due to water deficit. Second, there could be nutrient imbalances resulting in deficiencies or excesses of other ions. Third, there could be dehydrating effects of salts building up in the cell walls. A phenotype other than leaf injury is needed to distinguish between genotypes with different degrees of tissue tolerance.

5.3 Molecular Approaches to Achieve Salinity Tolerance

5.3.1 Gene Transformation

Significant advances in the field of molecular biology technology have been made during the past decade. The use of molecular techniques to selectively introduce desired genes may provide alternative ways to classical plant breeding to achieve salinity tolerance. These techniques will benefit the development of salinity-tolerant cultivars based on specific traits that are controlled by one gene, e.g. a transcription factor or an important ion channel. The work of Zhang and Blumwald (2001) shows the progress made by using molecular technology. The authors reported the development of a salinity tolerant transgenic tomato plant in which over-expression of the vacuolar Na^+/H^+ antiporter shows dramatic improvement of vegetative growth and of fruit yield. This antiporter is the only known transporter that would compartmentalise Na^+ in the vacuole, where Na^+ has little chance of toxic effect on metabolism, or to be transported to younger leaves and fruits. These studies indicate great potential for transgenic approaches.

Given the natural diversity that exists, and given the current suspicion of consumers to genetically-engineered crops, it might be more realistic to consider using the genes identified as perfect markers for naturally-occurring diversity, or linked molecular markers to physiological traits. The use of molecular markers in breeding programs is increasing rapidly as they have been shown to greatly improve the efficiency of the breeding programs. Marker-assisted selection is non-destructive and can provide information on the genotype of a single plant without exposing the plant to the stress. The technology is capable of handling large numbers of samples. In order to use marker-assisted selection in breeding programs, the markers must be closely linked to the trait, and work across different genetic backgrounds. QTL (Quantitative Trait Locus) mapping and marker-assisted selection is a technique that has many advantages over phenotypic screening as a selection tool. The efficiency of genetic mapping has improved greatly in recent years, with the advent of high-density maps incorporating microsatellite markers, which are overtaking the tedious RFLP markers, and unreliable RAPD markers.

QTLs for salinity tolerance have been described in several cereal species, including rice, barley and wheat. However, few studies have yielded robust markers that can be used across a range of germplasm, significant associations between the trait and the marker being confined to the populations in which they were derived.

We have been using the above approach to select for natural variation in durum (tetraploid) wheat. A particular accession, Line 149, was selected with exceptionally low rates of Na^+ accumulation in leaves, controlled by two dominant genes of major effect (Munns and James 2003). These have been named *Nax1* and *Nax2* (for Na+ exclusion). One gene retrieves Na^+ from the xylem in the roots, and enhances K^+ loading of the xylem, while the other retrieves Na^+ from the xylem in both the root and the shoots (James et al. 2006). Selected progeny of the original cross with the Australian durum cultivar Tamaroi were backcrossed into Tamaroi and other Australian cultivars, and are being evaluated in the field in saline and non-saline soil. Initial trials indicate a yield improvement of 10% in saline soil (R. Munns and R. James, unpublished data, 2006). The transfer of *Nax* genes from the durum wheat Line 149 into modern durum cultivars and into bread wheat has been assisted by molecular selection. *Nax1* was mapped to the long arm of chromosome 2A (Lindsay et al. 2004) and one very tightly linked marker, *gwm312*, is being used routinely to select low Na^+ progeny in our durum breeding program. *Nax2* has recently been mapped and a tightly linked marker is being used for selection of lines containing *Nax2*. In the field, the *Nax* genes reduce the Na^+ concentration in the leaves to very low concentrations, and also reduce the Na^+ in the grain (R. Munns and R. Hare, unpublished data 2006). The markers for *Nax1* and *Nax2* are being used to pyramid the *Nax* genes with traits associated with improved performance under drought (A.G. Condon and D. Mullan, unpublished data 2007).

6 Conclusions

Increases in the salinity tolerance of crops and pasture plants can overcome some of the impacts of salinity in areas not associated with rising water tables. Where there is a high water table, plants need to be tolerant of waterlogging as well as of salinity.

Solutions to salinity depend on lowering water tables, which means that plants should use as much water as possible, and use all the rain or irrigation water. Management of salinity includes planting perennials in areas identified as recharge areas. Salinity tolerance is not an important character for these plants, rather deep roots and efficient water extraction is more important.

Long-term solutions to salinity involve changed farming systems and a changed attitude to using water. Water-inefficient crops such as rice might give way to water-efficient crops. In catchments with rising water tables, solutions will also depend on social, economic, and political factors, and an acceptance that old ways have to change if agriculture is to be sustained.

References

Barrett-Lennard EG (2002) Restoration of saline land through revegetation. Agr Water Manage 53: 213–226.

Barrett-Lennard EG (2003) The interaction between waterlogging and salinity in higher plants: causes, consequences and implications. Plant Soil 253: 35–54.

Black AL, Brown PL, Halvorson AD, Siddoway FH (1981) Dryland cropping strategies for efficient water-use to control saline seeps in the Northern Great Plains, USA Agr Water Manage 4: 295–311.

Colmer TD, Munns R, Flowers TJ (2005) Improving salt tolerance of wheat and barley: future prospects. Aust J Exp Agr 45: 1425–1443.

Flowers TJ, Yeo AR (1986) Ion relations of plant under drought and salinity. Aust J Plant Physiol 13: 75–91.

Goyal SS, Sharma SK, Rains DW, Lauchli A (1999) Long term reuse of drainage waters of varying salinities for crop irrigation in a cotton-safflower rotation system in the San Joaquin Valley of California – A nine year study: I. Cotton (*Gossypium hirsutum* L.). J Crop Prod 2: 181–213.

Greenway H, Munns R (1980) Mechanisms of salt tolerance in nonhalophytes. Ann Rev Plant Physiol 31: 149–190.

Husain S, Munns R, Condon AG (2003) Effect of sodium exclusion trait on chlorophyll retention and growth of durum wheat in saline soil. Aust J Agr Res 54: 589–597.

James RA, Rivelli AR, Munns R, von Caemmerer S (2002) Factors affecting CO_2 assimilation, leaf injury and growth in salt-stressed durum wheat. Func Plant Biol 1393–1403.

James RA, Davenport R, Munns R (2006) Physiological characterisation of two genes for Na^+ exclusion in wheat: *Nax1* and *Nax2*. Plant Physiol 142: 1537–1547.

Kotuby-Amacher J, Koenig R, Kitchen B (2000) Salinity and plant tolerance. http://extension.usu.edu/htm/publications

Lindsay MP, Lagudah ES, Hare RA, Munns R (2004) A locus for sodium exclusion (*Nax1*), a trait for salt tolerance, mapped in durum wheat. Funct Plant Biol 31: 1105–1114

Maas EV, Hoffman GJ (1977) Crop salt tolerance - current assessment. J Irrig Drain Div Am Soc Civil Eng 103: 115–134.

Munns R (1993) Physiological processes limiting plant growth in saline soil: some dogmas and hypotheses. Plant Cell Environ 16: 15–24.

Munns R (2002) Comparative physiology of salt and water stress. Plant Cell Environ 25: 239–250.
Munns R (2005) Genes and salt tolerance: bringing them together. New Phytol 167: 645–663.
Munns R, James RA (2003) Screening methods for salinity tolerance: a case study with tetraploid wheat. Plant Soil 253: 201–218.
Munns R, Schachtman DP, Condon AG (1995) The significance of two-phase growth response to salinity in wheat and barley. Aust J Plant Physiol 22: 561–569.
Munns R, James RA, Läuchli A (2006) Approaches to increasing the salt tolerance of wheat and other cereals. J Exp Bot 57: 1025–1043.
Passioura JB, Munns R (2000) Rapid environmental changes that affect leaf water status induce transient surges or pauses in leaf expansion rate. Aust J Plant Physiol 27: 941–948.
Rengasamy P (2002) Transient salinity and subsoil constraints to dryland farming in Australian sodic soils: an overview. Aust J Exp Agr 42: 351–361.
Rhoades JD, Chanduvi F, Lesch S (1999) Soil salinity assessment: methods and interpretations of electrical conductivity measurements. Irrigation and Drainage Paper 57, Food and Agriculture Organization of the United Nations, Rome. ftp://ftp.fao.org/agl/aglw/docs/idp57.pdf
Stirzaker RJ (2003) When to turn the water off: scheduling micro-irrigation with a wetting front detector. Irrig Sci 22: 177–185.
Stirzaker RJ, Lefroy EC, Ellis TW (2002) An index for quantifying the trade-off between drainage and productivity in tree-crop mixtures. Agr Water Manage 53: 187–199.
Zhang HX, Blumwald E (2001) Transgenic salt-tolerant tomato plants accumulate salt in foliage but not in fruit. Nature Biotech 19: 765–768.

Chapter 12
Role of Vetiver Grass and Arbuscular Mycorrhizal Fungi in Improving Crops Against Abiotic Stresses

A.G. Khan

Abstract In the developing countries like Pakistan, not only good quality lands are degraded due to wind and water erosion, water-logging and salinity, loss of organic matter and biodiversity etc. but also are contaminated with heavy metals. Various physical, chemical, and hydrological approaches are being developed and used to remediate such lands. Recently, an alternative approach, Vetiver System, which involves the use of Vetiver grass, *Vetiveria zizinoides*, is being explored to restore degraded land. It is able to act as a natural barrier against erosion and pollution and produce massive odorous root system which can be used for the extraction of an essential oil of great economic importance. Most plants growing on degraded and heavy metals contaminated soils, including vetiver grass, possess arbuscular mycorrhizae, indicating that these arbuscular mycorrhizal fungi (AMF) have evolved a tolerance to HM and that they play an important role in the mycorrhizoremediation of stressed agricultural soils. The aim of this article is to discuss the possible role of Vetiver System and symbiotic arbuscular mycorrhizal fungi associated with its root system in improving crops against abiotic stresses in soils.

Keywords Vetiver grass • arbuscular mycorrhizas • salinity • trace element contamination • soil erosion • mycorrhizoremediation

A.G. Khan
Department of Microbiology, Quaid-i-Azam University, Islamabad, Pakistan
e-mail: a.khan@uws.edu.au

1 Biosphere Pollution

1.1 Causes

Biosphere pollution with heavy metals is accelerated dramatically during the last few decades due to mining, smelting, manufacturing, amendments of agricultural soils with agro-chemicals and soil sludge, etc. (Brazi et al. 1996). Problems associated with the contamination of soil and water such as animal welfare, health, fatalities and disruptions of natural ecosystems are well documented in literature (Sheppard et al. 1992). Heavy metals such as Pb, Cr, As, Cu, Cd, and Hg, being added to our soils through industrial, agricultural and domestic effluents, persist in soils and can either be adsorbed in soil particles or leached into ground water (Khan 2003, 2006a, b, 2008). Human exposure to these metals through ingestion of contaminated food or uptake of drinking water can lead to their accumulation in humans, plants and animals. The excessive use of these inputs has caused deteriorating effects on soils and environment (Garforth and Lawrence 1997; Chaudhry et al. 2006). According to Baier (1994) agriculture development at present has also many challenges to face like to maintain sustainable and progressive production increases; protect production resources and prevent their degradation. Whereas agriculture development would require practices that are socially acceptable, technically sound and appropriate, economically viable, environmentally non-damaging for achieving food security and improved quality of life for rural people (Baier 1994; Williams 2000; Chaudhry et al. 2006). With the increasing use of agrochemicals to maintain and improve soil fertility, unwanted elements such as cadmium into soils due to contaminated sources of fertilizers, especially

M. Ashraf et al. (eds.), *Salinity and Water Stress*,

in developing countries, are being introduced into agricultural soils, which, poses a potential threat to the food chain (Chaney and Oliver 1996). Mining and industrial operations also lead to significant challenges for the management of the natural environments during and after these activities. Increased public awareness of the environmental impact of such activities is demanding an interdisciplinary, inter-organizational, and international effort.

2 Remediation Techniques

2.1 *Non-biological Remediation Techniques*

Various physical and chemical techniques to de-contaminate soils have been undertaken during the last 25 years and millions of dollars being spent by governments all over the world (McEldowney et al. 1993). However, all of them are labor intensive and costly, and cannot be applied to thousands of hectares of land contaminated with inorganic heavy metals (Burns et al. 1996). These technologies results in rendering the soil biologically dead and useless for plant growth as they remove all flora, fauna and microbes including useful nitrogen fixing bacteria and P-enhancing mycorrhizal fungi.

2.2 *Biological Remediation Techniques*

Microbial bioremediation technology, well known for decontamination of organics (Flathman et al. 1994), is not available for large-scale biodegradation of inorganic heavy metals. The health hazards caused by the accumulation of toxic metals in the environment together with the high cost of removal and replacement of metal-polluted soil have prompted efforts to develop alternative and cheaper techniques to recover the degraded land.

Restoration of derelict land by establishing a plant cover is important before it poses serious health hazard by transferring the trace metals into the surroundings. Current research in this area now includes utilization of plants to remediate polluted soils and to facilitate improvement of soils structure in cases of severe erosion, the innovative technique being known as phytoreme-diation (Rao et al.1996; Chaudhry et al. 1998; Glass 2000; Khan et al. 2000; Pulford and Watson 2003). Phytoremediation includes a range of methods such as phytodegradation, phytostabilization, rhizofiltration, enhance rhizosphere biodegradation and phytoaccumulation (Table 12.1).

3 Plants for Phytoremediation

Plants that are used to extract heavy metals from contaminated soils have to be the most suitable for the purpose, i.e. tolerant to specific heavy metal, adapted to soil and climate, capable of high uptake of heavy metal(s), etc. Plants either take up one or two specific metals in high concentrations into their tissues (hyper-accumulators) with low biomass (Chaudhry et al. 1998; Brooks 1997), or extract low to average heavy metal (not metal specific) concentrations in their shoots with high biomass. Low biomass hyperaccumulators, generally, have a restricted root system (Ernst 1996). In contrast, non-accumulators, high biomass produ-

Table 12.1 Phytoremediation strategies to remove, transfer, stabilize, and/or degrade contaminants in soil, sediments and water (including mine tailing)

Phytoremediation strategy	Brief description	References
1. Phytodegradation (phytotransformation)	Enzyme-catalysed metabolism of contaminants within plant tissues	Black (1995)
2. Phytostabilization	Immobilization of contaminants at the interface of roots and soil by reducing their bioavailability	Chaudhry et al. (1998)
3. Rhizofiltration	Based on a combination of phytoextraction and phytostabilization of HM from polluted waters	Salt et al. (1995), Rice et al. (1997)
4. Enhanced rhizosphere biodegradation	Plant roots release substances that are nutrients for microbes which accelerate biological activity in rhizosphere	Chaudhry et al. (1998)
5. Phytoaccumulation (phytoextraction)	Uptake and translocation of contaminants by plant roots to shoot	Black (1995)

cing and tolerant plants have physiological adaptation mechanisms, which allow them to grow in contaminated soils better than others (Palazzo and Lee 1997). The tolerance and specific behaviour at the root level must be taken into consideration while selecting plants for phytoremediation (Keller et al. 2003; Khan 2003). Root system morphology allows some plants to be more efficient than others in nutrient uptake in infertile soil or stressed soil conditions (Fitter and Stickland 1991).

Recently Reeves (2003) have reviewed tropical hyperaccumulator of heavy metal plants and concluded that there is a lack of investigation for the occurrence of hyperaccumulator plant species. No botanical or biogeochemical exploration of trace metal tolerant and/or accumulating plant species has yet taken place in many parts of the world. Many plant species, which can accumulate high concentrations of trace elements, have been known for over a century. Renewed interest in the role of these hyperaccumulating plants in phyto-remediation has stimulated research in this area and several plant species or ecotypes, associated with heavy metal enriched soils, accumulate metals in the shoots. These plants can be used to clean up heavy metal contaminated sites by extracting metals from soils and accumulating them in aboveground biomass (Chaudhry et al. 1999; Khan et al. 2000). The metal enriched biomass can be harvested and smelted to recover the metal (phytomining). The plants used in phytoremediation, like all biological methods, do not remove 100% of the contaminant. Among the strategy to overcome this are the use of chelators to enhance metal solubility in soils (Salt et al. 1995) and breeding/genetically modifying phytoaccumulating plants (Raskin et al. 1994).

3.1 Vetiver Grass as an Ideal Plant for Phytoremediation

Vetiver grass (*Chrysopogon zizanioides*) has been known to be a useful plant for thousands of years and has been cultivated for the production of scented oil produced by its roots as well as for its ability to retain soil and prevent erosion (Maffei 2002). It is a perennial grass belonging to the family Poaceae (Gramineae), originally from India, growing wild or cultivated. Since late last century, it is being used by the sugar industry in West Indies, Fiji, and eastern African islands like Mauritius for its soil conservation properties.

Recently vetiver grass, due to its eco-friendly nature, found a new use for phytoremediation of contaminated sites. Vetiver grass, a xerophyte and a hydrophyte, is highly tolerant to various abiotic stresses such as droughts and water logging, frost, heat, extreme soil pH, sodicity, salinity, Al and Mn toxicity (Truong and Claridge 1996). It is suitable for the stabilization and rehabilitation and reclaiming of acid sulfate and trace metals contaminated soils, i.e. phytoremediation. In Australia, vetiver has been successfully used to stabilize mining overburden and highly saline, sodic, magnesic, and alkaline or acidic tailings of coal and gold mines (Truong 1999). This grass has been extensively used for land protection by mitigating soil erosion and water conservation, especially on very steep slopes, due to its faster root growth, i.e. root length may reach up to 3 m just in one year. Vetiver grass is regarded as a tool for environmental engineering (Mucciarelli et al. 1998) and as one of the most versatile crops of the third millennium (Maffei 2002). Truong (1999) furnished field observations relating to high tolerance levels of vetiver grass to adverse soil conditions, trace metal toxicities, and agrochemicals

Vetiver System (VS), is an effective, low-cost, and environmentally friendly technology to clean HM contaminated soils. VS is emerging as an alternative technology for rehabilitation of degraded, saline, or trace metal contaminated soils, and for purification of water polluted with trace elements, agrochemicals, and industrial-effluent disposals.

The success of phytoremedial efforts is not only dependant upon the choice of plant species but also their method of establishment (Bradshaw 1987). Among the plants involved in phytoremedial measures, Vetiver grass (*Chrysopogon zizaniodes*), should receive special attention. It is a densely tufted, awnless, wiry and glabrous plant occurring in large clumps as hydrophyte or xerophyte on vertisols through to red alfisols. It can grow on both acidic (pH 3) and alkaline (pH 11) soils, and is tolerant to high levels of various trace metals such as arsenic, cadmium, copper, chromium and nickel (Truong 1999). It produces up to 2 m high plant with a strong dense and mainly vertical root system often measuring >3 m, useful in soil erosion control (Greenfield 1988, 1989, 1993; Truong 2002). It is propagated vegetatively and is non-invasive (National Research Council, USA 1993). It is extremely resistant to insect pests and diseases and is widely used worldwide for soil and moisture conservation and soil restoration. It is immune to flooding, grazing, fires, and other hazards (Grimshaw and Helfer 1995). Maffei (2002) stated that this plant is one of the most versatile crops of the third millennium. On the request of the World Bank, the US Agency for International Development, and the US Conservation Services, the National Research Council of USA, evaluated the usefulness of the vetiver system and its implications. China is among the nations most active in studying vetiver system and massive projects have been started in several conservation areas, as soil and moisture conservation are China's national priorities. Vetiver was introduced in China in 1950s as a source of aromatic oil and when the oil prices dropped, the plant was abandoned. It was only in 1988, that its importance in soil and moisture conservation was realized and massive vetiver trial projects started by China's Ministry of Water Resources in many provinces. Chinese scientists and researchers also started their own vetiver networks to exchange their experiences and results of vetiver trials.

Recently, Shu et al. (2000) showed successful establishment and colonization of vetiver grass as pioneer plant species on Pb/Zn mine spoils in China and concluded that this plant should be considered as one of the plants to be used for mine site revegetation. Due to its unique morphological and physiological characteristics such as higher biomass; fast growth; higher metal tolerance, uptake, and accumulation; and strong ecological adaptability, the vetiver grass can play an important role in phytostabilizaation, phytoextraction, and phytofiltration of heavy metal contaminants (Mucciarelli et al. 1998; Khan et al. 2000; Lavania and Lavania 2000; Shu et al. 2000). Another benefit of using vetiver grass for phytoremediation is its foliage, which can be used as a mulch to improve physical properties of the soil.

4 Vetiver Grass and Mycorrhizoremediation

Most plants growing on HM contaminated soils possess arbuscular mycorrhizae (for references see Khan 2006a), indicating that these arbuscular mycorrhizal fungi (AMF) have evolved a tolerance to HM and that they play an important role in the mycorrhizoremediation of contaminated soils. Arbuscular mycorrhizal fungi are known to improve plant growth on nutrient-poor soils and enhance their uptake of P, Cu, Ni, Pb, and Zn (Khan et al. 2000; Zhu et al. 2001; Jamal et al. 2002). Other soil microbes such as plant growth promoting bacteria, phosphorus solubilizing bacteria and nitrogen fixing bacteria also have synergistic interactions with the AM fungi and benefit plants under stressed conditions (Khan 2008). Relatively few studies have focused on the effect of AMF and their associated microbes on metal remediation efforts.

Although vetiver grass is regarded to be a suitable pioneering candidate for phytoremediation of HM contaminated sites and for rehabilitation of abandoned metalliferous mine wastelands (Shu et al. 2002), no record of its mycorrhizal status exists in literature. The first record of occurrence of AM in vetiver grass was reported in vetiver plants growing in the South China Botanical Gardens in Guangzhou (Wong 2003). Recently, Wong et al. (2007) investigated the effect of AM fungi on growth and uptake of N, p, Zn, and Pb by vetiver as host and concluded that mycorrhizal colonization increased Pb and Zn uptake by plants under low soil metal concentration, whereas under higher concentrations, it decreased Pb and Zn uptake. These results showed that inoculation of the vetiver host plants with AMF protects them from the potential toxicity caused by increased uptake of Pb and Zn. This area merits greater attention in order to fully exploit the potentials of vetiver grass in controlling soil erosion, soil salinity, and mycorrhizoremediation of HM decontaminated soils. Overall, it is an ideal plant that can be used to combat soil degradation. The use of Vetiver System on agricultural lands is not being utilized at its fullest extent in the world in general and Pakistan in particular. It is anticipated that Vetiver technology will make the significant difference in ameliorating the environment, conserving natural resources, and improve crops against abiotic stresses caused by salinity, drought, heavy metal contamination, etc. (Khan 2006b).

References

Baier EG (1994). Gender, environment, population, education and sustainable development themes in agricultural education. FAO, Rome, Italy.

Black H (1995). Absorbing possibilities:: Phytoremediation. Environ Health Perspect 103: 1106–1118.

Bradshaw AD (1987). Reclamation of land and ecology of ecosystems. In: William RJ Gilpin ME, Aber JD (eds.) Restoration ecology. Cambridge University Press, Cambridge, pp 53–74.

Brazi F, Naidu R, McLaughlin MJ (1996). In: Naidu RS, Kookana RS, Olivers DP, Rogers S, McLaughlin MJ (eds.) Contaminants and the soil environment in the Australia Pacific Region. Kluwer, The Netherlands, p. 451.

Brooks RR (1997). Plants that accumulate heavy metals. CAB International, Wallingford, Oxon, UK.

Burns RG, Rogers S, McGhee I (1996). Remediation of inorganics and organics in industrial and urban contaminated soils. In: Naidu R, Kookana RS, Olivers DP, Ro McLaughlin MJ (eds.) Contaminants and the soil environment in the Australia Pacific region. Kluwer, The Netherlands, pp 361–410.

Chaney RL, Oliver DP (1996). Sources, potential adverse effects, and remediation of agricultural soil contaminants. In: Naidu R, Kookana RS, Oliver DP, Rogers S McLaughlin MJ (eds.) Contaminants and the soil environment in the Australia-Pacific region. Kluwer, Dordrecht, The Netherlands, pp 323–359.

Chaudhry K, Muhammad MS, Sharif I (2006). Alternate extension approaches to technology dissemination for sustainable agriculture in the Punjab, Pakistan. Int J Agr Biol 8: 836–839.

Chaudhry TM, Hayes WJ, Khan AG, Khoo CS (1998). Phytoremediation – Focusing on accumulator plants that remediation metal-contaminated soils. Aust J Ecotoxicol 41: 37–51.

Chaudhry TM, Hill L, Khan AG, Kuek C (1999). Colonization of iron and zinc-contaminated dumped filtercake waste by microbe, plants and associated mycorrhizae. In: Wong MH, Wong JWC, Baker AJM (eds.) Remediation and management of degraded land. CRC Press LLC, Boca Raton, FL, Chap 27, pp 275–283.

Ernst WHO (1996). Bioavailability of heavy metals and decontamination of soils by plants. Appl Geochem 11: 163–167.

Fitter AH, Stickland TR (1991). Architectural analysis of plant root system. 2. Influence of nutrient supply on architecture in contrasting plant species. New Phytol 118: 383–389.

Flathman PE, Jerger DE, Exner JH (1994). Bioremediation: Field experiences. CRC Press, Boca Raton, FL, p 548.

Garforth C, Lawrence A (1997). ODI Natural Resources Perspective No. 21.

Glass DJ (2000). Economic importance of phytoremediation. In: Raskin I, Ensley B (eds.) Phytoremediation of toxic metals: Using plants to clean up the Environment. Wiley, New York.

Greenfield JC (1988). Vetiver grass (*Vetiveria zizanioides*): A method for soil and water conservation. PR Press Services, New Delhi, India, pp 72.

Greenfield JC (1989). Vetiver grass (*Vetiveria zizanioides*): The ideal plant for vegetative soil and water conservation. The World Bank, Washington, DC.

Greenfield JC (1993). Vetiver grass: The hedge against erosion. 4th ed. The World Bank, Washington, DC.

Grimshaw RG, Helfer L (eds.) (1995). Vetiver grass for soil and water conservation, land rehabilitation, and embankment stabilization. World Bank Technical Paper No. 273. The World Bank, Washington, DC.

Jamal J, Ayub N, Usman M, Khan AG (2002). Arbuscular mycorrhizal fungi enhance zinc and nickel uptake from contaminated soil by soybean and lentil. Intl J Phytorem 4(3): 205–221.

Keller C, Hammer D, Kayser A (2003). Root development and heavy metal phytoextraction comparison of different species in the field. Plant Soil 249: 67–81.

Khan AG (2003). Vetiver grass as an ideal phytosymbiont for Glomalian fungi for ecological restoration of derelict land. In: Truong P, Hanping X (ed.) Proceedings of 3rd International Conference on Vetiver and Exhibition: Vetiver and Water, Guanzhou, China, October 2003. China Agricultural Press, Beijing, pp 466–474.

Khan AG (2006a). Review: Mycorrhizoremediation – an enhanced form of phytoremediation. J Zheiang Univ Sci B 7: 503–514.

Khan AG (2006b). Developing sustainable rural communities by reversing land degradatio through a miracle plant – Vetiver grass. In: Warren M, Yarwood R (eds.) Rural Futures Conference Proceedings: The RuralCitizen:Governance, Culture and Wellbeing in the 21st century, 5–7 April 2006. University of Plymouth, Plymouth, UK, pp 1–8.

Khan AG (2008). Microbial dynamics in the mycorrhizosphere with special reference to arbuscular mycorrhizae. In: Ahmad I, Pichtel J, Hayat S (eds.) Plant-bacteria interactions, strategies and techniques to promote plant growth. Wiley-VCH/ Verlag, Chap 13, Weinheim, pp 245–256.

Khan AG, Kuek C, Chaudhry TM (2000). Role of plants, mycorrhizae and phytochelators in heavy metal contaminated land remediation. Chemosphere 41: 197–207.

Lavania UC, Lavania S (2000). Vetiver grass technology for environmental technology and sustainable development. Curr Sci 78(8): 944–946.

Maffei M (ed.) (2002). Vetiveria: The genus Vetiveria. Taylor & Francis, London.

McEldowney PBA, Hardman DJ, Waite S (1993). Treatment technologies. In: McEldowney S, Hardman J, Waite S (eds.) Pollution, ecology and biotreatment. Longman, Singapore, pp 48–58.

Mucciarelli M, Bertea CM, Scannerini S (1998). Vetiveria zizanioides as a tool for environmental engineering. Acta Hortic 23: 337–347.

National Research Council, USA (1993). Vetiver grass, a thin green line against erosion. National Academy Press, Washington, DC.

Palazzo AJ, Lee CR (1997). Root growth and metal uptake of plants grown on zinc-contaminated soils as influenced by soil treatment and plant species. In: Extended abstracts of the 4th International Conference on the Biogeochemistry of Trace Elements, 23–26 June 1997, Berkeley, CA, pp 441–442.

Pulford ID, Watson C (2003). Phytoremediation of heavy metal-contaminated land by trees: A review. Environ Intel 29: 529–540.

Rao et al. 1996 PSC, Davis GB, Johnston CD (1996). Technologies for enhanced remediation of contaminated soils and aquifers: An overview, analysis and case studies. In: Naidu R, Kookana RS, Oliver DP, Rogers S, McLaughlin MJ (eds.) Contaminants and the soil environment in the Australia-Pacific region. Kluwer, Dordrecht, The Netherlands, pp 361–410.

Raskin I, Kumar PBAN, Dushenkov V, Salt DE (1994). Bioconcentration of heavy metals by plants. Curr Opin Biotechnol 5: 285–290.

Reeves R (2003). Tropical hyperaccumulators of metals and their potential for phytoextraction. Plant Soil 249: 57–65.

Rice PJ, Anderson TA, Anderson TA, Anhalt JC, Coats JR (1997). Phytoremediation of atrazine and metolachlro-contaminated water with submerged and floating aquatic plants.

Proceed. 12th Annual Conference on hazardous waste research, Kansas City, Missouri, May 19–22, 1997. Paper No. 52 (accessed via http://www.engg.ksu.edu/HSRC/97Proceed/proc97.html).

Salt DE, Blaylock M, Kumar PBAN, Dushenkov V, Ensley BD, Chet I, Raskin I (1995). Phytoremediation: A novel strategy for the removal of toxic metals from the environment using plants. Biotechnology 13: 468–474.

Shu WS, Xia HP, Zhang ZQ (2002). Use of Vetiver and three other grasses for revegetation of Pb/Zn mine tailings: Field experiment. Int Journal of Phytoremediation, 4(1): 47–57.

Truong P (1999). Vetiver grass technology for mine rehabilitation. Technical Bulletin No. 1999/2, PRVN/ORDPB, Bangkok, Thailand.

Truong P (2002). Vetiver grass technology. In: Maffei A (ed.) Vetiveria – The genus Vetiveria. Taylor & Francis, London, pp 114–132.

Truong P, Claridge J (1996). Effects of heavy metal toxicities on Vetiver growth. Thialand Vetiver Newtwork (TVN) Newsletter 15: 9.

Williams DL (2000). Students' knowledge of and expected impact from sustainable agriculture. J Agr Edu 41: (online).

Wong CC (2003). The role of mycorrhiae associated with *Vetiveria zizaniodes* and *Cyperus polystachyos* in the remediation of metals (lead and zinc) contaminated soils. M. Phil thesis, Hong Kong Baptist University, Hong Kong.

Wong CC, Kuek C, Khan AG, Wong MH (2007). The role of mycorrhizae associated with Vetiver for phytoremediating Pb/Zn contaminated soils: Greenhouse study. Restor Ecol 15(1): 60–67.

Zhu YG, Christie P, Laidlaw AS (2001). Uptake of Zn by arbuscular mycorrhizal white clover from zinc contaminated soil. Chemosphere 42: 193–199.

Chapter 13
Cell Membrane Stability (CMS): A Simple Technique to Check Salt Stress Alleviation Through Seed Priming with GA_3 in Canola

M. Jamil, M. Ashraf, S. Rehman, and E.S. Rha

Abstract Cell membrane stability (CMS) technique was used to assess whether salt tolerance could be improved in canola (*Brassica napus* L.) by soaking the seeds for 10h in distilled water (Control), 100, 150 and 200mg l^{-1} GA_3. The electrical conductivity (EC) values of the NaCl solution were 0 (control), 4.7, 9.4 and 14.1dS m^{-1} NaCl. Seed priming increased the final germination percentage and the germination rate ($1/t_{50}$, where t_{50} is the time to 50% of germination) under saline condition. Priming also alleviated the adverse effect of salt stress on canola in terms of fresh and dry weights of plants, roots and shoots. Similarly leaf area, leaf water contents (RWC) and chlorophyll contents (SPAD value) were significantly higher in plants raised from seeds primed with GA_3 as compared with those raised from seeds treated with distilled water. All pre-sowing seed treatments decreased electrolyte leakage of steep water as compared to that of non-primed seeds even after 24h of soaking. Similarly plants raised from seeds primed with GA_3, showed significantly lower cellular injury than seeds treated with distilled water.

Keywords Seed treatment • germination • gibberellic acid • salt stress • *Brassica napus* L.

M. Jamil
Department of Biotechnology and Genetic Engineering, Kohat University of Science and Technology (KUST) Kohat 26000, Pakistan

M. Ashraf
Department of Botany, University of Agriculture, Faisalabad, Pakistan

S. Rehman
Department of Botany, Kohat University of Science and Technology (KUST), Kohat 26000, Pakistan

E.S. Rha (✉)
College of Agriculture & Life Sciences, Sunchon National University, Suncheon 540–742, Republic of Korea
e-mail: euishik@sunchon.ac.kr

M. Ashraf et al. (eds.), *Salinity and Water Stress*,

1 Introduction

Numerous attempts have been made to improve the salt tolerance of crops by traditional breeding programmes, but commercial success has been very limited (Santa-Cruz et al. 2002). Pre-sowing seed treatment or seed priming is an easy technique and an alternative approach recently used to overcome salinity problems. Priming (osmo-conditioning) is one of the physiological methods, which improves seed performance and provides faster and synchronized germination (Ashraf and Foolad 2005). Pre-soaking or priming seeds of a number of crops has improved germination, seedling establishment and, in some cases, stimulated vegetative growth and hence crop yield (Amzallag et al. 1990; Kaur et al. 1998). Seed priming enhance many of the metabolic processes involved with the early phases of germination, and it has been noted that seedlings from primed seeds emerge faster, grow more vigorously, and perform better in adverse conditions (Desai et al. 1997).

Plant growth regulators (PGRs) have been found to play a key role in the integration of the responses expressed by plants under stress conditions (Amzallag et al. 1990). PGRs may also enhance germination and adaptation of plants to stress conditions (Banyal and Rai 1983). Gibberellic acid and kinetin have been reported to increase percentage germination and seedling growth in (name of crop species) (Kaur et al. 1998). Seed priming has been successfully demonstrated to improve germination and emergence in seeds of many crops, particularly seeds of vegetables (Heydecker and Coolbaer 1977). The beneficial effects of priming have also been demonstrated for many field crops such as canola, sugar beet, wheat, maize, soybean and sunflower (Zheng et al. 1996; Singh 1995; Ahmad et al. 1998; Parera and Cantliffe 1994; Khajeh-Hosseini et al. 2003; Sadeghian and Yavari 2004).

Cell membrane stability (CMS) is a technique that has often been used for screening against drought and salt tolerance in various crops such as wheat (Blum and Ebrecon 1981), maize (Premachandra et al. 1989), rice (Tripathy et al. 2000), wheat and wild relatives of wheat (Farooq and Azam 2002; Sairam et al. 2002). In the present study, we used CMS technique to assess whether salt tolerance could be improved in canola by soaking seed for 10h in distilled water (control), 100, 150 and 200 mg l^{-1} GA_3 at seed germination and subsequent vegetative growth. Secondly, to determine the effect of seed treatments on the relative water content and chlorophyll content in different plant leaves raised from seeds primed with GA_3.

2 Materials and Methods

Seeds of canola (*B. napus* L. Chinensis group cv Si jiu huang cai xin) were obtained from Institute of Soils and Fertilizers, Beijing, China. Seeds were surface sterilized in 5% sodium hypochlorite solution for 10 min, then rinsed with sterilized distilled water and air-dried at an ambient temperature of 25°C in the laboratory. The sterilized seeds were soaked for 10h in 100, 150 and 200 mg/l GA_3 or distilled water. After the pre-sowing treatment, all seed samples were rinsed with distilled water, and dried in an oven at 30°C for 2h to eliminate surface moisture.

2.1 Seed Water Uptake

Five grams of the seeds from each seed treatment were placed in Petri dishes containing 50 ml distilled water for 2, 4, 12 and 24h to determine water uptake of seeds necessary for germination. The water uptake was expressed as the percentage increase in moisture content on fresh weight basis.

2.2 Electrical Conductivity of Seed Leachate

Five grams seeds from each seed treatment were soaked in 50 ml at 25°C after washing in distilled water. Electrical conductivity of steep water was measured 2, 4, 12 and 24h after soaking using conductivity meter (Model CM-21P) and expressed as milliSiemens per meter.

2.3 Seeds Germination

Seeds were germinated in sterilized disposable Petri dishes (87 mm diameter, 15 mm height) containing two Whatman No. 2 filter papers soaked with 10 ml of distilled water or 4.7, 9.4 and 14.1 dS m^{-1} NaCl concentrations. Petri dishes were sealed with parafilm to prevent evaporation of water, thus minimizing the changes in concentration of the solutions. Seeds were incubated in growth chamber at 25°C. Five replicate dishes with 20 seeds per replicate were used for each treatment. Seeds were hand sorted to eliminate broken and small seeds. Seed germination was evaluated after every 12h up to 8 days and germination rate was calculated as $1/t_{50}$ (where t_{50} is the time to 50% of germination). After 36h, seeds started to germinate (seeds were considered to be germinated with the emergence of the radicle).

2.4 Glasshouse Experiment

A pot experiment was conducted in a glass-house supplied with natural sunlight to determine the influence of seed priming on vegetative growth. Seeds of canola were grown in plastic pots (16 cm diameter, 22 cm height) in sand culture. The sand was irrigated with full strength Hoagland's nutrient solution (Hoagland and Arnon 1950). All the pots were irrigated for 2 weeks with full strength Hoagland's nutrient solution. Salt treatments in this solution begun 15 days after the start of the experiment. The NaCl treatments used were 0 (control), 4.7, 9.4 and 14.1 dS m^{-1} in full strength Hoagland's nutrient solution. The average temperature for day/night was 25/15°C and photoperiod for the day/night cycle was 16/8h. The experiment was arranged in a completely randomized design with three replicates. Plants were harvested from each pot after 30 days of sowing. The plants were uprooted carefully, washed with distilled water and the fresh weights of Plants, shoots and roots were recorded. Plant samples were dried in an oven at 80°C to constant dry mass.

2.5 Leaf Area, Chlorophyll Content and Relative Water Contents (RWC)

Leaf areas of individual plant leaves were measured by using Area meter (AM-200, ADC Bio Scientific Ltd., England). Leaf chlorophyll content was measured using a hand-held chlorophyll content meter (CCM-200, Opti-Science, USA). For determination of RWC, fresh leaves were detached from each treatment, replicate and weighed immediately to record fresh weight (FW), followed by dipping in distilled water for 12h. The leaves were blotted to wipe off excess water, weighed to record fully turgid weight (TW), and subject to oven drying at 80°C for 24h to record the dry weight (DW). The RWC were determined by the equation proposes by Turner (1986) that is RWC = [FW – DW] × 100/[TW – DW].

2.6 Measurement of Cell Membrane Stability or Cellular Injury

CMS was determined by using fully expanded young leaf (4th leaf) was selected from each genotype, treatment and replication. Twenty pieces (1 cm diameter) cut from these leaves were submerged into distilled water contained in test tubes. The tubes were kept at 10°C in incubator for 24h, followed by warming at 25°C and measuring the electrical conductivity (C_1) of the contents. The leaf samples were then killed by autoclaving for 15 min at 120°C and electrical conductivity of the medium measured again (C_2). Cellular injury was determined by using the formula: (C_1/C_2) * 100 where C refers to conductivity one and two.

The experiment was designed by using a randomized complete block design with three replications. Analysis of variance was performed by using the Microsoft Excel version 5.0 (Middleton 1995). Means values for different plant characteristics were compared through LSD test (Li 1964).

3 Results

Water uptake of primed seeds was increased with increasing imbibitions period including all treatments and control (Fig. 13.1). Water uptake increased in all primed seeds than control after a long period soaking ranging from 2 to 24h. Maximum increase in water uptake was induced by 200 mg l^{-1} GA_3 on all measuring periods (Fig. 13.1).

Salt stress caused a significant reduction in germination percentage of both primed and non-primed seeds with increasing salt concentrations (Fig. 13.3A). There was considerable reduction in germination at highest

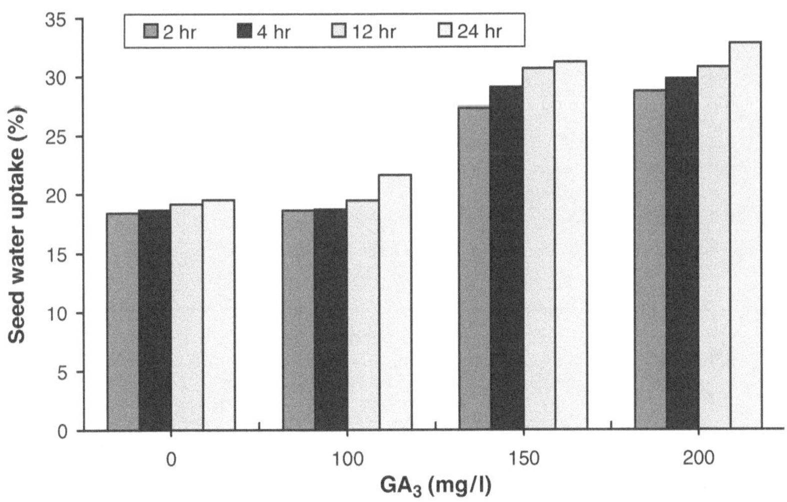

Fig. 13.1 Water uptake of control and primed seeds soaked for 2, 4, 12 and 24h in 50ml of distilled water

level of salinity (9.4 and 14.1 dS m^{-1} NaCl). However, priming increased germination percentage under salt stress. Among all priming treatments, 150 and 200 mg l^{-1} of GA_3 was the most effective in increasing germination under salt stress. Lowest germination was found in seeds treated with distilled water (Figs. 13.3A).

Germination rate of control and primed seeds under various conditions of salinity was expressed as a $1/t_{50}$ of the germination of seeds of the same population as in control. The germination response of the control and primed seeds under observation showed marked differences in the timing of initiation and completion of germination. Germination started within 36 h and was complete on the 8th day. Seed germination delayed as the level of salinity increased. However, priming treatments shortened the time to seed germination (Figs. 13.3B). Figure 3B also showed seeds primed with 150 and 200 mg l^{-1} GA_3 took less time to complete germination at high salt concentration as compared to other treatments.

Salt concentration of the growth medium caused a significant reduction in mean fresh and dry weights of plants, roots and shoots of both primed and non-primed seeds of canola with increasing salt stress (Figs. 13.4 and 5). However, higher fresh and dry weights were recorded from primed seeds as compared to control (Figs. 13.4 and 5). Plants raised from seeds treated with 200 mg l^{-1} GA_3 had the greater fresh and dry weights of plants, shoots and roots under both the salt and control treatments at 14.1 dS m^{-1} NaCl (Figs. 13.4 and 13.5).

As a general trend, increased NaCl salinity decreased total leaf area and leaf water content (Fig. 13.6). However, plants raised from GA_3 seeds, showed significantly higher leaf area and leaf water content as compared with those raised from seeds treated with distilled water (Fig. 13.6). Plants raised from seeds treated with 100 mg l^{-1} GA_3 had the greater leaf area and leaf water content at high salt concentration as compared to other treatments (Fig. 13.6).

Chlorophyll content increased significantly at all salinity levels (Fig. 13.7B). Plants raised from seeds treated with GA_3 had the higher chlorophyll content than in control (Fig. 13.7B). Plants raised from seeds treated with 200 mg l^{-1} GA_3 had the greater chlorophyll content at high salt concentration as compared to other treatments (Fig. 13.7B).

Pre-sowing treatments were effective in decreasing electrolyte conductivity of seed leachate (Fig. 13.2). Generally the electrolyte leakage was increased with increasing imbibitions period including all treatments and control. After a long period soaking ranging from 2 to 24 h all the seed treatments lowered down the electrolyte leakage than control. Maximum decrease in electrolyte leakage was induced by 200 mg l^{-1} GA_3 on all measuring periods. Similarly plants of canola, raised from GA_3 seeds, showed significantly lower cellular injury as compared with those raised from seeds treated with distilled water (Figs. 13.7a). Plants raised from seeds treated with 150 mg l^{-1} GA_3 had the lower cellular injury than other treatments (Fig. 13.7a).

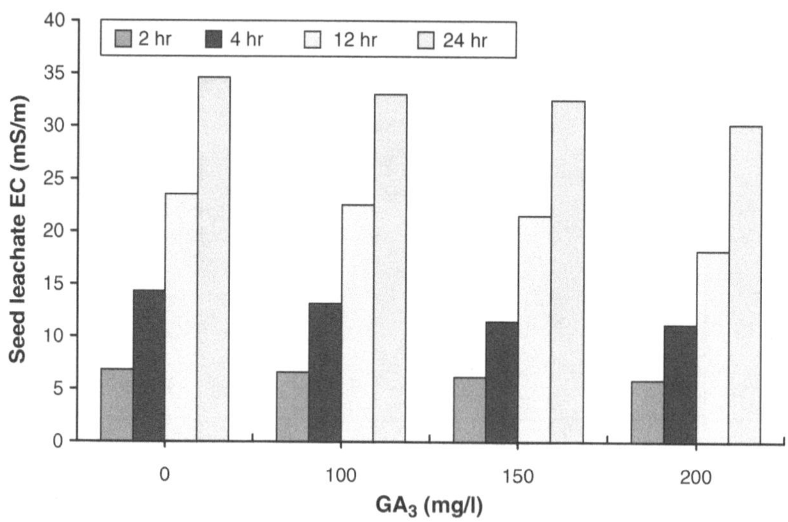

Fig. 13.2 Electrical conductivity of seed leachate of canola seeds primed with 0.0, 100, 150 and 200 mg l^{-1} GA_3

4 Discussion

Plant growth regulators have been found to play a key role in the integration of the responses expressed by plants under stress conditions (Amzallag et al. 1990). Gibberellic acid has been reported to increase percentage germination and seedling growth (Kaur et al. 1998). Primed seeds show enhanced performance under stress conditions. Seed priming shortened germination time but stress conditions delayed it considerably (Fig. 13.3b). Furthermore, priming resulted in increase of normal germination in all primed seeds (Fig. 13.3a). This could be explained by more rapid water uptake in primed seeds (Fig. 13.1) because germination in primed seeds started after 24 h. These results agree with Murillo-Amador et al. (2002) in cowpea, Demir and Van De Venter (1999) in watermelon, they suggested that salinity may influence germination by increasing the water uptake. The pre-sowing treatments cause initiation of the early metabolic processes and the re-drying of seeds arrest, but do not reverse, the initial stages of germination so that on the availability of suitable conditions, the time taken to germinate is reduced (Bewley and Black 1982). Priming induced activation of metabolic events (Hanson 1973) has been reported in seeds of various plant species. Akinola et al. (2000) reported that higher duration of exposure to seed treatment resulted in higher cumulative germination in wild sunflower and Caseiro et al. (2004) found that hydropriming was the most effective method for improving seed germination of

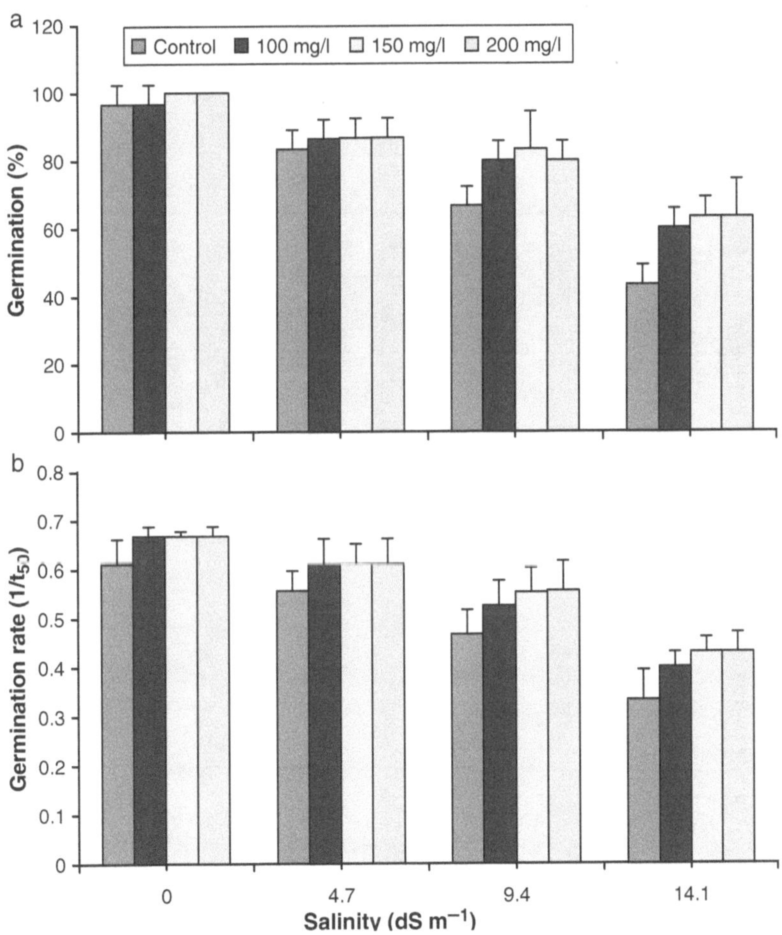

Fig. 13.3 Germination (**a**) and germination rate (**b**) of canola seeds primed with 0.0, 100, 150 and 200 mg l^{-1} GA_3 under various salt concentrations

onion, especially when the seeds were hydrated for 96h compared to 48h. The results are in line with the findings of Thornton and Powell (1992) in *Brassica* and Srinivasan et al. (1999) in mustard.

Higher fresh and dry weights of plants, shoots and roots were also recorded from primed seeds as compared to control (Figs. 13.4 and 13.5). Growth improvement with seed treatment was observed in seedling

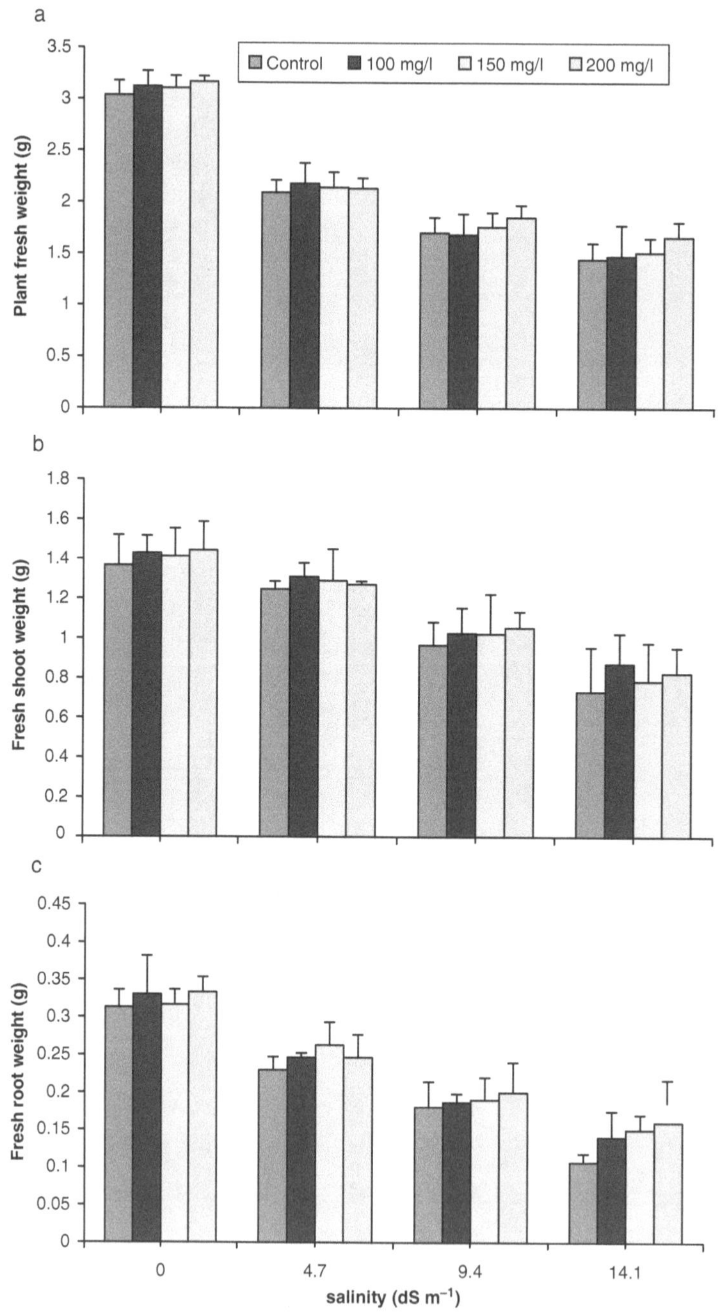

Fig. 13.4 Fresh weights of plants (A), shoots (B) and roots (C) of canola plants raised from seeds primed with 0.0, 100, 150 and 200 mg l^{-1} GA_3 under various salt concentrations

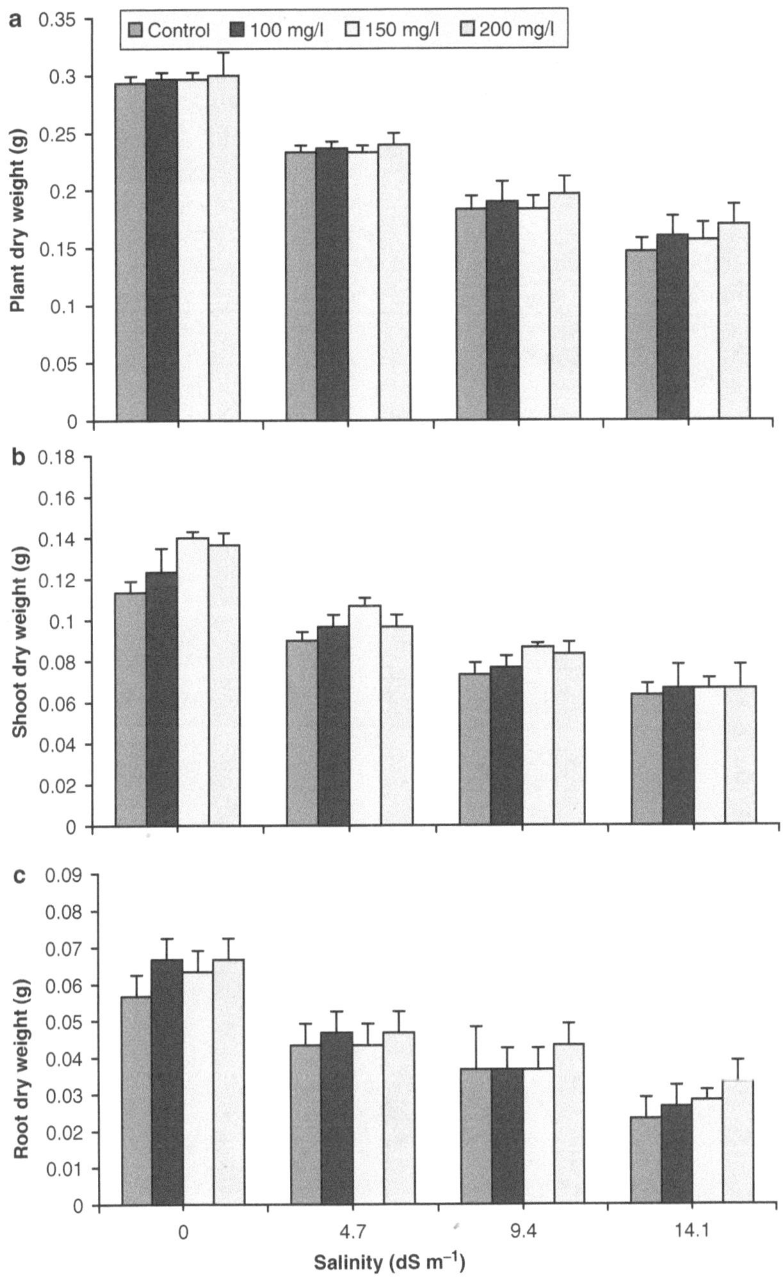

Fig. 13.5 Dry weights of plants (A), shoots (B) and roots (C) of canola plants raised from seeds primed with 0.0, 100, 150 and 200 mg l^{-1} GA_3 under various salt concentrations

growth due to faster emergence, giving seedlings a longer time to develop (Parera and Cantliffe 1994). Muhyaddin and Wiebe (1989) suggested that the enzymes are activated with an accompanying mobilization of reserve materials ending in transport of the reserve materials in the embryo by osmotic conditioning, and thus stronger seedlings are obtained as a result of embryo growth. The results of the present study are similar with those of Patel and Saxena (1994) who reported that fresh and dry weight of shoots and roots was increased in seedling raised from seed treated with Kinetin and GA_3 as compared to seed treated with NAA and Ethrel. There are numerous studies in the literature which exhibit the considerable effectiveness of seed priming on seed germination and later growth in different plant species under both saline and non-saline conditions e.g., wheat (Idris and Aslam 1975), *Acacia tortilis* and *Acacia coriacea* (Rehman et al. 1998) and under normal non-saline conditions e.g., maize (*Zea mays*), (Ashraf and Rauf, 2001) rice (*Oryza sativa*) and chickpea (*Cicer arietinum*) (Harris et al. 1999).

Similarly leaf area and chlorophyll contents were significantly higher in plants raised from seeds primed with GA_3 as compared with those raised from seeds treated with distilled water (Figs. 13.6a and 13.7b). Salt stress is known to reduce the life span of leaves. This causes accelerated senescence and, as a consequence, chlorophyll degradation (Yeo and Flowers 1984). However, in the present study, the chlorophyll content of leaves increased up to 14.1 dS m^{-1} salinity level (Fig. 13.7). GA_3 pre-treatments resulted in enhancing the chlorophyll content of the leaves. Although Misra et al. (1997) concluded that salt stress induced an increase in the chlorophyll content, which could be

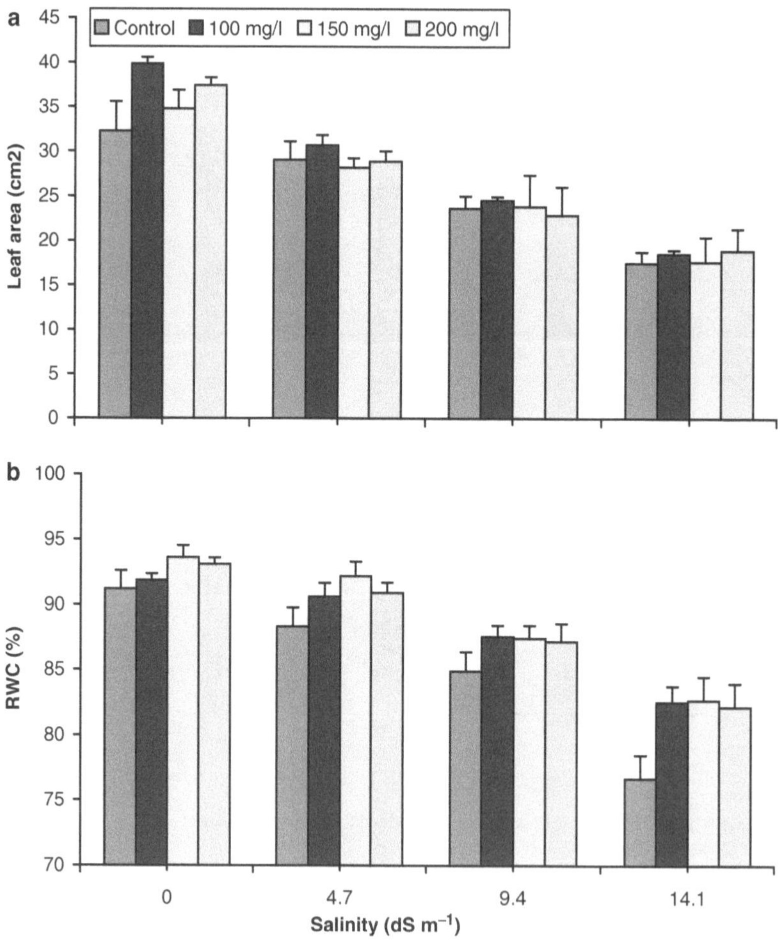

Fig. 13.6 Leaf area (**a**) and leaf water content (b) of canola plants raised from seeds primed with 0.0, 100, 150 and 200 mg l^{-1} GA_3 under various salt concentrations

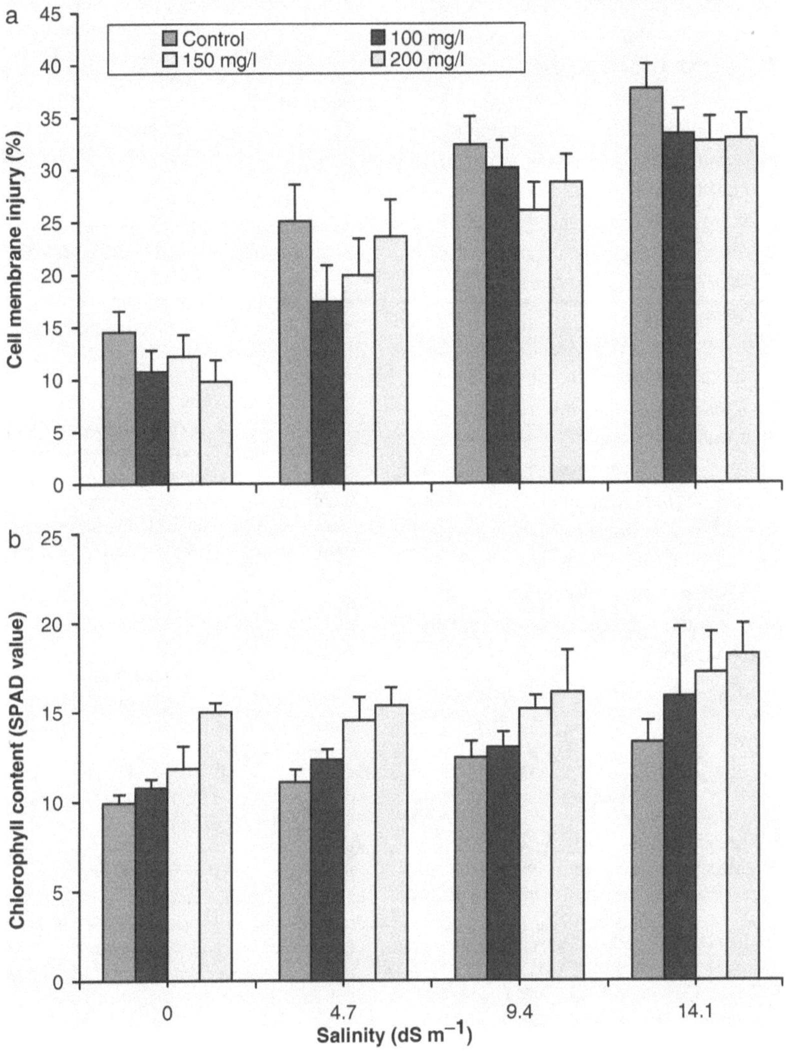

Fig. 13.7 Cell membrane injury (A) and chlorophyll content (B) of canola plants raised from seeds primed with 0.0, 100, 150 and 200 mg l^{-1} GA_3 under various salt concentrations

due to an increase in the number of chloroplasts in stressed leaves.

Seed leachate electrical conductivity is considered as an effective indicator of seed germination (Water and Blanchette 1983). All seed soaking treatments were effect in decrease the electrolyte conductivity of seed leachate, which shows membrane stability. Decreased leakage of solute in GA_3 treatments than control may be because of better membrane repair during hydration (Fu et al. 1988). Greater membrane integrity in primed seeds was reported by Rudrapal and Naukamura (1998) for the eggplant and radish and Afzal et al. (2002) for hybrid maize.

Maintaining integrity of cellular membranes under stress conditions is considered an important part of salinity tolerance mechanisms. This study showed that GA_3 reduced the amount of cellular injury in salt stressed canola plants indicating that GA_3 treatment has facilitated the maintenance of membrane functions under stress conditions. Results indicated, a progressive decrease in cellular injury and increased in RWC with increasing GA_3 treatments under salt stress (Figs. 13.6b and 7a). We believe that GA_3 absorbed by the seed was most likely distributed in the seedlings, which maintained the leaf water status and stabilized the cellular membranes. Supporting evidence was shown when salicylic acid reduced electrolyte leakage in corn leaf, rice leaf and cucumber hypocotyls under chilling stress (Kang and Saltveit 2001). These finding are in agreement with Kaya et al. (2001) who showed similar effects in cucumber and pepper.

It was concluded that seed priming induces physiological changes in plants, and these changes are clearly shown at advance stages of development. The results of this study shows that tolerance of salinity is maintained during the growth period, hence growth parameters remained significantly higher in plants raised from seeds primed with GA_3 as compared with those raised from seeds treated with distilled water. It was also concluded CMS can be used more effectively to check the enhancement in growth at seed germination and subsequent vegetative growth.

References

Afzal I, Basra SMA, Ahmad N, Cheema MA, Warraich EA, Khaliq A (2002) Effect of priming and growth regulator treatment on emergence and seedling growth of hybrid maize (*Zea mays*). Int. J. Agric. Biol. 4: 303–306.

Ahmad S, Anwar M, Ullah H (1998) Wheat seed pre-soaking for improved germination. J Agron Crop Sci 181: 125–127.

Akinola JO, Larbi A, Farinu GO, Odunsi AA (2000) Seed treatment methods and duration effects on germination of wild sunflower. Exp Agr 36: 63–69.

Amzallag GN, Lener HR, Poljakoff-Mayber A (1990) Exogenous ABA as a modulator of the response of sorghum to high salinity. J Exp Bot 541: 1529–1534.

Ashraf M, Foolad MR (2005) Pre-sowing seed treatment – a shotgun approach to improve germination, plant growth, and crop yield under saline and non-saline conditions. Adv Agron 88: 223–271.

Ashraf M, Rauf H (2001) Inducing salt tolerance in maize (*Zea mays* L.) through seed priming with chloride salts: growth and ion transport at early growth stages. Acta Physiol Plant 23: 407–414.

Banyal S, Rai VK (1983) Reversal of osmotic stress effects by gibberellic acid in *Brassica campestris*. Recovery of hypocotyls growth, protein and RNA levels in presence of GA. Physiol Plant 59: 111–114.

Bewley JD, Black M (1982) Physiology and Biochemistry of Seeds in Relation to Germination. Springer, Berlin, Germany.

Blum A, Ebrecon A (1981) Cell membrane stability as measure of drought and heat tolerance in wheat. Crop Sci. 21: 43–47.

Caseiro R, Bennett MA, Marcos-Filho J (2004) Comparison of three priming techniques for onion seed lots differing in initial seed quality. Seed Sci Technol 32: 365–375.

Demir I, Van De Venter HA (1999) The effect of priming treatments on the performance of watermelon (*Citrillus lanatus* (Thunb.) Matsum. & Nakai) seeds under temperature and osmotic stress. Seed Sci Technol 27: 871–875.

Desai BB, Kotecha PM, Salunkhe DK (1997) Seeds Handbook. Marcel Dekker, New York.

Farooq S, Azam F (2002) Co-existence of salt and drought tolerance in *Triticeae*. Hereditas 135: 205–210.

Fu JR, Lu XR, Chen Z, Zhang BZ, Ki ZS, Cai CY (1988) Osmoconditioning of peanut (*Arachis hypogaea* L.) seeds with PEG to improve vigor and some biochemical activities. Seed Sci. Technol. 16: 197–212.

Hanson AD (1973) The effects of imbibition drying treatments on wheat seeds. New Phytol 72: 1063–1073.

Harris D, Joshi A,Khan PA, Gothkar P, Sodhi PS (1999) On-farm seed priming in semi-arid agriculture: development and evaluation in maize, rice and chickpea in India using participatory methods. Exp Agr 35: 15–29.

Heydecker W, Coolbaer P (1977) Seed treatments for improved performance survey and attempted prognosis. Seed Sci Technol 5: 353–425.

Hoagland DR, Arnon DI (1950) The water culture method for growing plants without soil. Uni. Calif. Berkeley College. Agric. Exp. Stn. Circ. No. 347.

Idris M, Aslam M (1975) The effect of soaking and drying seeds before planting on the germination and growth of *Triticum vulgare* under saline and normal conditions. Can J Bot 53: 1328–1332.

Kang H, Saltveit ME (2001) Activity of enzymatic antioxidant defense systems in chilled and heat shocked cucumber seedling radicles. Physiol Plant 113: 548–556

Kaur S, Gupta AK, Kaur N (1998) Gibberellic acid and kinetin partially reverse the effect of water stress on germination and seedling growth. Plant Growth Regul 25: 29–33.

Kaya C, Kirnak H, Higgs D (2001) The effects of supplementary potassium and phosphorus on physiological development and mineral nutrition of cucumber and pepper cultivars grown at high salinity (NaCl). J Plant Nutr 24: 1457–1471.

Khajeh-Hosseini M, Powell AA, Bingham IJ (2003) The interaction between salinity stress and seed vigour during germination of soybean seeds. Seed Sci Technol 31: 715–725.

Khan AA (1993) Preplant physiological seed conditioning. Hortic Rev 13: 131–181.

Li CC (1964) Introduction to Experimental Statistics. McGraw-Hill, New York.

Middleton MR (1995) Data analysis using Microsoft Excel version 5.0. Duxbury press, Wadsworth Publishing, Belmont, CA.

Misra AN, Sahl SM, Misra M, Singh P, Meera T, Das N, Har M, Sahu P (1997) Sodium chloride induced changes in leaf growth, and pigment and protein contents in two rice cultivars. Biol Plant 39: 257–262.

Muhyaddin T, Wiebe HJ (1989) Effect of seed treatments with polyethylenglycol (PEG) on emergence of vegetable crops. Seed Sci Technol 17: 49–56.

Murillo-Amador B, Lopez-Aguilar R, Kaya C, Larrinaga-Mayoral J, Flores-Hernandez A (2002) Comparative effects of NaCl

and polyethylene glycol on germination, emergence and seedling growth of cowpea. J Agron Crop Sci 188: 235–247.
Parera CA, Cantliffe DJ (1994) Presowing seed priming. Hortic Rev 16: 109–139.
Patel I, Saxena OP (1994) Screening for PGRs for seed treatment in green gram and black gram. Indian J Plant Physiol 37: 206–208.
Premachandra GS, Saneoka H, Ogta S (1989) Nutrio-physiological evaluation of the polyethylene glycol test of cell membrane stability in maize. Crop Sci. 29: 1292–1297.
Rehman S, Harris PJC, Bourne WF (1998) Effect of pre-sowing treatment with calcium salts, potassium salts or water on germination and salt tolerance of *Acacia* seeds. J Plant Nutr 21: 277–285.
Rudrapal D, Naukamura S (1998) The effect of hydration dehydration pre-treatment on egg plant and radish seed viability and vigour. Seed Sci. Technol. 16: 123–130.
Sadeghian SY, Yavari N (2004) Effect of water-deficit stress on germination and early seedling growth in sugar beet. J Agron Crop Sci 190: 138–144.
Sairam RK, Rao KV, Srivastava GC (2002) Differential response of wheat genotypes to longterm salinity stress in relation to oxidative stress, antioxidant activity and osmolyte concentration. Plant Sci 163: 1037–1046.
Santa-Cruz A, Martinez-Rodriguez MM, Perez-Alfocea F, Romero-Aranda R, Bolarin M C (2002) The rootstock effect on the tomato salinity response depends on the shoot genotype. Plant Sci 162: 825–831.
Singh B G (1995) Effect of hydration-dehydration seed treatments on vigour and yield of sunflower. Indian J Plant Physiol 38: 66–68.
Srinivasan K, Saxena S, Singh BB (1999) Osmo and hydropriming of mustard seeds to improve vigour and some biochemical activities. Seed Sci Technol 27: 785–793.
Thornton JM, Powell AA (1992) Short term aerated hydration for the improved of seed quality on *Brassica oleracea*. Seed Sci Res 2: 41–49.
Tripathy JN, Zhang J, Robin S (2000) QTL for cell-membrane stability mapped in rice (*Oriza sativa* L.) under drought stress. Theor Appl Genet 100: 1197–1202.
Turner NC (1986) Crop water deficit: a decade of progress. Adv Agron 39: 1–51.
Waters Jr L, Blanchette BL (1983) Prediction of sweet maize field emergence by conductivity and cold tests. J. Amer. Soc. Hort. Sci. 108: 778–781.
Yeo AR, Flowers TJ (1984) Mechanisms of salinity resistance in rice and their role as physiological criteria in plant breeding. In: R.C. Staples (ed) Strategies for Crop Improvement. John Wiley and Sons Inc., New York, pp. 151–171.
Zheng GH, Gao YP, Wilen RW, Kirkland K, Gusta LV (1996) The potential of seed priming to enhance germination and yield of canola. Proc. Soils and Crops '96. Ext. Div. University of Saskatchewan, Saskatoon, pp. 318–327.

Chapter 14
Using Resources from the Model Plant *Arabidopsis thaliana* to Understand Effects of Abiotic Stress

M.G. Jones

Abstract Plants have to face many environmental stresses; all are of considerable agricultural importance as they reduce yield. To alleviated adverse effects of abiotic stresses on crop growth, additional cost inputs are required. There is interaction between the stress response pathways within the plant and the genetic, protein and physiological components involved in sensing and the response to stress in plants are starting to become known, but there are still a very large number of gaps in our knowledge. The complete sequencing and annotation of the genome of the model plant *Arabidopsis thaliana* (family Brassicaceae) has provided an unprecedented resource and opportunity for plant biologists, whatever their plant of interest. Selection of one plant species for intensive research has provided a new insight into plant biology that has advanced both fundamental and applied research. Greater understanding of the signaling and response pathways to abiotic stress within *Arabidopsis*, in particular the role of transcription factors, will provide leads that can be applied to the development of new, higher yielding varieties of crop species.

Keywords Abiotic stress • *Arabidopsis thaliana* • transcription factor

1 Introduction

Plants have a very wide range of mechanisms to adapt to abiotic stresses from the environment. These range from the slow growth of species in cold sub-arctic environments to very rapid life-cycles of ephemeral species triggered by rain in semi-deserts as well as a host of morphological adaptations to leaves, flowers, stems and roots (Bohnert et al. 1995). The seasonal life-cycles of most species are also adaptations to predictable annual changes in environment. Plants have therefore evolved strategies that ensure the dominance of particular species within stressful environments as well as mechanisms to adapt to seasonal or sudden changes. However, although these strategies ensure plant success within natural habitats, the demands of agriculture require growth cycles and yields that can be compromised severely by stress.

These abiotic stresses (Fig. 14.1) may be present throughout the life-cycle, such as unfavourable soil salinity, or can occur suddenly during the growing season, such as unusual heat. This problem is present throughout the world and affects profits in large-scale commercial farming as well as survival in regions with subsistence farming. The nature of abiotic stresses can be unexpected. For example, droughts occur regularly within some regions of the European Union, but low rainfall during 2006 in several Mediterranean and central European areas resulted in some of the worst droughts within the last 3 decades (see Table 14.1). If climate change results in more intense and prolonged lack of rainfall, current agricultural practices to mitigate drought stress may not be adequate.

The use of cultivars that maintain yield and quality in the face of abiotic stress is one important way to maintain productivity. Therefore, appropriate selection from existing cultivars and development of new ones are necessary. The numerous changes in plant growth and development pattern in response to stress offer opportunities for conventional breeding and genetic engineering strategies to develop varieties able to tolerate more abiotic stresses.

M.G. Jones (✉)
School of Biological Sciences, The Biosciences Building, The University of Liverpool, Liverpool, L69 7ZB, UK

M. Ashraf et al. (eds.), *Salinity and Water Stress*,

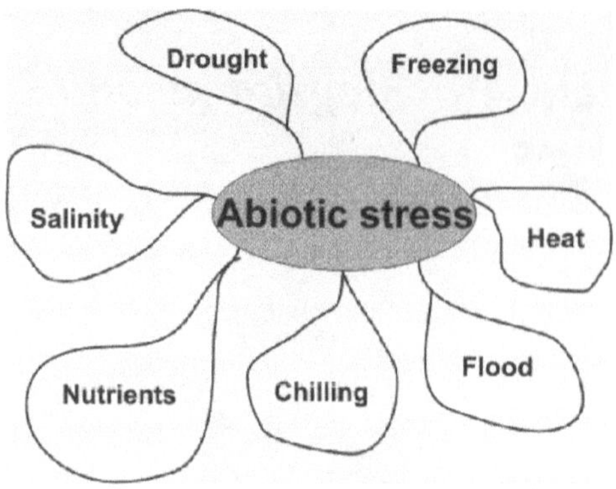

Fig. 14.1 A plant can experience either one or many abiotic stresses and must respond successfully to each for continued growth and reproduction

Many abiotic stresses affect the water status of the plant. These include drought, salinity, heat, chilling, flood and freezing. Some of the most obvious physiological effects are osmotic adjustments, primarily through ion exclusion and the synthesis of novel osmoprotectants (Bray 1997; Bartels and Sunkar 2005). However, stress causes changes to morphology and can accelerate the life-cycle in crop plants, indicating that the response encompasses many aspects of physiology, not only water relations. The signalling pathways that lead to these changes are of considerable interest, since it may be possible to modulate specific steps to maintain plant performance under stress.

Table 14.1 Rainfall balance for areas of the EU most affected by 2006 drought and comparison with the last 30 years (adapted from IP/06/1097 and MEMO/06/284, 'Crop yield forecasts for 2006 and drought effects', European Commission, Directorate General Joint Research Centre, 2006)

Country/region	Rainfall deficit (mm)	Ranking since 1975 (32 years)
Italy	−52.0	3rd worst year (after 1992, 2003)
Tuscany	−99.4	The worst year overall
Lazio	−89.4	The worst year overall
Sardinia	−56.5	2nd worst year (after 1983)
Spain	−28.6	8th worst year
Cataluna	−120.0	The worst year overall
Asturias	−120.0	3rd worst year (after 1995, 2003)
Extremadura	−39.5	10th worst year
Portugal	−44.6	7th worst year
Centro	−66.8	5th worst year
Alentejo	−44.9	7th worst year
Baltic countries	–	–
Estonia	−41.6	5th worst year
Latvia	−51.9	4th worst year
Lithuania	.60.6	3rd worst year (after 1979, 2000)
France	−41.5	3rd worst year (after 1976, 1996)
Provence-Alpes-Cote d'Azur	−104.9	The worst year overall
Languedoc-Roussillon	−133.2	The worst year overall
Midi Pyrénées	−101.3	2nd worst (after 1982)
Centre	−40.8	6th worst year
Pays de Loire	−59.9	3rd worst year
Bretagne	−73.2	2nd worst (after 1976)

2 *Arabidopsis thaliana* as a Model for Plant Stress Physiology

The study of plant stress has tended to focus, for very obvious reasons, on crop plants and on wild species that are highly adapted to resist abiotic stress. However, the first complete plant genome sequence was for the model plant *Arabidopsis thaliana* (Brassicaceae) and the very extensive information, data and genetic resources now available on this species makes it an important tool for stress research (Arabidopsis Genome Initiative 2000; The Arabidopsis Information Resource: TAIR www.arabidopsis.org). The value comes from the concerted international effort focused on this single species together with the conscious decision to make as much data as possible available to the entire scientific community. This has resulted in the availability of more information about this species than any other to date, although genome sequencing and hence post-genomic approaches in several other plant species is progressing, building on the experience with *Arabidopsis* (e.g. International Rice Genome Sequencing Project 2005; Plant Genome database: http://www.plantgdb.org/).

Although *A. thaliana* is a weed species without agricultural value and also does not exhibit unusual stress-tolerance, its value lies in its use for the discovery of gene and protein functions and for study of plant physiology, secondary metabolism, development and morphology. This can carried out more easily and rapidly in *Arabidopsis* and useful discoveries can then be applied to crop species through application of syntenic, orthologous and other evolutionary relationships.

The self-fertile, diploid, small plants have a generation time of around 6 weeks and can produce 20,000 seeds per plant. *A. thaliana* can be readily transformed using *Agrobacterium tumefaciens*, with the ability to transform developing seeds at a rate of 1% a particular advantage (Clough and Bent 1998). The complete genome sequence of *A. thaliana*, amounting to around 25,500 genes is curated and available on the TAIR website, which provides information on gene products, metabolism, genome maps, DNA and seed stocks and publications. A very extensive set of resources to exploit this information is also available via the links on the TAIR website. From the point of view of abiotic stress research, the value of large-scale microarray studies to examine the effects of stress on expression of the *Arabidopsis* genome became apparent soon after the genome was completed (e.g. Seki et al. 2002). These data have built up into extensive data-sets of publicly available microarray information (e.g. Genevestigator: https://www.genevestigator.ethz.ch/; TAIR: www.arabidopsis.org), including many sets from stress environments. Integration of transcriptomic, proteomic and metabolomic data has allowed models of metabolic and regulatory networks to be developed, providing new leads for understanding abiotic stress responses (e.g. Hirai et al. 2005; Yamaguchi-Shinozaki and Shinozaki 2006).

3 Transcription Factors and the Response to Abiotic Stress

Transcription factors (TFs) are proteins that control cell processes through binding to specific sequences in the promotors of target genes to activate or repress transcription. Approximately 15% of the genes within the *Arabidopsis* genome appear to be TFs (c. 1,500 genes) belonging to at least 37 gene families (Qu and Zhu 2006). One very interesting discovery from plant genome sequencing projects has been the way in which many families of TFs have expanded substantially, in comparison with animal TF families. This may well be related to the continual modulation of genome expression required by plants to adapt to their environment. Many TFs may therefore regulate gene expression in response to environmental factors, including stress.

Individual TFs function within multimeric protein structures, interacting with both specific DNA sequences and proteins. The interacting proteins may themselves also be TFs, or may have other roles. They are important in the response to signals from the environment through changing the sets of mRNA available for protein production (see Fig. 14.2). The full extent of these molecular structures remains to be identified, as are the details of how they receive and respond to signals, and is a very active area of research. The integration of data from microarray and other sources (e.g. Benedict et al. 2006; Zhang et al. 2006) will provide considerable insights into these processes.

The value of examining TFs in the response to stress was illustrated by the improved tolerance to drought,

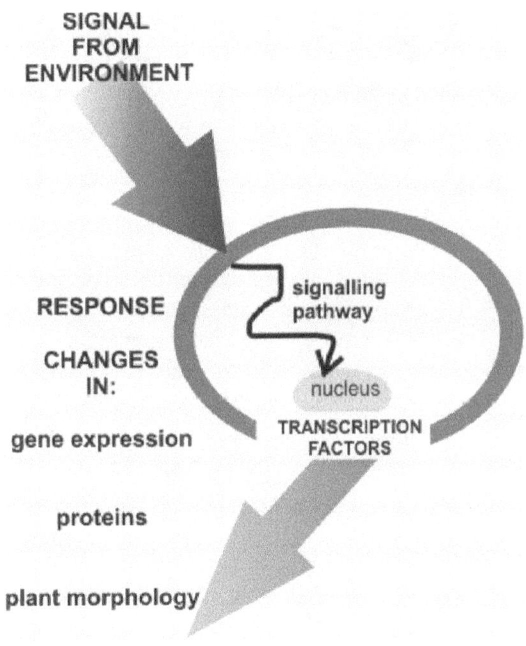

Fig. 14.2 Overview of the signaling pathway from abiotic stress imposed by the external environment to plant response. Perception of stress results in a signal being transmitted that results in a response that will include changes in gene expression and the cell's protein complement. Transcription factors will have a role in modulating gene expression

salinity and freezing conferred by over-expression of a single TF, DREB1A (Kasuga et al. 1999). This study also illustrated the importance of the promoter when using transgenes. Expression from a strong constitutive promoter resulted in growth retardation in normal conditions, while using a stress-inducible promoter resulted in minimal effects under normal growth conditions as well as increased tolerance to stress. The identity of many of the DNA motifs that act as promotors under stress conditions remain to be identified. Resources such as the PLACE database (Database of Plant Cis-acting Regulatory DNA Elements: http://www.dna.affrc.go.jp/PLACE/) are valuable compilations of information about these sequences. A more complete understanding of TF DNA-protein interactions will be necessary to guide the use of TFs for improved plant response to stress.

4 Prospects for the Future

Understanding of plant responses to abiotic stress is now at a very exciting stage. Decades of patient work has built up a substantial picture of responses and has also indicated productive strategies for improving plant performance. Post-genomic approaches provide new avenues, allowing views of genome, protein and cell function that were unimaginable a few years ago. The key components of signalling and response pathways are gradually being exposed. These will give further leads for developing plants with improved agricultural performance. Approaches pioneered within Arabidopsis will lead the way to develop new varieties of crop species as a result of collaboration between basic and applied plant science. Varieties developed using post-genomic insights will eventually appear in farmer's fields to provide improved agricultural productivity despite abiotic stress.

References

Arabidopsis Genome Initiative (2000) Analysis of the genome sequence of the flowering plant *Arabidopsis thaliana*. Nature 408: 796–815

Bartels D, Sunkar R (2005) Drought and salt tolerance in plants. Crit Rev Plant Sci 24: 23–58

Benedict C, Geisler M, Trygg J et al. (2006) Consensus by democracy. Using meta-analyses of microarray and genomic data to model the cold acclimation signaling pathway in *Arabidopsis*. Plant Physiol 141: 1219–1232

Bohnert HJ, Nelson DE, Jensen RG (1995) Adaptations to environmental stress. Plant Cell 7: 1099–1111

Bray EA (1997) Plant responses to water deficit. Trends Plant Sci 2: 48–54

Clough SJ, Bent AF (1998) Floral dip: a simplified method for *Agrobacterium*-mediated transformation of *Arabidopsis thaliana*. Plant J 16: 735–743

Hirai MY, Klein M, Fujikawa Y et al. (2005) Elucidation of gene-to-gene and metabolite-to-gene networks in *Arabidopsis* by integration of metabolomics and transcriptomics. J Biol Chem 280: 25590–25595

International Rice Genome Sequencing Project (2005) The map-based sequence of the rice genome. Nature 436: 793–800

Kasuga M, Liu Q, Miura S et al. (1999) Improving plant drought, salt, and freezing tolerance by gene transfer of a single stress-inducible transcription factor. Nature Biotech 17: 287–291

Qu L-J, Zhu Y-X (2006) Transcription factor families in *Arabidopsis*: major progress and outstanding issues for future research. Curr Opin Plant Biol 9: 544–549.

Seki M, Narusaka M, Ishida J et al. (2002) Monitoring the expression profiles of 7000 Arabidopsis genes under drought, cold and high-salinity stresses using a full-length cDNA microarray. Plant J 31: 279–292

Yamaguchi-Shinozaki K, Shinozaki K (2006) Transcriptional regulatory networks in cellular responses and tolerance to dehydration and cold stresses. Ann Rev Plant Biol 57: 781–803

Zhang T, Liu Y, Yang T et al. (2006) Diverse signals converge at MAPK cascades in plant. Plant Physiol Biochem 44: 274–283

Chapter 15
Improvement of Salt Tolerance Mechanisms of Barley Cultivated Under Salt Stress Using *Azospirillum brasilense*

M.N.A. Omar, M.E.H. Osman, W.A. Kasim, and I.A. Abd El-Daim

Abstract The present work was performed to study the improvement of the salt tolerance of two barley cultivars (Giza 123 and Giza 2000) which are known for their different tolerance to salt stress by using selected PGPR strain. The present study aimed to assess to what extent plant growth promoting rhizobacteria improve the salt tolerance to two barley cultivars (Giza 123 and Giza 2000) differing in salinity tolerance. The inoculant strain *Azospirillum brasilense* (NO40) used in the present study, isolated from hypersaline soil, was able to survive up to 1,800 mM NaCl in the basal media. A green house experiment was conducted to evaluate the effect of this inoculation on growth, photosynthetic pigments and capacity, proline and mineral contents of shoots and roots, some antioxidant and rhizosphere enzymatic activities and yield of the two barley cultivars grown under 250 and 350 mM NaCl. Salinity stress inhibited the growth and yield of both barley cultivars. Salt stress also reduced photosynthetic pigments, photosynthetic capacity, stomatal conductance, transpiration rate, accumulation of K, P, Mg, Ca and Fe in the shoots and roots. In addition, nitrogenase, dehydrogenase activities were also decre-ased in rhizosphere while catalase, peroxidase and superoxide dismutase activities were decreased in the leaves of both cultivars. However, the inoculation with *Azospirillum brasilense* significantly ameliorated the adverse effect of salinity on growth and yield. Furthermore, improvement in salt tolerance of both cultivars of barley was due to increased pigment contents, reduced accumulation of the osmoregulator proline, and reduced activities of antioxidant enzymes. PGPR-induced salt tolerance was more in slat sensitive cultivar Giza 123.

Keywords Antioxidant enzymes • *Azospirillum brasilense* • barley • dehydrogenase • nitrogenase • photosynthesis • proline • salinity • yield

1 Introduction

Salinity is an ever-increasing problem in many regions of the world, particularly in arid and semi-arid regions (Flowers 2004; Munns 2005; Ashraf and Foolad 2007). Salinity stress decreases crop growth and productivity due to salt-induced reduction in the photosynthetic activity (Ashraf 2004). However, generally salinity-induced reduction in growth and yield was due to salt-induced water stress, nutritional imbalance, specific ion toxicity, hormonal imbalance, and generation of reactive oxygen species, which may cause membrane destabilization (Hasegawa et al. 2000; Parida and Das 2005; Ashraf 1994, 2004; Tester and Davenport 2003; Flowers 2004; Munns 2002, 2005). Antioxidant enzyme activities are usually affected by salinity and used as indicators of oxidative stress in plants (Mittler 2002). To protect against oxidative stress, plant cells produce both antioxidant enzymes such as superoxide dismutase (SOD), peroxidase (POX) and catalase (CAT), and non-enzymatic antioxidants such as ascorbate, glutathione and tocopherol (Mittler 2002; Del Rio et al. 2003). The presence of high concentration of salt in the growth medium often results in accumulation of low molecular mass compounds such as proline and glycine betaine, commonly termed compatible solutes, which do not interfere with the normal biochemical

M.N.A Omar (✉) and I.A. Abd El-Daim
Microbiology Department, Soils, Water and Environment Research Institute, Agricultural Research Centre, Giza, Egypt
e-mail: Nabilomarster@gmail.com

M.E.H. Osman and W.A. Kasim
Botany Department, Faculty of Science, Tanta University, Tanta, Egypt

M. Ashraf et al. (eds.), *Salinity and Water Stress*,

reactions (Zhifang and Loescher 2003; Hasegawa et al. 2000; Ashraf and Foolad 2007).

Plant growth promoting rhizobacteria (PGPR) are a group of bacteria that can actively colonize plant roots and improve plant growth and yield by direct and indirect mechanisms (Noel et al. 1996). They are of two general types: those that form a symbiotic relationship with the plant such as the nitrogen-fixing *Rhizobium* spp. and those that are free-living (Glick 1995). They produce plant growth promoting compounds including phytohormones, auxins, cytokinins and gibberellins (Garcia de Salamone et al. 2001). The effectiveness of the plant's rhizosphere colonization by PGPR depends on numerous factors in the agro-ecosystem; among these are the soil type, climatic conditions and fertilization level which play an important role in root colonization (Dobbelaere et al. 2001). In addition, soil pH and moisture are crucial for ultimate attachment and spread of the microbes (Burr et al. 1978).

Azospirillum strains form different types of association with diverse vegetable species. Initially it was believed that they were found only in the rhizosphere, but later on they were isolated from the soil. Certain endophytic strains are able to colonize internally the plant, supplying it more efficiently with nitrogen (Ramos et al. 2002; Fischer et al. 2003).

The aim of this work was to improve salt tolerance of two barely cultivars (Giza 123 as a salt sensitive, and Giza 2000 as a salt tolerant) as one of the most important crop plants, using the relatively cheap and easy method of inoculation with a strain of the plant growth promoting rhizobacteria *Azospirillum brasilense* NO40, and to explore the possible mechanism(s) which lead to that improvement.

2 Materials and Methods

2.1 Selection of Soil and Bacterial Strain

The soil used in this study consisted of 69.25%, 20.4% and 10.35% of sand, silt and clay, respectively. Soil pH was 7.53 and the EC was 0.69 mmohs. Its mineral composition was: 0.9151, 0.511, 0.7, 0.131, 0.436, 0.331 and 0.005 g/100 g soil of N, P, K, Na, Ca, Mg and Fe. A basal fertilizer (N, P and K) was applied according to the recommendations of the Egyptian Ministry of Agriculture. A screening experiment of strains belonging to four different species showed that *Azospirillum brasilense* (NO40) was the most salt tolerant strain and was selected for the colonization of barley cultivated under salt stress.

2.2 Plant Growth

Grains of the two barley cultivars (*Hordeum vulgare*) Giza 123 (salt sensitive) and Giza 2000 (salt tolerant) were obtained from Barley Department, Agricultural Research Center, Giza, Egypt. Grains were soaked in water for 24 h before coating with *Azospirillum brasilense*. Inoculation was performed through mixing grains with the appreciate amount of bacterial strain *Azospirillum brasilense* (NO40). A single inoculation grain harbored 10^6 bacteria on its surface. Liquid inoculation was replicated three times: (1) at sowing date, (2) two months after sowing, and (3) three months after sowing. Eight coated or uncoated grains were sown in each plastic pot and irrigated every other day with 250 or 350 mM NaCl, or with tap water as control. Each treatment was represented by six plastic pots. The pots were drained with water once a week. The seedlings were left to grow for 30 days at 25 ± 2°C in a relative humidity of 65% and 16 h photoperiod. From each treatment, six seedlings were taken to measure the growth criteria (length, fresh and dry weights); 15 seedlings were dried at 50°C for analysis of proline, nitrogen and minerals. Fifteen seedlings were kept frozen in liquid nitrogen and used in the measurements of photosynthetic pigments and enzyme activities.

The photosynthetic pigments, chlorophyll-a, chlorophyll-b and carotenoids were measured in fresh leaves according to Metzner et al. (1965). Stomatal conductance, photosynthetic and transpiration rates were measured using infra red gas analysis system by a clipping single leaf in a Parkinson leaf chamber of a portable ADC-LCA4 system (The Analytical Development Company Ltd, Hoddesdon, Herts, UK) at a photon flux density of about 450 μmol m^{-2} s^{-1} (PAR).

Antioxidant enzymes (catalase and peroxidase) activities were determined according to Kato and Shimizu (1987). Proline contents were measured in dried shoots and roots according to Bates et al. (1973). Total nitrogen contents were measured in dried shoots and roots according to Cottenie et al. (1982). Mineral contents of shoots and roots (P, K, Na, Ca, Mg and Fe) were measured using atomic absorption spectrophotometer. Nitrogenase and dehydrogenase activities in the rhizosphere were assayed according to the methods of Hardy et al. (1973) as modified by Somasegaran and Hoben (1985) and Thalmann (1967), respectively.

Another set of six pots with 10 coated or uncoated grains were left to grow under the same conditions for five months until yield. Criteria used for measuring productivity were spike weight (g), spike length (cm), weight of 1,000 grains (g), number of grains/spike, weight of grains/spike (g) and weight of grains/plant (g).

2.3 *Statistical Analysis*

Data based on replicates were subjected to three-way ANOVA analysis to determine the significance of treatment differences using SPSS v. 10 for Windows. Comparison of the main effects was performed using the Least Significant Difference (LSD) test.

3 Results

3.1 *Selection of Salt Tolerant PGPR Strain*

Results present in Table 15.1 show that *Azospirillum brasilense* (NO40) and *Azospirillum lipoferum* (4B) were able to survive at 1,800 mM NaCl in the basal medium. The effect of different NaCl concentrations on log number, cells dry weight and nitrogen contents of both strains (data not shown) favored the selection of *Azospirillum brasilense* (NO40) for the inoculation of barley grains.

3.2 *Effects of Inoculation with* Azospirillum brasilense

3.2.1 Enzymatic Activities in the Rhizosphere

Nitrogenase and dehydrogenase activities were significantly inhibited with increasing salinity in the rhizosphere of all un-inoculated plants during seedling (30 days) and flowering (two months) stages of both cultivars (Table 15.2). Activities of both enzymes were significantly increased in the rhizosphere of all inoculated plants. The salt sensitive cultivar Giza 123 showed much higher response to inoculation than the salt tolerant cultivar Giza 2000.

3.2.2 Plant Growth

Table 15.3 shows that shoot height, root depth, water contents and fresh and dry weights of shoots and roots of un-inoculated 30-day old seedlings were significantly decreased with increasing NaCl levels. Giza 123 was more affected by NaCl in all un-inoculated seedlings compared with Giza 2000. *A. brasilense* inoculant reversed the inhibitory effect of salinity on all of the measured growth criteria of both cultivars but the response of Giza 123 was much greater.

3.2.3 Pigment Contents, Photosynthetic and Transpiration Rates

Table 15.4 shows that chl-a, chl-b, carotenoids, photosynthetic and transpiration rates and stomatal conductance were decreased with increasing salinity in un-inoculated 30-day old seedlings. The effect was more remarkable in Giza 123 than in Giza 2000. Inoculation with *A. brasilense* significantly alleviated this inhibitory effect of NaCl in all inoculated seedlings compared with their un-inoculated counterparts.

Table 15.1 Effect of NaCl concentrations on the growth of some PGPR strains (+ve = growth, −ve = no growth)

Strain	Control	80 mM	175 mM	350 mM	900 mM	1,800 mM
Azospirillum brasilense (NO40)	+ve	+ve	+ve	+ve	+ve	+ve
Azospirillum lipoferum (4B)	+ve	+ve	+ve	+ve	+ve	+ve
Azotobacter chroococcum	+ve	+ve	+ve	+ve	−ve	−ve
Bacillus polymyxa	+ve	+ve	+ve	−ve	−ve	−ve

Table 15.2 Effect of bacterial inoculation with *Azospirillum brasilense* (NO40) on nitrogenase (µmol C_2H_4/g soil/h) and dehydrogenase (µg TPF/g soil/day) activities in the rhizosphere of two cultivars of *Hordeum vulgare* (Giza 123 and Giza 2000) cultivated under 250 and 350 mM NaCl (Nf: no nitrogen fixation)

			Nitrogenase activity nanomoles C_2H_4/g soil/hr			Dehydrogenase activity µg TPF/g soil/day		
		Treatments	Seedling	Flowering	Yield	Seedling	Flowering	Yield
Un Inoculated	Giza 123	Control	5.10	2.86	Nf	5.85	7.23	4.85
		250 mM NaCl	2.30	2.41	Nf	4.13	5.55	3.32
		350 mM NaCl	1.88	Nf	Nf	4.25	4.81	3.36
	Giza 2000	Control	4.35	Nf	Nf	5.1	6.94	4.77
		250 mM NaCl	1.95	Nf	Nf	3.96	4.92	4.39
		350 mM NaCl	1.72	Nf	Nf	3.87	4.44	4.56
Inoculated	Giza 123	Control	14.32	12.29	2.52	19.23	29.53	6.61
		250 mM NaCl	35.64	30.15	Nf	38.12	44.12	5.14
		350 mM NaCl	22.85	19.45	Nf	26.88	40.61	4.31
	Giza 2000	Control	12.26	Nf	Nf	15.94	20.73	4.95
		250 mM NaCl	19.90	15.37	Nf	21.33	25.68	4.22
		350 mM NaCl	17.52	Nf	Nf	18.53	22.17	4.12
	Native soil			2.3902			4.69	
	LSD 0.01		0.3685	0.3876	0.2776	1.0356	1.2192	1.1386

Results of 4-way ANOVA
F values: ** = significant at $p \leq 0.01$; * = significant at $p \leq 0.05$

Treatments	Nitrogenase activity	Dehydrogenase activity
Cultivar	34,465**	2,184**
Inoculation	104,584**	25,904**
Cultivar × inoculation	25,376**	2,054**
Salinity	4,600**	345.4*
Cultivar × salinity	2,645**	125.2**
Inoculation × salinity	7,162**	687.1**
Cultivar × inoculation × salinity	2,542*	190.8*
Stage	54,778**	7,525**
Cultivar × stage	12,560**	642.3**
Inoculation × stage	30,324**	6,400**
Cultivar × inoculation × stage	8,725**	426.9**
Salinity × stage	1,691**	138.5**
Cultivar × salinity × stage	1,017*	56.61*
Inoculation × salinity × stage	3,022*	172.2**
Cultivar × inoculation × salinity × stage	1,124**	53.67**

Table 15.3 Effect of the inoculation with *Azospirillum brasilense* (NO40) on growth criteria of two cultivars (Giza 123 and Giza 2000) of *Hordeum vulgare* grown under 250 mM and 350 mM NaCl salinity stress

			Shoot				Root			
		Treatments	Height (cm)	Water content (g)	Fresh wt (g)	Dry wt (g)	Depth (cm)	Water content (g)	Fresh wt (g)	Dry wt (g)
Un Inoculated	Giza 123	Control	16.6	0.6662	0.732	0.0658	7.8	0.503	0.5308	0.0278
		250 mM NaCl	8.9	0.2178	0.253	0.0352	7.1	0.1471	0.1621	0.0150
		350 mM NaCl	8.6	0.1349	0.163	0.0281	5.9	0.116	0.1295	0.0135
	Giza 2000	Control	18.9	0.5539	0.626	0.0721	8.06	0.465	0.5143	0.0493
		250 mM NaCl	11.6	0.2818	0.320	0.0382	7	0.1703	0.1875	0.0172
		350 mM NaCl	11.4	0.3063	0.342	0.0357	6.06	0.134	0.1495	0.0155
Inoculated	Giza 123	Control	17.9	0.8544	0.929	0.0746	8.8	0.527	0.5595	0.0325
		250 mM NaCl	11.6	0.2849	0.336	0.0511	8.7	0.331	0.3698	0.0388
		350 mM NaCl	10.4	0.2873	0.322	0.0347	8.1	0.219	0.2427	0.0237
	Giza 2000	Control	20.8	0.7517	0.830	0.0783	10.7	0.505	0.5639	0.0589
		250 mM NaCl	13.9	0.4816	0.526	0.0444	9.4	0.267	0.2893	0.0223
		350 mM NaCl	13.5	0.4249	0.471	0.0461	7.4	0.241	0.2583	0.0173
		LSD 0.01	1.065	0.046	0.047	0.0055	0.94	0.049	0.048	0.0028

Results of 3-way ANOVA

F values: ** = significant at $p \leq 0.01$; * = significant at $p \leq 0.05$; ns = non-significant

	Shoot				Root			
Treatments	Height	Water content	Fr wt	Dry wt	Depth	Water content	Fr wt	Dry wt
Cultivar	359.5**	83.9**	93.7**	37.7**	8.92*	7.89**	ns	197.5**
Inoculation	197.2**	721.9**	758.8**	115.9**	216.65**	150.2**	196.8**	531.6**
Salinity	1,143**	1,805.2**	2,036.5**	942.6**	77.64**	880**	1,085.9**	
Cultivar X inoculation	ns	ns	ns	7.34*	ns	5.99*	9.43**	82.67**
Cultivar X salinity	ns	173.19**	161.6**	23.43**	5.92*	12.69**	9.43**	707.8**
Inoculation X salinity	4.75*	11.9**	10.57**	ns	ns	27.88**	32**	40.12**
Cultivar X inoculation X salinity	ns	14.85 **	12.35**	9.05**	5.49*	5.31*	7.34**	77.1**

Table 15.4 Effect of inoculation with *Azospirillum brasilense* (NO40) on photosynthetic rate, transpiration rate, stomatal conductance, chl-a, chl-b and carotenoids of seedlings of two cultivars (Giza 123 and Giza 2000) of *Hordeum vulgare* cultivated under 250 mM and 350 mM NaCl salinity stress

		Treatments	Photosynthetic rate (µmol m^{-2} s^{1})	Transpiration rate (mmol m^{-2} s^{-1})	Stomatal conductance (mol m^{-2} s^{-1})	Pigment content (mg/g dry wt) Chl-a	Chl-b	Carotenoids
Un Inoculated	Giza 123	Control	0.17	0.28	0.01	30.158	19.714	16.226
		250 mM NaCl	0.095	0.205	0.01	22.665	19.597	12.939
		350 mM NaCl	0.065	0.25	0.01	15.611	14.718	9.984
	Giza 2000	Control	0.215	0.34	0.01	39.412	25.256	20.285
		250 mM NaCl	0.17	0.29	0.01	33.869	22.181	18.622
		350 mM NaCl	0.08	0.27	0.015	33.412	21.336	18.200
Inoculated	Giza 123	Control	1.035	0.27	0.01	33.260	24.171	19.263
		250 mM NaCl	0.125	0.3	0.02	25.035	15.329	17.494
		350 mM NaCl	0.115	0.23	0.00	24.271	13.521	16.703
	Giza 2000	Control	0.275	0.355	0.2	41.508	26.520	20.513
		250 mM NaCl	0.19	0.33	0.1	36.875	26.297	19.780
		350 mM NaCl	0.09	0.28	0.00	34.306	25.481	18.135
		LSD 0.01	0.138	0.094	0.113	1.587	1.023	1.14

Results of 3-way ANOVA:
F values: ** = significant at $p \leq 0.01$; * = significant at $p \leq 0.05$; ns = non-significant

Treatments	Photosynthetic rate	Transpira-tion rate	Stomatal conductance	Chl-a	Chl-b	Carotenoids
Cultivar	1,835.54**	44.985**	8.747**	2,440.28**	1,928.87**	470.109**
Inoculation	3,188.37**	4.237**	2.527*	266.75**	102.14**	236.153**
Salinity	4,632.14**	15.709**	ns	555.85**	492.63**	159.2**
Cultivar × inoculation	2,305.83**	ns	8.057**	41.68**	100.16**	145.34**
Cultivar × salinity	1,814.51**	ns	ns	43.52**	80.60**	16.749**
Inoculation × salinity	2,239.14**	5.603*	3.443*	5.10*	31.37**	17.206**
Cultivar × inoculation × salinity	1,018.07**	24.152*	ns	25.30**	126.53**	5.959**

3.2.4 Mineral Contents

As shown in Tables 15.5 and 15.6, Na contents of shoots and roots were significantly increased with increasing salinity with and without inoculation. This increase was less in case of inoculation and this was more pronounced in Giza 123 than in Giza 2000. On the other hand, K, P, Mg, Ca and Fe contents of shoots and roots were significantly decreased in shoots and roots in both barley cultivars under salinity stress with or without inoculation. Bacterial inoculation inhibited this decrease, especially in Giza 2000.

3.2.5 Total Nitrogen, Proline and Antioxidant Enzymes

Nitrogen contents of shoots and roots of 30-day old un-inoculated seedlings of both barley cultivars were significantly reduced with increasing salinity (Fig. 15.1). The same trend was obtained on inoculation with *A. brasilense* but the reductions were significantly lower than in the corresponding un-inoculated seedlings.

Proline contents in shoots and roots of both cultivars increased significantly with increased salinity with or without bacterial inoculation (Fig. 15.2). In the salt sensitive cultivar Giza 123, the inoculation increased proline accumulation in shoots but decreased it in roots compared with their corresponding un-inoculated counterparts. In the salt tolerant cultivar Giza 2000, inoculation reduced the proline accumulation in shoots and roots.

Figure 15.3 shows that increased salinity resulted in a significant increase in catalase and peroxidase activities in all salt-stressed leaves of both cultivars with and without inoculation, although inoculation lowered the magnitudes of increase.

Table 15.5 Effect of inoculation with *Azospirillum brasilense* (NO40) on shoots mineral contents of seedlings of two cultivars (Giza 123 and Giza 2000) of *Hordeum vulgare* grown under 250 and 350 mM NaCl salinity stress

			Shoots minerals contents (g/100 g dry weight)					
		Treatments	Na^{+}	K^{+}	P^{+3}	Mg^{+2}	Ca^{+2}	Fe^{+2}
Un Inoculated	Giza 123	Control	0.465	2.751	0.425	0.325	0.4575	0.325
		250 mM NaCl	1.151	1.375	0.325	0.155	0.3975	0. 169
		350 mM NaCl	1.568	1.211	0. 291	0 145	0.3525	0. 138
	Giza 2000	Control	0.425	2.775	0.475	0.375	0.4725	0.352
		250 mM NaCl	1.0725	1.975	0.341	0.195	0.4201	0.165
		350 mM NaCl	1.125	1.252	0.321	0.183	0.3825	0.152
Inoculated	Giza 123	Control	0. 415	2.815	0.463	0.397	0.4625	0.332
		250 mM NaCl	0.864	1.694	0.342	0 169	0.3841	0. 175
		350 mM NaCl	1.055	1.326	0.320	0.152	0.3924	0. 140
	Giza 2000	Control	0.4275	2.825	0.445	0.389	0.4755	0.345
		250 mM NaCl	1.0525	1.982	0.332	0.197	0.4322	0.175
		350 mM NaCl	1.1115	1.394	0.329	0.188	0.3925	0.164
LSD 0.01			0.62	0.641	0.197	0.158	0.097	0.063

Results of 3-way ANOVA
(F values: ** = significant at $p \leq 0.01$; * = significant at $p \leq 0.05$; ns = non-significant)

Treatments	Na^{+}	K^{+}	P^{+3}	Mg^{+2}	Ca^{+2}	Fe^{+2}
Cultivar	ns	5.521*	ns	ns	ns	9.288**
Inoculation	6.266**	14.617**	ns	ns	ns	ns
Salinity	33.24**	67.957**	12.931**	46.33**	12.272**	225.77**
Cultivar X inoculation	8.623**	8.889**	7.4151**	ns	ns	ns
Cultivar X salinity	ns	Ns	ns	ns	ns	ns
Inoculation X salinity	ns	Ns	ns	ns	ns	ns
Cultivar X inoculation X salinity	ns	Ns	ns	ns	ns	ns

Table 15.6 Effect of inoculation with *Azospirillum brasilense* (NO40) on roots mineral contents of seedlings of two cultivars (Giza 123 and Giza 2000) of *Hordeum vulgare* grown under 250 and 350 mM NaCl salinity stress

Treatments			Roots minerals contents (g/100 g dry weight)					
			Na^+	K^+	P^{+3}	Mg^{+2}	Ca^{+2}	Fe^{+2}
Un Inoculated	Giza 123	Control	0.0616	1.7866	0.3933	0.431	0.0603	0.024
		250 mM NaCl	0.2196	1.0733	0.1833	0.2766	0.0273	0.0196
		350 mM NaCl	0.2603	0.54	0.0933	0.18	0.021	0.018
	Giza 2000	Control	0.0596	1.8866	0.4666	0.3866	0.0616	0.027
		250 mM NaCl	0.1873	1.15	0.2666	0.3333	0.0373	0.0246
		350 mM NaCl	0.1956	0.63	0.18	0.2333	0.0253	0.0206
Inoculated	Giza 123	Control	0.0573	2.1966	0.4266	0.47	0.062	0.028
		250 mM NaCl	0.1856	1.1233	0.23	0.2633	0.029	0.0203
		350 mM NaCl	0.1786	0.61	0.13	0.2166	0.0243	0.0183
	Giza 2000	Control	0.0543	1.89	0.48	0.4433	0.0646	0.028
		250 mM NaCl	0.1583	1.14	0.2965	0.345	0.0396	0.0256
		350 mM NaCl	0.189	0.6466	0.2233	0.2366	0.0294	0.0213
LSD 0.01			0.018	0.185	0.032	0.0067	0.0486	0.0025

Results of 3-way ANOVA
(F values: ** = significant at $p \leq 0.01$; * = significant at $p \leq 0.05$; ns = non-significant)

Treatments	Na^+	K^+	P^{+3}	Mg^{+2}	Ca^{+2}	Fe^{+2}
Cultivar	116.62**	10.99**	49.613**	14.993**	8.4866**	11.932**
Inoculation	48.34**	ns	250.25**	16.229**	35.756**	73.285**
Salinity	1,226.5**	818.11**	1,227.7**	289.17**	534.66**	126.22**
Cultivar X inoculation	31.36**	10.19**	ns	ns	ns	ns
Cultivar X salinity	23.69**	4.679**	ns	7.0815*	ns	ns
Inoculation X salinity	9.65**	3.811*	ns	13.457**	6.1512**	8.3909**
Cultivar X inoculation X salinity	27.3 9**	4.619*	ns	ns	ns	ns

3.2.6 Grain Yield

Salinity induced significant decreases in number of spikes/plant, spike length, spike weight, number of grains/spike, weight of grains/spike, grain weight/plant and weight of 1,000 grains (Table 15.7). Reductions were greater in the salt-sensitive cultivar Giza 123 than in the salt-tolerant cultivar Giza 2000. Bacterial inoculation alleviated the reduction in yield parameters especially in the salt-tolerant cultivar. As shown in Fig. 15.4 (A and B), the inoculation is correlated positively with the increase in the dry matter accumulation, even at the high salinity level of 350 mM NaCl.

4 Discussion

Salinity stress significantly reduced the productivity of the salt-sensitive (Giza 123) and the salt-tolerant (Giza 2000) cultivars of barley (*Hordeum vulgare*) as an ultimate outcome of the salt-induced reduction in stomatal conductance, transpiration rate, accumulation of the osmoregulator proline, availability of soil minerals, photosynthetic pigments and capacity, antioxidant enzyme activities and, consequently, accumulation of dry matter in the plants' shoots and roots. Such effects were comparatively more pronounced in the sensitive than in the tolerant cultivar. On the other hand, the plant growth promoting rhizobacteria (PGPR) *Azospirillum brasilense* (strain NO40) ameliorated the deleterious effects of salinity on these parameters and improved crop productivity.

Results of the present experiment might be easily understood in the light of the interpretations put forward by numerous authors. Thus, the inhibitory effect of salinity on crop yield might be caused by several reasons which include delaying the emergence of panicle and flowering (Khatun et al. 1995), decreasing seed set through reduced pollen viability (Khatun and Flowers

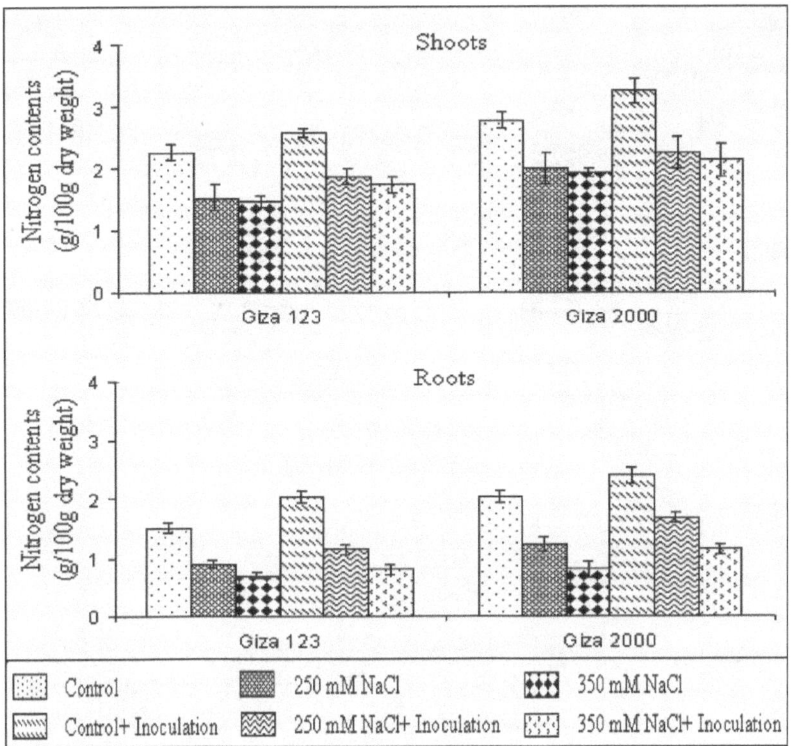

Results of 3-way ANOVA:

(F values: ** = significant at $p \leq 0.01$; F values: * = significant at $p \leq 0.05$; n.s. = non-significant)

Treatments	Shoots	Roots
Cultivar	104.1666 **	144.230 **
Inoculation	48.1661 **	122.0769 **
Salinity	100.0416 **	466.845 **
Cultivar x Inoculation	ns	ns
Cultivar x Salinity	ns	3.769 **
Inoculation X Salinity	ns	4.354 **
Cultivar x Inoculation x Salinity	ns	4.216 **

Fig. 15.1 Effect of bacterial inoculation with *Azospirillum brasilense* (NO40) on shoot and root nitrogen contents (g/100 g dry weight) of two cultivars of *Hordeum vulgare* (Giza 123 and Giza 2000) cultivated under 250 and 350 mM NaCl

1995; Khatun et al. 1995) and suppression of germination which was mainly due to the osmotic stress (Heenan et al. 1988). Reduction in seedling growth and loss of standing yield due to salinity were implicated as causative factors for losses in crop production (Scardaci et al. 1996; Shannon et al. 1998). Effects of salinity on plant growth were attributed to increasing stiffness of cell walls, probably due to altered cell wall structure (Sweet et al. 1990), reduced rates of new cell production (Kinraide 1999; Shabala et al. 2000), a decrease in cell wall extensibility and/or to ion toxicity (Van Volkenburgh and Boyer 1985), the reduction in light interception due to the reduced leaf area (Marcelis and Hooijdonk 1999), and osmotic effects, nutritional deficiency as well as mineral disorders (Katerji

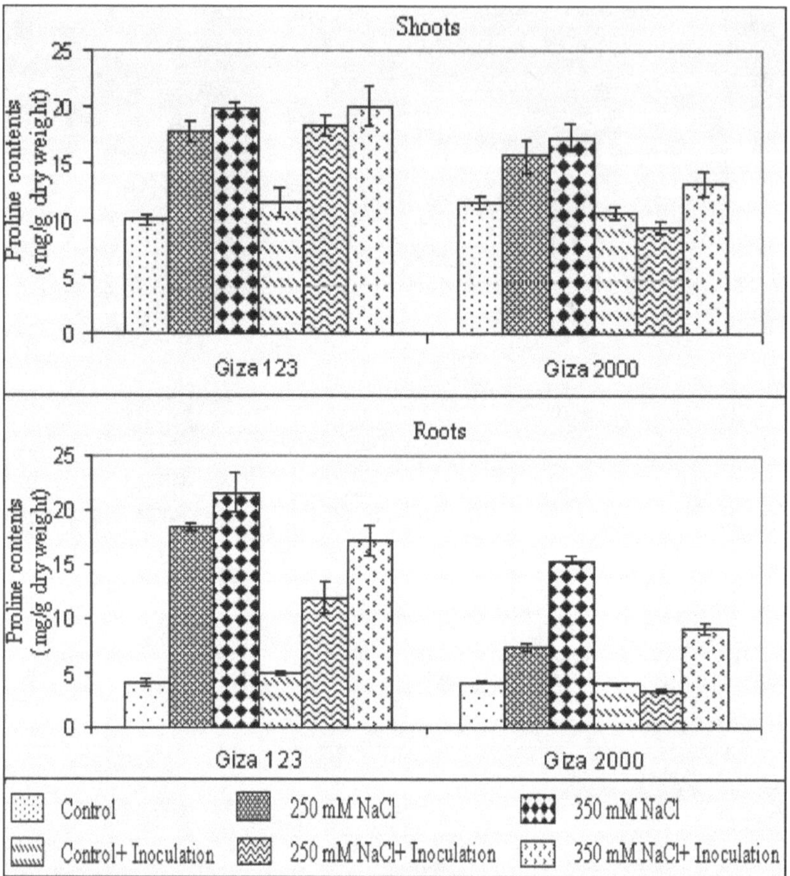

Results of 3-way ANOVA:

(F values: ** = significant at $p \leq 0.01$; * = significant at $p \leq 0.05$; n.s. = non-significant)

Treatments	Shoots		Roots	
Cultivar	5.904	*	323.637	**
Inoculation	16.413	**	909.146	**
Salinity	106.424	**	989.647	**
Cultivar X inoculation	234.67	**	87.706	**
Cultivar X salinity	ns		128.430	**
Inoculation X salinity	29.965	**	262.774	**
Cultivar X inoculation X salinity	25.755	**	91.935	**

Fig. 15.2 Effect of bacterial inoculation with *Azospirillum brasilense* (NO40) on shoot and root proline contents (mg/g dry weight) of two cultivars of *Hordeum vulgare* (Giza 123 and Giza 2000) cultivated under 250 and 350 mM NaCl

et al. 1998; Hoorn et al. 2002). Furthermore, salinity was shown to reduce the biosynthesis of the photosynthetic pigments (Kasim and Hamada 2003). This reduction in pigment content affected the strength of binding forces in the pigment-protein complex in the chloroplast structures which consequently resulted in structural changes in photosynthetic system (Strogonov et al. 1970). Salinity stress also caused changes in the integrity and composition of chloroplast membranes (Quaratacci and Narvari-Izzo 1992). GuiBin and FuLiang (2004) and Martinez-Ballesta

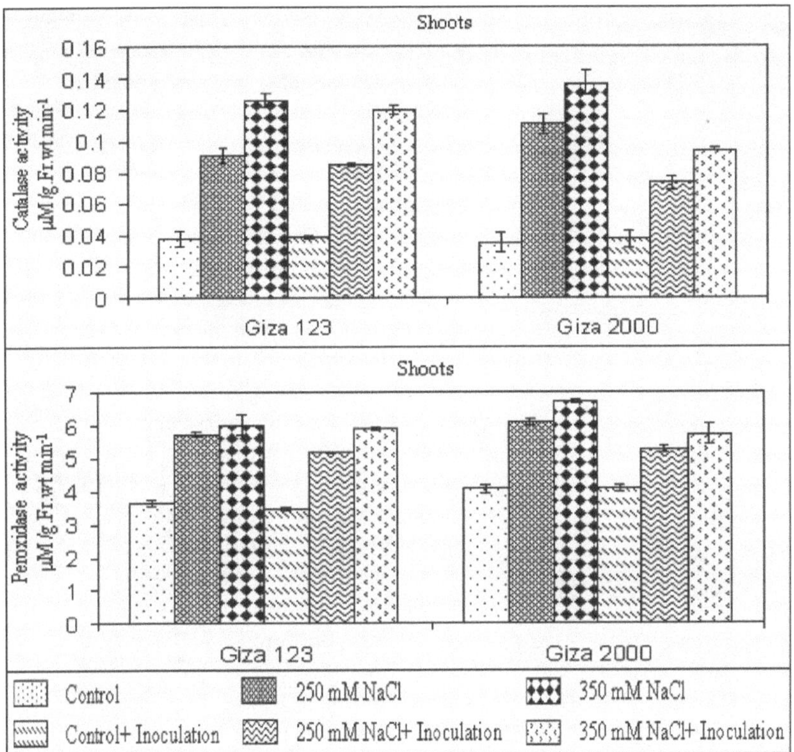

Results of 3-way ANOVA:

(F values: ** = significant at $p \leq 0.01$; * = significant at $p \leq 0.05$; n.s. = non- significant)

Treatments	Catalase	Peroxidase
Cultivar	ns	103.9504 **
Inoculation	178.331 **	174.2449 **
Salinity	1901.03 **	1676.026 **
Cultivar X inoculation	111.635 **	22.89268 **
Cultivar X salinity	7.54831 **	6.802789 **
Inoculation X salinity	53.8429 **	29.6637 **
Cultivar X inoculation X salinity	31.7439 **	21.8918 **

Fig. 15.3 Effect of inoculation with *Azospirillum* (NO40) on catalase and peroxidase activities in shoots of two cultivars (Giza 123 and Giza 2000) of *Hordeum vulgare* cultivated under 250 and 350 mM NaCl salinity stress

et al. (2004) suggested that the decrease in photosynthetic and transpiration rates under salt stress may be attributed to stomatal limitations caused by stomatal closure induced by drought resulting from the decrease in plant water contents as well as non-stomatal limitations caused by metabolic factors such as decrease in chlorophyll content or mineral disorders. De la Rosa and Maiti (1995) suggested that the inhibitory effect of salt stress on chlorophyll content may be due to the synthesis of nitrogen compounds such as proline which consumes large amount of nitrogen. This observation is in agreement with the results obtained in the present study which showed that the increase of chlorophyll contents in the inoculated salt-stressed seedlings of both cultivars was accompanied with a decrease in proline contents under the same treatments. The proposed functions of proline under stress

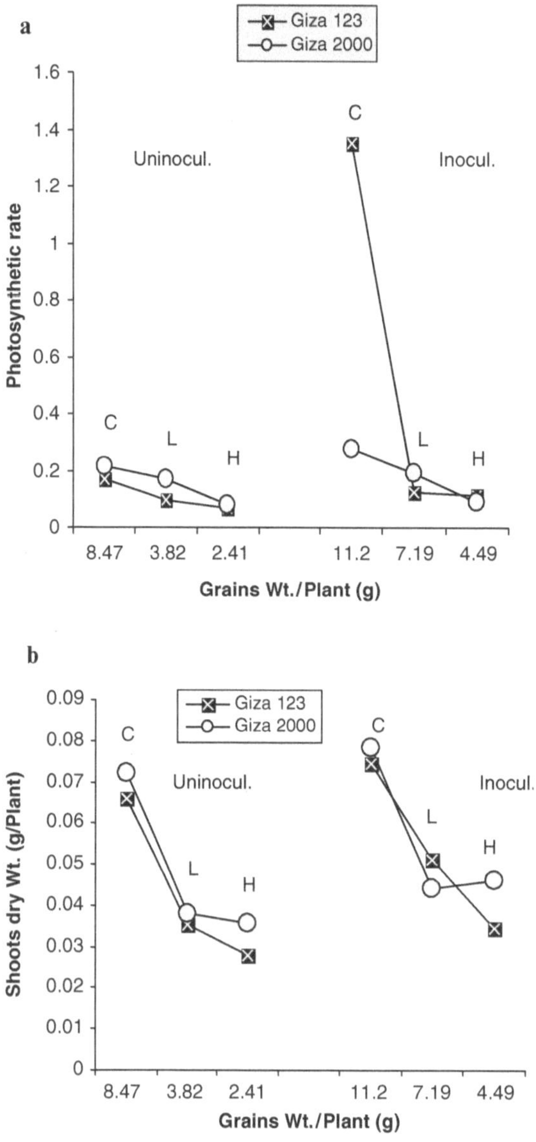

Fig. 15.4 Effect of bacterial inoculation with *Azospirillum brasilense* (NO40) on the relation between the photosynthetic rate in μmol m^{-2} s^{-1} (**a**), shoot dry weight in g/plant (**b**) and the weight of grain yield/plant of two cultivars of *Hordeum vulgare* (Giza 123 and Giza 2000) cultivated under control (**c**), 250 mM NaCl (L) and 350 mM NaCl (H) with and without inoculation

conditions include its role as a solute for intercellular osmotic adjustment, protection of enzymes and membranes, and acting as a reservoir of energy and nitrogen for utilization during exposure to salinity (Bandurska 1993; Perez-Alfocea et al. 1993; Rai et al. 2003; Silveira et al. 2003). Zhu (2001) and Sairam and Tyagi (2004) demonstrated that both osmotic and ionic effects of NaCl salinity caused stomatal closures which reduced the CO_2/O_2 ratio in the leaves of plants and inhibited CO_2 fixation. These conditions increased the rate of reactive oxygen species (ROS) formation *via* enhanced leakage of electrons to oxygen in the chloroplast; ROS are responsible for a secondary oxidative stress that can damage cellular structure and metabolism.

Plant growth promoting rhizobacteria were used for agricultural purposes because they stimulated plant growth through the production of plant growth regulators and the increase of nitrogen fixation as the major mechanisms of action for the enhancement of plant growth (Warembourg et al. 1987; Lucy et al. 2004). Furthermore, according to Glick (1995) rhizosphere bacteria can promote plant growth either directly or indirectly. Direct mechanisms of plant growth promotion may involve the synthesis of substances by the bacterium or facilitation of the uptake of nutrients from the environment (Glick et al. 1999). These direct growth promoting mechanisms may take place through the following processes: (1) nitrogen fixation, (2) solubilization of phosphorus, (3) sequestering of iron by production of siderophores, and/or (4) production of several phytohormones such as auxins, cytokinins, gibberellins and indoleacetic acid (Kloepper et al. 1988; Glick 1995; Glick et al. 1999). The indirect promotion of plant growth occurred when PGPR lessened or prevented the deleterious effects of pathogens on plants by the production of substances inhibitory to the pathogen, or by increasing the natural resistance of the host (Handelsman and Stabb 1996; Nehl et al. 1996; Cartieaux et al. 2003). Many *Azospirillum* strains produced several phytohormones such as indoleacetic acid, isobutyric acid and cytokinins which promoted plant growth (Omay et al. 1993).

The enhanced photosynthesis due to increased chlorophyll contents is a known response of plants to the inoculation with *A. brasilense* (Bambal et al. 1998; Omar et al. 2000; Panwar and Singh 2000). Bashan et al. (2006) found that inoculation of wheat cultivated under salt stress with *A. brasilense* resulted in the enhanced production of photosynthetic pigments, which in turn led to a significant increase in photosynthetic rate and CO_2 fixation. Bashan et al. (2006) reported that inoculation of wheat with *Azospirillum* under salt stress enhanced the production of auxiliary

Table 15.7 Effect of the inoculation with *Azospirillum brasilense* (NO40) on some yield parameters of two cultivars (Giza 123 and Giza 2000) of *Hordeum vulgare* cultivated under 250 and 350 mM NaCl salinity stress

		Treatments	No. of spikes/plant	Spike length (cm)	Spike wt (g)	No. of grains/spike	Grains wt/spike (g)	Grains wt/plant (g)	Weight of 1,000 grains (g)
Un Inoculated	Giza 123	Control	4.66	17.68	1.715	38	1.7617	8.2	47.82
		250 mM NaCl	3.33	16.965	1.417	29	1.1946	3.97	39.51
		350 mM NaCl	2.75	16.65	0.761	25.5	0.8863	2.43	34.31
	Giza 2000	Control	5	18.35	1.815	41.5	2.0032	10.016	49.45
		250 mM NaCl	3.55	17.45	1.605	34	1.6433	5.83	47.91
		350 mM NaCl	3	16.655	0.831	30	1.3945	4.18	44.82
Inoculated	Giza 123	Control	5.33	18.725	1.905	42.5	2.1282	11.34	49.45
		250 mM NaCl	4.33	17.825	1.665	35.5	1.7106	7.4	46.75
		350 mM NaCl	3.66	17.035	0.932	31.5	1.2406	4.54	38.95
	Giza 2000	Control	5.66	18.955	1.885	44	2.1963	12.43	49.81
		250 mM NaCl	4.66	18.1	1.655	39	1.8842	8.78	49.12
		350 mM NaCl	4	17.01	0.991	32	1.4807	5.92	45.85
LSD 0.01			1.97	0.23	0.143	3.03	0.136	0.622	1.205

Results of 3-way ANOVA:
(F values: ** = significant at $p \leq 0.01$; * = significant at $p \leq 0.05$; ns = non-significant)

Treatments	No. of spikes/plant	Spike length	Spike wt	No. of grains/spike	Grains wt/spike	Grains wt/plant	Weight of 1,000 grains
Cultivar	ns	65.8**	9.46**	48.3**	198.3**	407**	198.3**
Inoculation	10.66**	372**	50.1**	99.1**	216.9**	158.4**	216.9**
Salinity	15.6**	742**	746**	237**	503**	947**	503**
Cultivar × inoculation	ns	11.3**	6.82**	7.94**	36.11**	7.91*	36.11**
Cultivar × salinity	ns	18.03**	Ns	ns	10.75**	6.32*	10.75**
Inoculation × salinity	ns	17.6**	Ns	ns	5.37**	Ns	5.37**
Cultivar × inoculation × salinity	ns	ns	Ns	ns	ns	ns	ns

photoprotective pigments such as violaxanthin, zeaxanthin, antheroxanthin, lutein, neoxanthin, and β-carotene which may protect chl-a and chl-b from oxidation during exposure to salt stress. They attributed the effect of inoculation to possible interactions between some complex mechanisms such as hormonal effect, N_2 fixation, proton extrusion and/or mineral uptake. Moreover, PGPR strains were shown to produce bacterial exopolysaccharides which bind with some cations including Na (Geddie and Sutherland 1993; Han and Lee 2005). This notion is further supported by the findings of Ashraf et al. (2004) that increasing the population density of PGPR in the root zone could decrease the Na available for plant uptake, thus helping to alleviate salt stress in plants. Results of the present study and their possible interpretations strongly point to the feasibility of using PGPR in improving the productivity of crops grown under salinity stress.

References

Ashraf M (1994) Breeding for salinity tolerance in plants. Crit Rev Plant Sci 13:17–42.

Ashraf M (2004) Impact evaluation of water resources development in the common areas of small dams. Pakistan Council of Research in Water Resources, Research Report 5–2004.

Ashraf M, Foolad MR (2007) Roles of glycine betaine and proline in improving plant abiotic stress resistance. Environ Exp Bot 59: 206–216.

Ashraf M, Berge SH, Mahmood OT (2004) Inoculating wheat seedlings with exopolysaccharide-producing bacteria restricts sodium uptake and stimulates plant growth under salt stress. Biol Fert Soils 40: 157–162.

Bambal AS, Verma RM, Panchbhai DM, Mahorkar VK, Khankhane RN (1998) Effect of biofertilizers and nitrogen levels on growth and yield of cauliflower (*Brassica oleracea* var. *botrytis*). Orissa J Hortic 26: 14–17.

Bandurska H (1993) In vivo and in vitro effect of proline on nitrate activity under osmotic stress in barley. Acta Physiol Plant 15: 83–88.

Bashan Y, Bustillos JJ, Leyva LA, Hernandez JP, Bacilio M (2006) Increase in auxiliary photoprotective photosynthetic pigments in wheat seedlings induced by *Azospirillum brasilense*. Biol Fert Soils 42: 279–285.

Bates LS, Waldren RP, Tear ID (1973) Rapid determination of free proline for water stress studies. Plant Soil 39: 205–207.

Burr TJ, Schroth MN, Suslow T (1978) Increased potato yields by treatment of seed pieces with specific strains of *Pseudomonas fluorescens* and *Pseudomonas putida*. Phytopathology 68: 1377–1383.

Cartieaux FP, Nussaume L, Robaglia C (2003) Tales from the underground: molecular plant-rhizobacteria interactions. Plant Cell Environ 26: 189–199.

Cottenie A, Verloo M, Velghe M, Camerlynck R (1982) Chemical analysis of plant and soil. Laboratory of Analytical and Agrochemistry, State University, Ghent, Belgium.

De la Rosa IM, Maiti RK (1995) Biochemical mechanism in glossy *Sorghum* lines resistance to salinity stress. J Plant Physiol 146: 515–519.

Del Rio LA, Corpas FJ, Sandalio LM, Palma JM, Barroso JB (2003) Plant peroxisomes, reactive oxygen metabolism and nitric oxide. IUBMB Life 55: 71–81.

Dobbelaere S, Croonenborghs A, Thys A, Ptacek D, Vanderleyden J, Dutto P, Labandera-Gonzales C, Caballero-Mellado J, Aguirre JF, Kapulnik Y, Brener S, Burdman S, Kadouri AD, Sarig S, Okon Y (2001) Responses of agronomically important crops to inoculation with *Azospirillum* sp. Aust J Plant Physiol 28: 871–879.

Fischer SE, Miguel MJ, Mori GB (2003) Effect of root exudates on the exopolysaccharide composition and the lipopolysaccharide profile of *Azospirillum brasilense* Cd under saline stress. FEMS Microbiol Lett 219: 53–62.

Flowers TJ (2004) Improving crop salt tolerance. J Exp Bot 55: 307–319.

Garcia de Salamone IE, Hynes RK, Nelson LM (2001) Cytokinin production by plant growth promoting rhizibacteria and selected mutants. Can J Microbiol 47: 404–411.

Geddie JL, Sutherland IW (1993) Uptake of metals by bacterial polysaccharides. J Appl Bacteriol 74: 467–472.

Glick BR (1995) The enhancement of plant growth by free-living bacteria. Can J Microbiol 41: 109–117.

Glick BR, Patten CL, Holguin G, Penrose DM (1999) Biochemical and genetic mechanisms used by plant growth promoting bacteria. Imperial College Press, London, UK, p. 267.

GuiBin W, FuLiang C (2004) Effects of soil, water and salt contents on photosynthetic characteristics. J Nanjing Forestry University (Natural Sciences Edition) 28(3): 14–18.

Han HS, Lee KD (2005) Plant growth promoting rhizobacteria: effect on antioxidant status, photosynthesis, mineral uptake and growth of lettuce under soil salinity. Res J Agr Biol Sci 1: 210–215.

Handelsman J, Stabb EV (1996) Biocontrol of soil-borne plant pathogens. Plant Cell 8: 1855–1869.

Hardy RWF, Burns RC, Holsten RO (1973) Application of the acetylene–ethylene assay for measurement of nitrogen fixation. Soil Boil Biochem 5: 47–81.

Hasegawa PM, Bressan RA, Zhu JK, Bohnert HJ (2000) Plant cellular and molecular responses to high salinity. Annu Rev Plant Physiol Plant Mol Biol 51: 463–499.

Heenan DP, Lewin LG, McCaffery DW (1988) Salinity tolerance in rice varieties at different growth stages. Aust J Exp Agr 28: 343–349.

Hoorn W, Katerji N, Hamady A, Mastroilli M (2002) Effect of salinity on yield and nitrogen uptake of four grain legumes and on biological nitrogen contribution from the soil. Agr Water Manage 51: 87–98.

Kasim WA, Hamada EAM (2003) Effect of salinity stress on some metabolites, α- and β- amylase activities and protein patterns in *Eruca sativa* seedlings. Egypt J Biotechnol 14: 126–141.

Katerji N, Hoorn J, Hamady A, Mastroilli M (1998) Response of tomatoes, a crop of indeterminate growth, to soil salinity. Agr Water Manage 38: 59–68.

Kato M, Shimizu S (1987) Chlorophyll metabolism in higher plants. VII. Chlorophyll degradation in senescing tobacco leaves: phenolic-dependent peroxidative degradation. Can J Bot 65: 729–735.

Khatun S, Flowers TJ (1995) Effects of salinity on seed set in rice. Plant Cell Environ 18: 61–67.

Khatun S, Rizzo CA, Flowers TJ (1995) Genotypic variation in the effect of salinity on fertility in rice. Plant Soil 173: 239–250.

Kinraide TB (1999) Interactions among Ca^{+2}, Na^{+} and K^{+} in salinity toxicity: Quantitative resolution of multiple toxic and ameliorative effects. J Exp Bot 338: 1496–1505.

Kloepper JW, Hume DJ, Scher FM, Singleton C, Tipping B, Laliberté M, Frauley K, Kutchaw T, Simonson C, Lifshitz R, Zaleska I, Lee L (1988) Plant growth-promoting rhizobacteria on canola (rapeseed). Plant Dis 72: 42–45.

Lucy M, Reed E, Glick BR (2004) Applications of free living plant growth-promoting rhizobacteria. Antonie van Leeuwenhoek 86: 1–25.

Marcelis LFM, Hooijdonk JV (1999) Effect of salinity on growth, water use and nutrient use in radish (*Raphanus sativus* L.). Plant Soil 215: 57–64.

Martinez-Ballesta MC, Martinez V, Carvajal M (2004) Osmotic adjustment, water relations and gas exchange in pepper plants grown under NaCl or KCl. Environ Exp Bot 52: 161–174.

Metzner H, Rau H, Senger H (1965) Untersuchungen zur synchronisierbarkeit einzelner pigmentmangel-mutanten von *Chlorella*. Planta 65: 186–194.

Mittler R (2002) Oxidative stress, antioxidants and stress tolerance. Trends Plant Sci 7: 405–410.

Munns R (2002) Salinity, growth and phytohormones. In: Läuchli A, Lüttge U (eds). Salinity: environment - plants - molecules. Kluwer, Dordrecht, pp. 271–290.

Munns R (2005) Genes and salt tolerance: bringing them together. New Phytol 167: 645–663.

Nehl DB, Allen SJ, Brown JF (1996) Deleterious rhizosphere bacteria: an integrating perspective. Appl Soil Ecol 5: 1–20.

Noel TC, Sheng C, Yost CK, Pharis RP, Hynes MF (1996) *Rhizobium leguminosarum* as a plant growth-promoting rhizobacterium: direct growth promotion of canola and lettuce. Can J Microbiol 42: 279–283.

Omar MN, Fang P, Jia XM (2000) Effect of inoculation with *Azospirillum brasilense* NO40 isolated from Egyptian soils on rice growth in China. Egypt J Agr Res 78: 1005–1014.

Omay SH, Schmidt WA, Martin P (1993) Indoleacetic acid production by the rhizosphere bacterium *Azospirillum brasilense* Cd under *in vitro* conditions. Can J Microbiol 39: 187–192.

Panwar JDS, Singh O (2000) Response of *Azospirillum* and *Bacillus* on growth and yield of wheat under field conditions. Indian J Plant Physiol 5: 108–110.

Parida SK, Das AB (2005) Salt tolerance and salinity effects on plants. Ecotoxicol Environ Safety 60: 324–349.

Perez-Alfocea F, Estan MT, Santa Cruz A, Bolarin MC (1993) Effects of salinity on nitrate, total nitrogen, soluble protein and free amino acid levels in tomato plants. J Hortic Sci 68: 1021–1027.

Quaratacci MF, Navari-Izzo FR (1992) Water stress and free radical mediated changes in sunflower seedlings. J Plant Physiol 139: 621–625.

Rai SP, Luthra R, Kumar S (2003) Salt-tolerant mutants in glycophytic salinity responses (GRS) genes in *Catharanthus roseus*. Theor Appl Genet 106: 221–230.

Ramos HJO, Roncato-Maccari LDB, Souza EM, Soares-Ramos JRL, Hungria M, Pedrosa FO (2002) Monitoring *Azospirillum*-wheat interactions using the gfp and gusA genes constitutively expressed from a new broad-host range vector. J Biotechnol 97: 243–252.

Sairam RK, Tyagi A (2004) Physiology and molecular biology of salinity stress tolerance in plants. Curr Sci 86: 707–421.

Scardaci SC, Eke AU, Hill JE, Shannon MC, Rhoades JD (1996) Water and soil salinity studies on California rice. Rice Publication 2. Cooperative Extension, University of California, Colusa, CA.

Shabala S, Babourina O, Newman I (2000) Ion specific mechanisms of osmoregulation in bean mesophyll cells. J Exp Bot 51: 1243–1253.

Shannon MC, Rhoades JD, Draper JH, Scardaci SC, Spyres MD (1998) Assessment of salt tolerance in rice cultivars in response to salinity problems in California. Crop Sci 38: 394–398.

Silveira JA, Viegas Rde A, da Rocha IM, Moreira AC, Moreira Rde A, Oliveira JT (2003) Proline accumulation and glutamine synthetase activity are increased by salt-induced proteolysis in cashew leaves. J Plant Physiol 160: 115–123.

Somasegaran P, Hoben HG (1985) Methods in legume-Rhizobium technology, NIFTAL Project and MIRCEN. Department of Agronomy and Soil Science, College of Tropical Agriculture and Human Recourse, University of Hawaii, Honolulu, HI, pp. 320–327.

Strogonov BP, Kabanov VV, Shevajakova NI, Lapine LP, Komizerko EI, Propov BA, Dostonova RK, Prykod'Ko LS (1970) Structure and function of plant cell in saline habitats. Vauka, Moscow (in Russian).

Sweet WJ, Morrison JC, Labavitch JM, Matthews MA (1990) Altered synthesis and composition of cell wall of grapevines during expression and growth inhibiting water deficits. Plant Cell Physiol 31: 407–414.

Tester M, Davenport R (2003) Na^{+} tolerance and Na^{+} transport in higher plants. Ann Bot 91: 503–527.

Thalmann A (1967) Über die mikrobielle aktivitat und Ihr Beziehung zu Fruchtbar merkmalen einiger Acherbeden ünter besonderer Berüksichtigung der dehydrogenase aktivität (TTC redakion). Ph.D. dissertation, Giesen, Germany.

Van Volkenburgh E, Boyer JS (1985) Inhibitory effects of water deficits on maize leaf elongation. Plant Physiol 77: 190–194.

Warembourg FR, Dressen R, Vlassak K, Afont F (1987) Peculiar effect of *Azospirillum* inoculation on growth and nitrogen balance of winter wheat. Biol Fert Soil 4: 55–59.

Zhifang G, Loescher WH (2003) Expression of a celery mannose 6-reductase in *Arabidopsis thaliana* enhances salt tolerance and induces biosynthesis of both mannitol and a glucosyl-mannitol dimer. Plant Cell Environ 26: 275–283.

Zhu JK (2001) Plant salt tolerance. Trends Plant Sci 6: 66–71.

Chapter 16
Genetic Resources for Some Wheat Abiotic Stress Tolerances

A. Mujeeb-Kazi, A. Gul, I. Ahmad, M. Farooq, Y. Rauf, A.-ur Rahman, and H. Riaz

Abstract Breeding for abiotic stress tolerances has a similar input necessity given to the incorporation of genetic diversity that combats biotic stress constraints. The major difference is that abiotic stress tolerance genetic transfers are expected to be more durable since in the absence of pathogenic influences the traits are considered static entities. Wheat production scenarios have generated the need to breed cultivars with tolerance to drought, salinity, heat, waterlogging, plus cold, aluminum, some micronutrients, boron and copper efficiency to a lesser degree. Sources of allelic diversity exploited exist in the conventional wheat germplasm, landrace cultivars or in its relative species distributed across the Triticeae gene pools. Harnessing the above diversity has protocol specificity that opens doors for wheat improvement programs across various phases as influenced by genetic introgression simplicity or complexity mediated by novel breeding techniques, achievement of homozygosity and use of molecular tools.

Keywords Abiotic stresses • gene pools • genetic diversity • wheat breeding • alien introgression • synthetic wheats

1 Introduction

Some years ago, Valkoun (2001) estimated *ex situ* gene bank holdings to be around 8 million accessions of which about 3% were wild wheat relatives. These are separated into three gene pools primary, secondary and tertiary (Jiang et al. 1994) comprising of both annual and perennial species (Harlan and De Wet 1971; Dewey 1984). Among the 325 Triticeae species, there are approximately 250 perennials. The huge germplasm collection provides an invaluable resource for wheat breeding goals set for high yield outputs around numerous biotic and abiotic stresses. This allelic diversity will be crucial to harness for meeting the global production estimates of 720 million tones by 2025 to sustain the needs of the projected global population of around 8.2 billion (Mujeeb-Kazi and Rajaram 2002). Current national production trends in Pakistan where 8.3 million hectares are planted to wheat are 23.5 million tones at 2.6 t/ha and short-term projects of the next few years are set at 4.5% per annum to give by 2011 an output of 26.41 million tones at 2.9 t/ha. The strategies of achieving the above global and national targets have been separated into a management area and pertinent to this presentation the germplasm development area. For achieving the latter, resource diversity, priorities, technical integration, efficient technologies emanating from genetic resources are essential. Further, having balances of resistances/tolerances spread across biotic and abiotic stresses constraints are necessary for achieving production targets. The emphasis of this presentation is to address the exploitation of genetic resources for development of wheat germplasm that fits the farmers needs relative to some key abiotic stresses. The main focus is on salinity and drought but in addition heat, waterlogging, and minor but important abiotic factors are also considered. The key sources of variability at a priority are the primary gene pool diploid D genome donor accessions to wheat: *Aegilops tauschii* and some sources from the tertiary gene pool possessing high potential values. Integrated aspects with the diversity introgression shall elucidate association of new technologies and multidisciplinary

A. Mujeeb-Kazi (✉) and A. Gul
National Institute of Biotechnology and Genetic Engineering (NIBGE), Faisalabad, Pakistan
e-mail: m.kazi@cgair.org

I. Ahmad, M. Farooq, Y. Rauf, A.-ur Rahman, and H. Riaz
National Agricultural Research Center (NARC), Islamabad, Pakistan

specialties that facilitate exploitation of molecular tools hopefully to add to breeding efficiency.

2 The Triticeae Gene Pools

2.1 *Distribution of Genetic Diversity*

Though genetic diversity can be induced, for more controlled well directed incorporation of diversity naturally present in the annual and perennial Triticeae species maintains its priority. This natural diversity resides in the conventional wheat germplasm, and in closely or distantly related alien species sources. The species resources are distributed within gene pools and genetic transfers can be realized for wheat improvement from these pools over short- or long-term time frames. The gene pools are structured upon the genomic constitution of the species, and are comprised of three groups: primary, secondary and tertiary.

The primary gene pool species include the hexaploid land races, cultivated tetraploids, wild *Triticum dicoccoides*, and diploid donors of the A and D genomes to durum/bread wheats. Genetic transfers from these two genomes occur as a consequence of direct hybridization and homologous recombination with breeding protocols contributing different backcrossing and selection strategies. Some cross combinations require embryo rescue, but no cytogenetic manipulation procedures are necessary. The secondary gene pool is formed of the polyploidy *Triticum* plus *Aegilops* species which share one genome with the three genomes (A, B, and D) of wheat. The diploid species of the Sitopsis sections are included in this pool, and hybrid products within this gene pool demonstrate reduced chromosome pairing. Gene transfers occur as a consequence of direct crosses, breeding protocols, homologous exchange between the related genome or through use of special manipulation strategies amongst the non-homologous genome. Embryo rescue use a complementary aid for obtaining hybrids. Diploid and polyploidy species are members of the tertiary gene pool. Their genomes are non-homologous. Hence, genetic transfers require special techniques that assist homoeologous exchanges facilitated by irradiation or callus culture mediated translocation induction. Diploid and polyploidy species with genomes that are non-homologous to wheat reside in the tertiary gene pool. Homologous exchanges cannot affect genetic transfers, but genomic homoeology of these species does permit transfer of genes by somewhat complex protocols.

2.2 *Utilization of Gene Pool Diversity*

For practical end-products to be obtained, some transfer pre-requisites that encompass all the three gene pool species span from hybrid production, embryo rescue, plant regeneration, cytological diagnostics, breeding methodology, and stress screening, culminating in stability of the advanced derivatives as contributed by induced homozygosity protocols (Mujeeb-Kazi and Riera-Lizarazu 1996; Mujeeb-Kazi et al. 2006). Based upon these pre-requisites and genetic transfer ease, the primary gene pool diversity holds priority significance for wheat improvement. The species of the diploid A and D genomes contribute novel genes, allow direct recombinational exchanges with their respective genome partners to facilitate both durum and bread wheat improvement over a relatively "short-term" time frame, than what is provided by the secondary or tertiary gene pool species. Setting the above as a base, is illustrated the strategy and outcome of exploiting the diversity of the three gene pools and their accessions.

2.3 *Transfer Pre-requisites Across Gene Pools*

Some pre-requisites for achieving gene transfers from the annual/perennial Triticeae across all three gene pools are related with varying degrees of complexity that involves hybrid production, embryo rescue, plant regeneration, hybrid validation through multiple diagnostic protocols, breeding methodology, biotic/abiotic stress screening, and stability of the wheat/alien derivatives achieved by use of sexual homozygosity inducing protocols. In general genetic transfers from the primary pool are rapid and products emerge over a shorter time frame versus the complex and lengthy outputs from the distant tertiary pool species. However, to ensure durability, tapping beneficial genetic diversity across all pools is advantageous. The Fig. 16.1 schematic (modified from Mujeeb-Kazi et al. 2006, 2008) elucidates the role of annual grass species in the evolution scheme of wheat, and endorses the significance of the use of the three diploid (A, B, D) accessions for harnessing additional genetic diversity for rapid wheat improvement.

2.4 *Interspecific Hybridization and Abiotic Stresses*

2.4.1 Use of the Primary Gene Pool Diversity

With the excitement generated by the earlier report of wheat/barley hybridization by Farrer (1904), accompanied by the outputs of Kruse (1967, 1969, 1973) interest in wide crosses spread swiftly with innovative research options also emerging (Bates and Deyoe 1973) that were short lived. It was the classical approaches validated by cytogenetics that set the solid foundation for alien introgression efforts beyond 1978. Earlier efforts were focused on the complex tertiary gene pool resources (Islam et al. 1978; Mujeeb et al. 1978; Sharma and Gill 1983) but when the chances of quality practical outputs from this route were evaluated, research efforts saw a strategy shift towards the primary gene pool targeting on the D genome as the priority resource complimented by other diploid progenitors (Alonso and Kimber 1984; Gill and Raupp 1987; Valkoun et al. 1990; Cox et al. 1991; Mujeeb-Kazi and Hettell 1995; Mujeeb-Kazi et al. 1996). Though direct crosses are an elite protocol for introgressing D genomic diversity into recipient bread wheat cultivars swiftly (Gill and Raupp 1987) major effort was simultaneously given to using *Ae. tauschii* accessions via 'bridge crosses' for wheat improvement since they in addition to introgressing the D genome diversity also allow for the A and B genome diverse alleles to be utilized (Mujeeb-Kazi 2003, 2006; Mujeeb-Kazi et al. 2008; Trethowan and Mujeeb-Kazi 2008). In general the major abiotic stresses recognized to limit crop productivity are salinity and drought. The former is of concern in irrigated input oriented wheat cultivation environments. Drought is significant in marginal and rain-fed areas of wheat cultivation. A third stress (heat) is showing importance in production areas under influence of cropping systems characterized in Pakistan by rice-wheat or cotton wheat that causes late wheat plantings to occur creating vulnerability for terminal heat stress. The major focus in this presentation shall be on the first two stresses addressed via two hybridization categories, the first "interspecific" covered by *Ae. tauschii*.

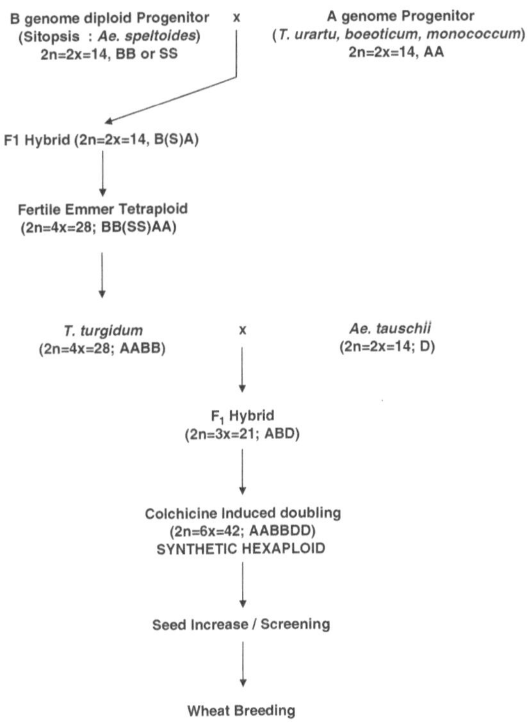

Fig. 16.1 Origin of wheat showing role of diploid progenitors and their further usage in wheat improvement

2.4.2 Genetic Diversity of the Salinity Status in the Triticeae

Ae. tauschii and Salinity Tolerance

Due to the pressing need to meet projected wheat production goals, strategies giving short-term benefits are preferred. Hence interspecific hybridization was given a priority with most emphasis assigned to *Ae. tauschii* (syn. *T. tauschii* (Coss.) Schmalh., syn. *Ae. squarrosa* auct. non L., $2n = 2x = 14$, DD) because of its close genetic relationship to the D genome of wheat. *Ae. tauschii*; unequivocally accepted as the D genome donor to *T. aestivum* (Kimber and Feldman 1987); is further attributed with a wide range of resistances/tolerances to biotic/abiotic stresses (Valkoun et al. 1990; Cox et al. 1991). One mechanism – of the three that exist – is exploiting the *T. turgidum x Ae. tauschii* hybrids ($2n = 3x = 21$, ABD) leading to the generation of synthetic hexaploid wheats ($2n = 6x = 42$, AABBDD) upon colchicine treatment or by spontaneous induction. The synthetic germplasm generated through using elite durum wheats and numerous accessions of *Ae. tauschii* became the storehouse of allelic diversity that made available a rich reservoir of novel genes for wheat improvement encompassing a wide array of target objectives (Mujeeb-Kazi et al. 2008). This output enabled researchers to form synthetic sub-set groups around various objectives and one such group was categorized as the Elite 1 set of 95 synthetic entries. Screening this elite 1 set for salinity tolerance unravelled the enormous potential of the synthetic germplasm for the stress as all materials exhibited a positive response (Pritchard et al. 2002) measured through *in vitro* evaluation using the K:Na screen (Gorham et al. 1987; Shah et al. 1987). The test basis was associated to the contribution of the D genome chromosome 4D and a brief detail follows to elucidate the significance and correlation of the test parameter i.e. K:Na.

Chromosomal association of the salinity trait was provided from analyses of the D genome chromosome substitution stocks for the A and B genome chromosomes of *T. turgidum* cv. Langdon and D genome based synthetic hexaploid germplasm (Gorham et al. 1987; Shah et al. 1987). These studies demonstrated that the D genome (*Ae. squarrosa* L. 2n = 2x = 14; syn. *Ae. tauschii*) contained a trait for enhanced K^+:Na^+ discrimination located on chromosome 4D. Use of Chinese Spring ditelosomic lines provided confirmation that chromosome 4D and specifically its long arm (4DL) possessed the discrimination trait. A locus for this trait, the *Knal* locus, has been mapped to a region on the distal third of chromosome 4DL (Dubcovsky et al. 1996). The trait operated over a wide salinity range and appeared constitutive (Gorham et al. 1987).

Subsequent contributions suggested that a consensus existed amongst researchers as to the tolerance diversity of *Ae. tauschii*. The D genomes significant effects on K^+:Na^+ discrimination in salt-stressed hexaploid wheat, its desirable variation for salinity tolerance in gravel tank nutrient cultures between a E.C. 30 to 40 dS/m^2 (Farooq et al. 1989), and the K/Na diversity of this diploid D genome donor to bread wheat (Gorham 1990) provided an impetus to assign a high priority for its usage. Hence the emphasis was placed on the exploitation of *Ae. tauschii* via the bridge crossing route (Mujeeb-Kazi et al. 1996) that involved production, screening and exploitation of synthetic hexaploids (*T. turgidum*/*Ae. tauschii* 2n = 6x = 42, AABBDD) for bread wheat improvement. The exploitation of synthetic hexaploid wheats (SH) has over the last decade gained greater priority globally (Coghlan 2006; Simonite 2006) for multiple stresses that limit wheat production. Specifically for salinity testing under hydroculture with synthetics exhibiting superior K:Na discrimination over their durum parents volatile SH utilization programs for salinity have taken off (Colmer et al. 2006; Mujeeb-Kazi et al. 2004; Mujeeb-Kazi 2006; Reynolds et al. 2005; Trethowan and Mujeeb-Kazi 2008). The data trends are consistent with earlier projections made by Shah et al. (1987) and Gorham (1990). These findings allowed for conclusions to be drawn indicating that durum cultivars had equivalent values for K and Na thus yielding a ratio of 1.0 or close to this value. When such durums were hybridized with *Ae. tauschii* accessions lower Na content and high K gave K:Na ratios of higher than 1.0 allowing for conclusions to be made that the D genome progenitor was contributing to the observed K:Na discrimination trend. The extensive screening of durum germplasm subsequently (Munns and James 2003; Munns et al. 2006) led to the identification of tetraploid wheat germplasm with K:Na ratios that deviated from the established perception that all durums were susceptible. The A genome contributed; specifically chromosome 2A. Thus opened another door for enriching bread wheats with this new durum wheat attribute through gene pyramiding. James et al. (2006) reported that durum wheats (line 149) contained two novel major genes (Nax_1 and Nax_2) that excluded Na^+ from leaf blades describing the *modus operandi*. Nax1 was mapped to the distal region of chromosome 2AL. The second gene independent of Nax_1 was suggested to contribute to the full expression of the Na^+ exclusion trait (Lindsay et al. 2004) and was named Nax_2. *Triticum monococcum* (2n = 2x = 14, AA; C68–101) was the original source of both *Nax* genes. Both genes restrict Na^+ transport from roots to shoots resulting in enhanced K^+:Na^+ ratios in the leaf blade. The mechanism of *Nax2* was similar to the *Knal* locus in bread wheat (*T. aestivum*). *Nax1* differs from *Nax2* in its capacity to remove Na^+ from the xylem in the lower part of the leaf and root, which differs from the action in bread wheat. *Nax2* functioned only in the root. The recent review by Munns and Tester (2008) elucidates the details of salinity tolerance mechanisms as they relate to wheat breeding.

In order to realize the practical benefits from the potential of the above genetic resources for wheat improvement one requires a breeding strategy that can effectively harness the potent diversity identified in the A and D genome accessions. Simultaneously one must be cognizant of the fact that abiotic tolerance transfers alone will not lead to varietal releases with maximized yield performance since the realization of yield target is interlinked with prevalent biotic stress factors and thus a holistic integrated approach must exist to capture the scattered significant scientific research contributions.

The Conventional Germplasm

Triticum aestivum L. cv. Kharchia 65, a saline/sodic tolerant cultivar, has been exploited by growers in Rajasthan, India for several decades. Controlled testing in saline media unequivocally attests to the cultivars saline/sodic tolerance potential (Mujeeb-Kazi et al. 1993). Attempts to incorporate this cultivars stress tolerance genetic diversity for wheat improvement have been made, and the singular success was accredited to the KRL1-4 wheat cultivar release in India. Later followed KRL-19. Another salt tolerant *T. aestivum* cultivar selected from LU26 was LU26S in Pakistan. Additional inputs to this conventional gene pool of Kharchia 65, KRL 1-4, KRL19 and LU26S are some collections of promising cultivars and land races acquired from various programs and countries. These are Chinese Spring, SNH-9, WH-157, Sakha 8, Shorawaki and Pasban-90, Pericu, Calafia, Cochimi, Mepuchi and S-24. Standard in vitro tests around protocols of Gorham et al. (1987) and Shah ct al. (1987) supported the tolerance attributes of the above genetic resource (Mujeeb-Kazi and Diaz-de-Leon 2002)

Susceptible cultivars in the above group of diverse germplasms tested for tolerance were a durum (PDW34) and a bread wheat cultivar, Oasis. The CIMMYT Wheat Wide Crossing program developed this testor set in order to establish germplasm evaluation uniformity among researchers. Undoubtedly versus time additional promising entries will be added to the current set and a revised list of germplasm could be formulated. The set is comprised of land races (Kharchia 65 and Shorawaki), wheat cultivars with good agronomic phenotypes coupled with high yield and those with high tolerance but poor plant type (Chinese Spring). There is abundant diversity just within this conventional set to improve our elite wheat varieties for saline/sodic conditions. When other genomic diversity options from exotic sources are integrated the outputs definitely stand out to be phenomenal and highly recommended.

Some Salinity Screening Methodologies

In general the methodologies are separated into field evaluations and *in vitro* testing. Both procedures provide information that shed light on genetic diversity amongst the germplasms. We present here the procedure used by Mujeeb-Kazi and Diaz-de-Leon (2002). Greater depth is in Munns (2007).

Field Screening Using Sea-Water Dilutions

Germplasm screening for salinity response was conducted under field conditions in La Paz, Baja California Sur, Mexico. The normal well-water for irrigation with an electrical conductivity (EC) level of 4.5 dS m^{-1} served as the control concentration. Sea water was in close proximity to the field screening site. Mixing sea water with the normal well-water provided the necessary EC levels selected for the field evaluation. These levels ranged between 8.0 to 20.0 dS m^{-1} and were kept at 8.0, 12.0, 16.0 and 20.0 dS m^{-1} with the option to vary or extend them beyond 20.0 dS m^{-1}. Plots were separated from each other on all sides by black plastic line dividers for avoiding error via seepage. The plots were individually flood irrigated according to the treatment category with 200 l twice a week. Electrical conductivity (EC) of the irrigation water was measured, and soil samples randomly taken from each plot after 24 hours. Determination of the soil EC followed the established extraction procedure with steps of soil sampling/plot done at random points. After 1 week of plant growth all plots were fertilized with 15 g of urea per week up to 8 weeks. The urea application made each week was with each irrigation. The protocol was related to the sandy soil conditions at the experimentation site and was associated with adult plant performance of the germplasm built around phenology parameters and yield. The testing situation would vary according to natural soil test conditions, and further could be modified to suit adult plant testing in cement blocks, or in potted soil where the concentrations could be more stringently controlled. Contrary to these precision outputs that provide a crucial test sieve assessed at plant maturity, the ultimate test would be under natural conditions that have allowed cultivars like Kharchia 65, LU26S, Shorawaki, Pasban 90, KRL 19 to be cultivated.

In Vitro Screening

(a) These protocols utilize the hydroponics system (Gorham 1990) where NaCl levels from 50 up to 200 mol m^{-3} are used as discrimination levels; 50 mol m^{-3} being preferred. Growth parameter assessments are made, and the procedure is strengthened by K:Na discrimination analyses.
(b) In addition, two other strategies have been used where growth analyses are made by 15 days from the start of the experimentation. This allows one to take a first selective cut of the germplasm, and use these initial selections for the field sea-water dilution or related study. These strategies have utilized
    (i) Saline agar media based upon the standard Murashige-Skoog media culture with variable EC levels that stimulate the field sea water dilution study ranges.
    (ii) Use the same sea water dilution ranges for doing a Petri-dish germination/growth study, complemented by Hoagland's nutrition solution as a seedling growth variable if necessary.

In general the above evaluatory screenings have been associated with inferences that support results with the test materials salt tolerance ability. Caution however needs to be in place that such tests fit a uni-directional system where only salinity tolerance is the limiting constraint. If a complex field system also involves sodicity then the testing should be accordingly adjusted in order to provide bi-directional test data that addresses the dual saline-sodic constraint. This dual test necessity apart from aiding the interspecific and conventional programs would also extend to the diversity present in the species within the tertiary gene pool and categorized under 'intergeneric hybridization'.

2.5 Intergeneric Hybridization: Use of Secondary and Tertiary Gene Pools

Harnessing the genetic diversity of the resources residing in the secondary and tertiary gene pools is a complex and cumbersome task with practical beneficial outputs taking considerable time (Mujeeb-Kazi 2003, 2005, 2006). Gauging the potential of the perennial alien species removed from the primary gene pool have been based upon diverse artificially controlled testing procedures for the various biotic and abiotic stress parameters. This has been the easier facet from which projections are made. Practical outputs however are beset with complexities and time-consuming. Colmer et al. (2006) reviewed the use of wild relatives to improve salt tolerance in wheat and addressing the germplasm range covered several tertiary gene pool resources. These were the E and J genome tall wheatgrasses, *Thinopyrum ponticum*, and *Th. junceum*. These perennial species comprised of 2 diploids ($2n = 2x = 14$, commonly EE and JJ); *Th. bessarabicum* and *Th. elongatum*. Also a potent source was one decaploid ($2n = 10x = 70$) *Th. elongatum* syn. *Th. ponticum* syn *Elytrigia ponticum* that had gained significant importance as a salinity tolerant grass (Gorham et al. 1985; McGuire and Dvorak 1981) and a hexaploid ($2n = 6x = 42$, J1J1J2J2EE) *Th. junceum*. In general, hybridization of such perennial Triticeae species with bread wheats, production of amphiploids, cytogenetic advances leading to alien chromosome additions/substitutions/translocations/introgressions are essential before positive conclusion of tolerant genetic transfers can be made. Such germplasm developmental procedures have identified chromosome contributions to salt tolerance for *Th. bessarabicum* and *Th. elongatum* (Forster et al. 1988; Dvorak et al. 1988). Wang (1989) produced an amphiploid involving the above two *Thinopyrum* diploids that furthered research incentives for pyramiding genes in order to facilitate the transfer of potent cumulative salinity tolerance genes to wheat. This amphiploid was hybridized with *Triticum*, and cytogenetically documented. Viable end products, however, have still to be realized. The developed base germplasms have not yet made a significant impact on improving wheat cultivation on saline lands.

In addition to the species identified by Colmer et al. (2006) with proven practicality for field cultivation has been the wheat derivative reported by Wang et al. (2003a, b) derived from *Th. junceiforme* via *Ph1b* inhibition using the Ph^{I} inhibitor genetic stock derived from *Ae. speltoides*.

Due to the tremendous genomic distance between wheat and the perennial Triticeae species, cytogenetic manipulation complexity, and distribution of the salinity tolerance trait control mechanism/s on more than one alien chromosome (Dvorak et al. 1988), research successes are anticipated to emerge over a long time frame. We feel that there is a possibility for desired results to be obtained faster through the incorporation of novel methodology shifts associated with *Ph* locus manipulation coupled with the integration of molecular diagnostics and sexually induced polyhaploidy avenues (Gill and Gill 1996; Qu et al. 1998). The use of the Ph^I genetic stock is also seen as a unique avenue (Chen et al. 1994). These mechanisms allow alien introgressions to be ideally incorporated in homoeologous sites in wheat and render germplasm the crucial homozygosity necessary to address polygenic traits and soil heterogeneity.

In a recent research paper, Colmer et al. (2006) reviewed the usage of wild relatives for improving salt tolerance in wheat. The status and potential of various Triticeae resources was elucidated comprising of durums, bread wheats, the diploid wheat progenitors, wild tetraploids, synthetic hexaploids including the D genome and other combinations, *Aegilops* species with unrelated genomes of wheat, tertiary pool species with diverse genomes (E and J) with their previous ploidy levels, and *Hordeum* species (I, H and X genomes) giving explicit evidence that an enormous genetic resource was available for wheat improvement. Nevertheless the cumulative findings of numerous researchers over the past two decades have clearly suggested that to achieve success by exploiting the alien species, research structuring must be based upon ability to harness alien diversity with positive potential for the trait via homologous exchanges or homoeologous exchanges mediated by cytogenetic manipulation strategies (Mujeeb-Kazi et al. 2008). There must also be cognizance that conventional germplasm must be involved in such programs as crucial building blocks in order to deliver practical outputs possessing potent pyramided genes from diverse sources. Homoeologous transfers are complex since the ideal exchanges from alien sources need to be placed in the best recipient chromosome so as to compensate adequately and yield maximum practical benefits. The complexity is further enhanced when a trait is under polygenic control and extended over several alien chromosomes as has been seen for addition line response of *Th. bessarabicum* (Mujeeb-Kazi and Diaz de Leon 2002) and *Th. elongatum* (Dvorak et al. 1988) diploids. The multiple chromosome control of *Th. bessarabicum* is contradictory to the earlier observation that associated chromosome 5J with salinity tolerance (Forster et al. 1987, 1988). These two sources however are the most promising to tap from the tertiary gene pool and a strategy (Mujeeb-Kazi 2003) that has potency to give rapid returns should be pursued rigorously. The basis is around the *Ph1b* manipulation system and the schematic of Fig. 16.2 demonstrates how genetic exchanges can be maximized between the full complement of homoeologous chromosomes of wheat and *Th. bessarabicum* similarly covering *Th. elongatum*. The system offers flexibility to work at the amphiploid level, or even at individual disomic alien chromosome addition level. It has the capacity to maximize exchanges between

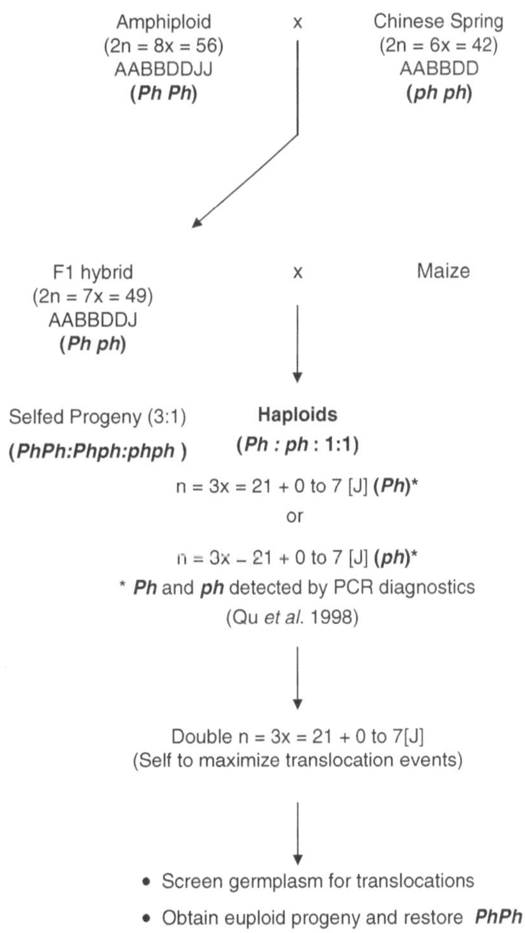

Fig. 16.2 Schematic elucidating the recurrent backcrossing steps that will be used to transfer the existent wheat/alien chromosome translocations into agronomically superior wheat cultivars

wheat/alien chromosomes by repeated selfings of the backcrossed materials under the *phph* structure and restore the *Ph1bPh1b* mechanism yielding euploids that are 42 chromosomes carrying the alien exchange as a translocation with the stability for stress testing.

Managing stressed soils requires attention and coupled with this is the genetic improvement angle. The ample diversity that is available needs to be stringently assessed and that most suited in terms of giving rapid returns exploited for wheat improvement. The course that we feel is prudent to adopt for bread wheat improvement for salinity is to exploit conventional cultivar variation > land races > wheats diploid progenitors > alien distant resources with a preference for alien diploids around established homoeologous relationships. In the case that durum wheat improvement is a target, A and B genomic potency should have a preference over all other options. However, since a durum cultivar Capelli exists with a chromosome 5B deletion location *(ph1c ph1c)* homoeologous transfers of need can be realized. The extensive practical attributes present in the various accessions of *Ae. tauschii* also expressed in their derived synthetic stocks (Mujeeb-Kazi et al. 2008) and crossed derivatives with bread wheat provide an excellent means to exploit this D genomic variability to enrich durum wheats via D to A genome homoeologous exchanges mediated by the *ph1c ph1c* locus in Capelli.

2.6 Genetic Diversity for Drought Tolerance

Productivity in marginal environments is a challenge that has emerged over the past decade and is a global situation. Like salinity tolerance wheats must possess the appropriate genetic structure to be successfully cultivated and yield under reduced irrigation conditions. There are the conventional germplasms and in addition with growing emphasis on close relative genetic diversity utilization novel strategies have emerged. Of these, the strategy that exploits the D genome diploid progenitor accessions of *Ae. tauschii* holds priority with this group of researchers. This focus however does not overlook the value and need of the conventional germplasm that forms the mainstream wheat breeding efforts globally. The leading cultivars are widely recognized and over the years have emerged additional genetic derivatives that have led to significant contributions towards yield and performance in rain-fed systems. The germplasms are recognized as derivatives from the winter X spring wheat cross combination diversity known as the famous T1BL.1RS translocations that possess various biotic stress resistance genes, are widely adapted, give significant yield advantage per hectare and form the bulk entry numbers of promising lines under reduced irrigation. The practical impact of this translocation germplasm has been the subject of many reports (Rajaram et al. 1983; Jahan et al. 1990; Ter-Kuile 1991; Villareal et al. 1997, 1998; Mujeeb-Kazi et al. 1999a, b; 2000, 2001a, b, c; Warburton et al. 2002) and globally such wheats are cultivated on more than 50 million hectares. The origin of the T1BL.1RS translocation event was fortuitous and its spread via breeding into numerous varieties deployed globally a function of its performance ability but over time the resistance genes (*Lr26, Sr31, Yr9, Pm8*) on the 1RS chromosome arm have lost their resistance impact though still such varieties figure significantly in international nurseries and national wheat varietal trials. The search for other potent sources of diversity that address the reduced water usage scenario emerged at an appropriate time and has moved significantly forward. The germplasm making this contribution was *Ae. tauschii* and the take-off location was from CIMMYT, Mexico that built this arsenal of usable diversity for wheat improvement (Mujeeb-Kazi and Hettell 1995).

2.6.1 Contribution of Synthetic Wheats to Drought Tolerance

Synthetic wheats were developed by crossing elite tetraploid durum wheat cultivars (*T. turgidum*, $2n = 4x = 28$, AABB) with diploid *Aegilops tauschii* ($2n = 2x = 14$, DD) accessions. The outcome was the production of approximately 1,150 spring and winter habit synthetic wheats that have become a valuable source of new genetic diversity for bread wheat and durum wheat improvement. During the utilization phase of this output various sub-sets were formed that suited breeder needs for conducting recombination breeding programs covering biotic and abiotic stresses. The initial sub-set that formed the nucleus to address drought breeding was composed of 23 synthetic hexaploids. Microsatellite markers facilitated the identification of DNA polymorphisms within this set and allowed for targeted utilization of the most genetically diverse synthetics for wheat improvement and for QTL mapping options via the development of molecular mapping populations.

2.6.2 Contribution of Synthetics to Parentage in CIMMYT's Breeding Program for Marginal Areas

The breadwheat stress-breeding program at CIMMYT has steadily increased the contribution of synthetic wheat parentage in the breeding program over the past 7 years. Since 1997 until 2005 of all breeding materials in CIMMYT over 50% had synthetic wheat somewhere in their parentage.

2.6.3 Drought Screening at Ciudad Obregon, Mexico and Its Relationship to Global Yield Performances

The drought screening site near Ciudad Obregon is arid, and wheat is grown under irrigation. Drought stress screening is conducted using a combination of gravity and drip irrigation methods to generate controlled moisture stress scenarios. Genotype x year interactions under moisture stress are low (Trethowan et al. 2003a) and the relevance of germplasm selected at this site, under limited and optimal irrigation, to global wheat growing environments has been demonstrated (Trethowan et al. 2001a, 2003b). When germplasm selected at this site using one or two gravity irrigations is tested globally, significant rates of improvement in productivity have been observed (Trethowan et al. 2002). The soils at this location have been carefully characterized for biotic stresses (nematodes, root rots) and other abiotic stresses (micronutrient imbalances) thereby ensuring that the observed differentiation of genotypes is due to water and not other confounding factors.

2.6.4 Contribution of Synthetics to Improved Productivity Under Moisture Stress

In 1996, the highest yielding line under drought in the CIMMYT wheat breeding program was Baviacora. The best new advanced lines at that time yielded 95% off this yield standard. This drought stress standard was outclassed in 2000 and replaced by a Baviacora derivative called Weebill 1. Weebill 1 was on average 11% higher yielding than Baviacora, based on 29 head-to-head comparisons conducted across 2 years. Many new advanced lines were higher yielding than Weebill 1 and they combined variability for drought stress tolerance from traditional (germplasm adapted to marginal environments globally) and synthetic sources thereby demonstrating the superiority of a few of these new synthetic derivatives over Weebill 1 and Pastor, a line adapted to dry environment globally, under drought stress at Ciudad Obregon, Mexico. The advantage that many synthetic derivatives have over the bread wheat and durum wheat parents has become evident over the last decade. Unfortunately, many of the early durum wheats used to make the primary synthetics that were later crossed to develop these derivatives are no longer available. However, the advantage of the synthetic derivatives over their adapted bread wheat parents is clear.

2.6.5 Performance of Synthetic Derivatives Under Different Moisture Stress Scenarios

The timing and intensity of drought stress is varied at Obregon, Mexico to identify genotypes that perform well across a range of stress levels while maintaining the ability to yield well should moisture conditions improve. The performance and contributions from this strategy has led to the identification of synthetic germplasm that is broadly adapted to stress. This important new variability is now being introgressed into the gene pool targeted to marginal environments globally.

2.6.6 Contribution of Synthetics to Maintaining Seed Weight Under Drought Stress

Wheat plants grown under moisture or heat stress often produce shriveled seed. This reduces the yield and market value of the wheat crop. Maintenance of seed weight under moisture stress is thus a key adaptive mechanism that influences performance under stress. Significant new variability for this character has been found in various primary synthetics. These selected primary synthetics have larger seed size (measured as the weight of 1,000 seeds) than Baviacora (recognized as having large seed) under stress-free conditions. All synthetics tested had higher seed weight than the check under all conditions. A synthetic (#1) was able to maintain its seed weight best under both drought and heat stress with very little observable difference among treatments. This important new variability is being introgressed and adds another dimension to the utility/potential of primary synthetics in wheat breeding.

2.6.7 Contribution of Synthetics to the Simultaneous Improvement of Grain Yield and Protein Content

Historically it has been very difficult to breed for both high yield and high protein content, as these characteristics are negatively associated. However, if sufficient variability for both characteristics is present in a crossing program, lines that optimize both yield and protein can be found at very low frequency (Trethowan et al. 2001b). Synthetic wheats offer significant new variability for protein sub-unit compositions. When crossed to high yielding, drought tolerant elite bread wheat parents, it was possible to find lines with high yield and high protein. The two synthetic derivatives studied had yields equivalent to that of the elite drought tolerant check cultivar, Weebill; but had up to 20% more grain protein.

2.6.8 Conclusion

Drought is a major wheat production constraint and screening the synthetic wheats under reduced irrigation near Ciudad Obregon in Northwestern Mexico has yielded enormous desirable diversity that has been harnessed for improving drought tolerance in bread wheat over the past decade. Incorporation of these synthetics into the breeding program has led to advanced free-threshing derivatives that are significantly superior to recognized benchmark cultivars Weebill 1 and Pastor. Notable attributes of the improved germplasm are its unique early canopy growth, deep root system, maintenance of high 1,000-kernel weight, germination from greater planting depth, high yield and desirable protein quality. F1 doubled haploid based (Mujeeb-Kazi et al. 2006) molecular mapping populations have been developed (Mujeeb-Kazi et al. 2008) from molecularly diverse drought tolerant synthetic hexaploid/drought susceptible bread wheat (Opata) combinations. With ample seed increase having been achieved the populations are ideal candidates for Phenotyping and genotyping leading to QTL mapping. The precaution taken ensured that the tolerant synthetic wheats were molecularly diverse from each other and also possessed DNA polymorphisms from the drought susceptible bread wheat CIMMYT variety 'Opata'. Researchers in some locations desire to develop mapping populations around their local wheat varieties which is fine as long as the variety is drought susceptible and possesses DNA polymorphism compared to the tolerant parent. Drought tolerance breeding has emanated from the drought tolerant synthetic hexaploid trait donors and this has become the basis of having advanced wheat derivatives possessing 50% contribution of the resource in their pedigrees within CIMMYTs program. Not exploited is the potential of direct crosses (Alonso and Kimber 1984; Gill and Raupp 1987) where tolerant *Ae. tauschii* accessions as identified from the synthetic hexaploid pedigrees of the five mapping populations (Mujeeb-Kazi et al. 2008) are crossed onto recipient elite bread wheat cultivars with poor drought tolerance.

In general, the better performance of synthetic wheats and their cross derivatives is associated with root distribution which was linked to a greater investment in root biomass at depth in the soil profile, rather than a greater root dry weight (Reynolds et al. 2007). Summarizing the key traits under moisture stress (Reynolds and Trethowan 2007) synthetic derivatives were 24% higher yielders, possessed 57% more biomass, had a 46% lower root:root ratio, and were 41% more water-use efficient than their recurrent parents with the ability to maintain seed weight under drought/heat stress (Trethowan et al. 2005).

2.7 *Some Other Abiotic Stresses Influencing Wheat Production*

Synthetic hexaploid wheats have been studied under various stress parameters and diversity for numerous constraints that limit wheat productivity globally has been identified in the primary form or also within their advanced cross derivatives. Heat tolerance (entire cycle or terminal) has been positively studied (Yang et al. 2002) alongwith low temperature/frost tolerance (Maes et al. 2001). Terminal heat tolerance is a major constraint where the cropping systems like rice/wheat or cotton/wheat force the wheat planting to be delayed thus exposing the crop near its maturity stage to high temperatures. Primary synthetics tolerant to waterlogging conditions have been identified and registered (Villareal et al. 2001) with others having the potential of being 15% better for zinc efficiency than the zinc efficient modern cultivars (Genc and McDonald 2004). Further synthetics involving *T. dicoccum*/*Ae. tauschii* possessing high levels of iron and zinc have also been identified (Ortiz-Monasterio et al. 2007) and are a value for human nutrition research efforts. Boron toxicity limits wheat productivity in some environments (Torun et al. 2001) and synthetics tolerant to boron have been identified (Dreccer et al. 2003). A elaborate review of the vast potential of synthetics has been recently published or is in the printing process and elucidates greater details (Mujeeb-Kazi et al. 2008; Trethowan and Mujeeb-Kazi 2008).

Two abiotic stress constraints where synthetics have yet to find promise for are copper efficiency and aluminum tolerance.

2.7.1 Copper Efficiency

Evaluation of copper (Cu) deficiency symptoms has been conducted at Cu levels ranging from 00 to 4.0 mg/pot (Graham 1978) with an ultimate measure of grain yield. The reproductive phase is effected more than the vegetative (Graham 1975). *Secale cereale* L. cv. Imperial was identified as a Cu efficient source. Chromosome 5R from *S. cereale* disomically added to *Triticum aestivum* cv. Chinese Spring (seed source: ER Sears) was observed to be responsive and subsequently the 5RL rye arm was positively associated. CIMMYT obtained the T5AS/5RL translocation line from TE Miller (AFRC-IPSR, Cambridge) and used it as the male parent in crosses onto some CIMMYT spring bread wheat cultivars for generating the 5A, 5AS/5RL heterozygote F1 combinations. Subsequent backcrossing (BC) of these F1s with their respective recurrent parents up to BC8 followed by an eventual selfing of the best heterozygote BC derivative resulting in elite *T. aestivum* near isogenic lines with the T5AS/5RL translocation homozygote. The T5AS/5RL heterozygote was identified at each BC through the morphological presence of the hairy peduncle (*hp*) marker mapped on 5RL, controlled by a dominant gene and effective in the hemizygous stage also. Differential C-banding checks on the heterozygote BC derivatives were integrated to ensure adequate advance accuracy. The translocation germplasms utility is located in small pockets of which Kenya is one site. The translocation has been transferred to elite spring bread wheats and durum wheats (Mujeeb-Kazi and Cortez 2001) and remains at this stage more a scientific curiosity.

2.7.2 Copper Toxicity

Since Cu is bound strongly to soil particles, occurrence of toxicity is rare. However, in acid soils over-fertilization and use of fungicide applications increases Cu toxicity. Though alternative remedial solutions exist, a mix with genetical inputs is advantageous.). *Th. bessarabicum* has been identified as a potent alien source to contribute tolerant genes Manyowa and Miller (1991) with *S. cereale* chromosome 2R. Disomic chromosome additions 2J and 5J (also $2E^b$ and $5E^b$) were the tolerant contributing disomics (Manyowa and Miller 1991). The potent 5J/6J ($T5E^bL/6E^bL$) may well be a positive contributor because of $5E^bL$. *Th. repens* has also been associated as a potential source; however, its hexaploid status may complicate its usage. Rapid outputs could be expected from the *Th. bessarabicum* germplasm with rye serving as a secondary source.

2.7.3 Aluminum Tolerance

Like other stress attributes genetic diversity for aluminum tolerance Al^{3+} is also distributed within conventional wheat germplasm and in some progenitor sources. Having screening protocols that permit *in vitro* tests to be conducted accurately are a big plus for making germplasm generation advances readily and a simple one as developed for aluminum tolerance testing has been an asset.

Screening

Direct observation of wheat seedling roots under Al^{3+} stress was developed as a selection system by Polle et al. (1978), a methodology with transportation ease (Lopez-Cesati et al. 1986). The process is based on the fact that Al^{3+} tolerance in wheat is largely a function of its exclusion from the roots. Root tip growth, after immersion of the roots in a nutrient solution containing 46 ppm Al^{3+}, followed by: (i) staining of the roots with 0.2% aqueous hematoxylin solution, (ii) observing continued root re-growth, and (iii) scoring on a 1 to 3 scale covered the screening needs of the Triticeae germplasm for tolerance.

Based upon the above screening schedule experiments were conducted extensively in CIMMYT, MEXICO over at least 3 decades using: (i) conventional germplasm, (ii) alien species with their wheat amphiploids, (iii) some *S. cereale* cultivars and (iv) an elite 1 sub-set of synthetic hexaploid wheats with their durum wheat parents. The Al^{3+} test levels were 0 and 46 ppm for (i) and (ii) and (iv), and 0, 46 and 70 to 95 ppm for the *S. cereale* cultivars. Root regrowth scores after hematoxylin staining allowed for the estimation of the germplasms tolerance to Al^{3+} at various concentration levels.

Genetic and Wide Hybridization Studies

Conventional Genetic Studies

The T1BL.1RS bread wheat variety Glennson 81 is highly susceptible to Al at 46 ppm, cv. Chinese Spring is medium tolerant, with Maringa and CNT-1 being highly tolerant. Genes controlling Al^{3+} tolerance in wheat have been reported to range from one to the additive effect of two or more, dominant in action, located genomically and chromosomally with variability across cultivars (Manyowa and Miller 1991). Since a monosomic series is now available in cv. Glennson 81, highly susceptible to Al^{3+}, a monosomic study could be designed for an in-depth understanding of the tolerance mechanism. The information would be helpful as *in vitro* tests with the cultivar 'Alondra S' (T1BL.1RS cultivar) and its field performance plus phosphate influence have offered interesting variations.

Wide Hybridization Studies

The superior tolerance within *S. cereale* germplasm and *Ae. variabilis* accessions offer a potent means to exploit these sources for Al^{3+} tolerance breeding. There has not been any advance in this area since the report of Mujeeb-Kazi et al. (1993). However with the new technologies falling in place around *Ph1b* manipulation and use of doubled haploidy with molecular cytology and markers making and detecting alien transfers from both the above alien sources should be possible. Alien species with the N genome have also been reported to possess Al^{3+} tolerance (Berzonsky and Kimber 1989). Though one could utilize *T. ventricosum* (DDNN) or *T. rectum* (UUMMNN) exploiting *T. uniaristata* (NN) is preferred due to its diploid status.

The fastest practical outputs are offshoots of breeding programs that exploit diversity that is prevalent in the primary gene pool resources of the Triticeae. In the case of various other stress constraints the D genome diploid donor accessions gave ample usable diversity and thus a combination of conventional plus the variation within *Ae. tauschii* would have a plus for Al^{3+} tolerance also. Over the past near 2 decades of exploiting the *Ae. tauschii* resource Al^{3+} is the only stress for which no tolerance has been identified in the derived synthetic hexaploid wheats; a possible link with the diploid ancestral progenitor grasses growing locations (habitat) not possessing the toxicity constraint.

Conclusions

Wheat production encompasses the performance of high yielding varieties that are allelicly enriched to combat location dependent key biotic and abiotic stresses. For successful durable outputs varieties need to possess dual biotic and abiotic stress factor resistances/tolerances. Breeding wheats for abiotic tolerances has a built in advantage that tolerant varieties once produced are free of direct pathogen influence on the trait and hence have the potential to stay in cultivation over extended time durations. They get discarded after biotic stress susceptibility attacks the variety which happens when pathogen effectivity shifts set in. Global scenarios of wheat production environments are variable and appropriate genetic packages are structured by breeders that inevitably have the rust resistances forming a solid base of each variety via major and minor genes to ensure durability of resistance around a high yield potential that is paramount to all wheat improvement programs. The abiotic stress constraints that play a key role in wheat improvement generally are drought and salinity. To these has been added heat tolerance that in some countries like Pakistan is influenced by national cropping systems causing wheat to be planted late and thus getting exposed to the damaging terminal heat exposure. An appropriate wheat breeding strategy exercised by programs with global focus is to have genetic balance of appropriate major/minor genes for key biotic stresses and build in abiotic target requirements that can pay quick dividends. There are ample resources available for improving wheats around various traits as researchers have generated enormous data from their basic studies. Utilization of such identified diversity however requires stringent selection of appropriate strategies and also having a total grip that the outputs of the efforts must result in quality products that are high yielding, durable, with the sustainability measured through tonnes per hectare on farmers fields. What is the ideal route to take for breeding salt tolerant wheats? Does one exploit the durum based diversity (Munns) via the pentaploid protocol to improve both bread and durum wheats, is there a need to capture the durum based salt tolerance attribute and pyramid other genetic sources onto this base, or use the D genome diversity via synthetics (Mujeeb-Kazi et al. 2008) or by direct hybridization (Gill and Raupp 1987), or capture the benefits of the tertiary gene pool diploids (Mujeeb-Kazi 2006), or identify allelic variations in the D genome diploid progenitor, pyramid them (Mujeeb-Kazi and Hettell 1995) and exploit the cumulative package for maximizing returns reflected in more secure production systems. One also needs to be cognizant that what is readily available and apparently easily workable should also receive top-priority; the conventional germplasm that has potent salinity diversity spread across elite cultivars and landraces. This conventional germplasm as a foundation has the potential for stacking of other genes onto it; the choice of those other genes however, must be stringently set/guided by the practicality time frame governed by global needs for food and population increase trends. The aspect not addressed here but worthy of having in mind is the potential of exploiting transgenes that have the capacity to short-cut breeding complexities of both biotic and abiotic production targets. Instead of unidirectional efforts these sophisticated technologies will in our perception generate greater impact if they operate in tandem with the conventional tools as the output target is to secure global wheat productivity against the changing scenarios of various stress constraints.

References

Alonso LC, Kimber G (1984) Use of restitution nuclei to introduce alien genetic variation into hexaploid wheat. Z. Pflanzenzuchtung 90:185–189.

Bates LS, Deyoe CW (1973) Wide hybridization and cereal improvement. Econ Bot 27:401–412.

Berzonsky WA, Kimber G (1989) The tolerance to aluminum of *Triticum* N-genome amphiploids. Plant Breed 103:37–42.

Chen PD, Tsujimoto H, Gill BS (1994) Transfer of the Ph^{I} genes promoting homoeologous pairing from *Aegilops speltoides* to common wheat. Theor Appl Genet 88:97–101.

Coghlan A (2006) Synthetic wheat offers hope to the world. New Scientist Print Edition; February 11, 2006.

Colmer TD, Flowers TJ, Munns R (2006) Use of wild relatives to improve salt tolerance in wheat. J Exp Bot 57:1059–1078.

Cox TS, Harrell LG, Chen P, Gill BS (1991) Reproductive behaviour of hexaploid/diploid wheat hybrids. Plant Breed 107:105–118.

Dewey DR (1984) The genomic system of classification as a guide to intergeneric hybridization with the perennial Triticeae. In: Gustafson JP (ed) Gene manipulation in plant improvement. 16th Stadler Genetics Symposium, Plenum Press, New York, pp. 209–279.

Dreccer MF, Ogbonnaya F, Borgogone G, Wilson J (2003) Boron tolerance is present in primary synthetic wheats. In: Proceedings of the 10th International Wheat Genetics Symposium, Paestum, Italy, 1–6 September 2003, pp. 1130–1132.

Dubcovsky J, Santa-Maria G, Epstein E, Luo M-C, Dvorak J (1996) Mapping of the K/Na discrimination locus *Knal* in wheat. Theor Appl Genet 92:448–454.

Dvorak J, Edge M, Ross K (1988) On the evolution of the adaptation of *Lophopyrum elongatum* to growth in saline environments. Proc Natl Acad Sci USA 85:3805–3809.

Farooq S, Niazi MLK, Iqbal N, Shah TM (1989) Salt tolerance potential of wild resources of the tribe Triticeae. II. Screening of species of the genus Aegilops. Plant Soil 119:255–260.

Farrer W (1904) Some notes on the wheat Bobs: its peculiarities, economic value and origin. Agr Gaz NSW 15:849–854.

Forster BP, Gorham J, Miller TE (1987) Salt tolerance of an amphiploid between *Triticum aestivum* and *Agropyron junceum*. Plant Breed 98:1–8.

Forster BP, Miller TE, Law CN (1988) Salt tolerance of two wheat-*Agropyron junceum* disomic addition lines. Genome 30:559–564.

Genc Y, McDonald GK (2004) The potential of synthetic hexaploid wheats to improve zinc efficiency in modern bread wheat. Plant Soil 262:23–32.

Gill BS, Raupp WJ (1987) Direct gene transfers from *Aegilops squarrosa* L. to hexaploid wheat. Crop Science 27:445–450.

Gill KS, Gill BS (1996) A PCR- based screening assay of *Ph1*, the chromosome pairing regulator gene of wheat. Crop Sci 36:719–722.

Gorham J (1990) Salt tolerance in the Triticeae: ion discrimination in synthetic hexaploid wheats. J Exp Bot 41:623–627.

Gorham J, McDonnell E, Budrewicz E, Wyn Jones RG (1985) Salt tolerance in the Triticeae: growth and solute accumulation in leaves of *Thinopyrum bessarabicum*. J Exp Bot 36:1021–1031.

Gorham J, Hardy C, Wyn Jones RG, Joppa LR, Law CN (1987) Chromosomal location of a K/Na discrimination character in the D genome of wheat. Theor Appl Genet 74:584–588.

Graham RD (1975) Male sterility in wheat plants deficient in copper. Nature (London) 254:514–515.

Graham RD (1978) Tolerance of Triticale, wheat and rye to copper deficiency. Nature (London) 271:542–543.

Harlan J, de Wet JMJ (1971) Towards a rational classification of cultivated plants. Taxon 20:509–517.

Islam AKMR, Shepherd KW, Sparrow DHB (1978) Production and characterization of wheat-barley addition lines. In: Proceedings of the 5th International Wheat Genetics Symposium, New Delhi, India, pp. 365–371.

Jahan Q, Ter-Kuile N, Hashmi N, Aslam M, Vahidy AA, Mujeeb-Kazi A (1990) The status of 1B/1R translocation chromosome in some released wheat varieties of Pakistan. Pak J Bot 22:1–10.

James RA, Davenport RJ, Munns R (2006) Physiological characterization of two genes for Na^+ exclusion in durum wheat, *Nax1* and *Nax2*. Plant Physiol 142:1537–1547.

Jiang J, Friebe B, Gill BS (1994) Recent advances in alien gene transfer in wheat. Euphytica 73:199–212.

Kimber G, Feldman M (1987) Wild wheat: an introduction, Special Report 353. College of Agriculture, University of Missouri, Columbia, pp. 1–131.

Kruse A (1967) Intergeneric hybrids between *Hordeum vulgare* L. ssp. *distichum* (v Pallas 2n = 14) and Secale cereale L. (v Petkus 2n = 14). In: Royal Veterinary and Agricultural College Yearbook 1967, Copenhagen, pp. 82–92.

Kruse A (1969) Intergeneric hybrids between *Triticum aestivum* L. (v Koga ll 2n = 42) and *Avena sativa* L (v Stal 2n = 42) with pseudogamous seed formation. In: Royal Veterinary and Agricultural College Yearbook 1967, Copenhagen. pp. 188–200.

Kruse A (1973) *Hordeum* x *Triticum* hybrids. Hereditas 73:157–161.

Lindsay MP, Lagudah ES, Hare RA, Munns R (2004) A locos for sodium exclusion (*Nax1*), a trait for salt tolerance, mapped in durum wheat. Funct Plant Biol 31:1105–1114.

Lopez-Cesati J, Villegas E, Rajaram S (1986) CIMMYT's laboratory method for screening wheat seedling tolerance to aluminum. In: Kohli MM, Rajaram S (eds) Wheat breeding for acid soils. Review of Brazilian/CIMMYT collaboration, 1974–1986, CIMMYT, Mexico DF, Mexico, pp. 59–65.

Maes B, Trethowan RM, Reynolds MP, van Ginkel M, Skovmand B (2001) The influence of glume pubescence on spikelet temperature of wheat under freezing conditions. Aust J Plant Physiol 28:141–148

Manyowa NM, Miller TE (1991) The genetics of tolerance to high mineral concentrations in the tribe Triticeae – a review and update. Euphytica 57:175–185.

McGuire PE, Dvorak J (1981) High salt tolerance potential in wheat grasses. Crop Sci 21:702–705.

Mujeeb-Kazi A (2003) New genetic stocks for durum and bread wheat improvement. In: Tenth International Wheat Genetics Symposium, Paestum, Italy, pp. 772–774.

Mujeeb-Kazi A (2005) Wide Crosses for durum wheat improvement. In: Roya C, Nachit MN, DiFonzo N, Araus JL, Pfeiffer WH, Slafer GA (eds) Durum wheat breeding: current approaches and future strategies. Food Products Press, An imprint of The Haworth Press, Inc. New York, London, Oxford, pp. 703–743.

Mujeeb-Kazi A (2006) Utilization of genetic resources for bread wheat improvement. In: Singh RJ, Jauhar PP (eds) CRC Series pp. 61–97.

Mujeeb-Kazi A, Diaz de Leon JL (2002) Conventional and alien genetic diversity for salt tolerant wheats: focus on current status and new germplasm development. In: Ahmad R, Malik KA (eds) Prospects for saline agriculture, Kluwer, Dordrecht, pp. 69–82.

Mujeeb-Kazi A, Hettel GP (1995) Utilizing wild grass biodiversity in wheat improvement: 15 years of wide cross research at CIMMYT, pp. 1–140.

Mujeeb-Kazi A, Rajaram S (2002) Transferring alien genes from related species and genera for wheat improvement. In: Bread Wheat Improvement and Production, FAO, Rome, pp. 199–215.

Mujeeb-Kazi A, Riera-Lizarazu O (1996) Polyhaploid production in the Triticeae by sexual hybridization. In: Jain SM et al. (ed) *In vitro* haploid production in higher plants, pp. 275–296.

Mujeeb-Kazi A, Thomas JB, Rodriguez R, Waters RF, Bates LS (1978) Chromosme instability in hybrids of *Hordeum vulgare* L. with *Triticum turgidum* and *T. aestivum*. J Hered 69: 170–182.

Mujeeb-Kazi A, Gorham J, Lopez-Cesati J (1993) Use of wild Triticeae relatives for stress tolerance. In: International Journal of Crop Science, Crop Science Society of America, pp. 549–554.

Mujeeb-Kazi A, Rosas V, Roldán S (1996) Conservation of the genetic variation of *Triticum tauschii* (Coss.) Schmalh. (*Aegilops squarrosa* auct. non L.) in synthetic hexaploid wheats (*T. turgidum* L. s. lat. X *T. tauschii*; 2n = 6x = 42, AABBDD) and its potential utilization for wheat improvement. Genet Res Crop Evol 43: 129–134.

Mujeeb-Kazi A, Cortes A, Rosas V, William MDHM, Delgado R (1999a) Some *Triticum turgidum* L. cultivars possessing the chromosome T1BL.1RS substitution. Cereal Res Commun 27:373–376.

Mujeeb-Kazi A, Cortes A, Rosas V, William MDHM, Delgado R (1999b) Development of near-isogenic sets of derivatives with T1BL.1RS or 1B chromosome substitutions in bread wheat. Wheat Inf Serv 89:13–16.

Mujeeb-Kazi A, William MDHM, Villareal RL, Cortés A, Rosas V, Delgado R (2000) Registration of 11 new isogenic T1BL.1RS chromosome translocation and 11 extracted chromosome 1B lines in *Triticum turgidum* L. cv. 'Altar 84'. Crop Sci 40:588–589.

Mujeeb-Kazi A, William MDHM, Villareal RL, Cortés A, Rosas V, Delgado R (2001a) Registration of 10 isogenic chromosome 1B and 10 T1BL.1RS chromosome translocation bread wheat germplasms. Crop Sci 41:280–281.

Mujeeb-Kazi A, Cortes A, Rosas V, Cano S, Delgado R (2001b) Registration of 17 isogenic chromosome 1B and 17 T1BL.1RS chromosome translocation bread wheat germplasms. Crop Sci 41:596–597.

Mujeeb-Kazi A, Cortes A, Rosas V, Cano S, Delgado R (2001c) Registration of six isogenic T1BL.1RS chromosome translocation and six chromosome 1B durum germplasms. Crop Sci 41:595–596.

Mujeeb-Kazi A, Delgado R, Cortes A, Cano S, Rosas V, Sanchez J (2004) Progress in exploiting *Aegilops tauschii* for wheat improvement. Ann Wheat Newslett 50:79–88.

Mujeeb-Kazi A, Gul A, Ahmad J, Mirza JI (2006) A simplified and effective protocol for production of bread wheat haploids (n = 3x = 21, ABD) with some application areas in wheat improvement. Pak J Bot 38:393–406.

Mujeeb-Kazi A, Gul A, Farooq M, Rizwan S, Ahmad I (2008) Rebirth of synthetic hexaploids with global implications for wheat improvement. Aust J Agr Res 59:391–398.

Munns R (2007) Utilizing genetic resources to enhance productivity of salt-prone land. In: CAB reviews in Agriculture, Veterinary Science, Nutrition and Natural Resources, pp. 1–11.

Munns R, James RA (2003) Screening methods for salinity tolerance of wheat and other cereals. Plant Soil 253:201–218.

Munns M, Tester, M (2008) Mechanisms of salinity tolerance. Annu Rev Plant Biol 59:651–681.

Munns R, James RJ, Lauchli A (2006) Approaches to increasing the salt tolerance of wheat and other cereals. J Exp Bot 47:1025–1043.

Ortiz-Monasterio JI, Palacios N, Meng E, Pixley K, Trethowan RM, Pena RJ (2007) Enhancing the mineral and vitamin content of wheat and maize through plant breeding. J Cereal Sci 46:293–307.

Polle E, Konzak CF, Kittrick JA (1978) Rapid screening of wheat for tolerance to aluminum in breeding varieties better adapted to acid soils. Technical Series Bulletin No. 21, Office of Agriculture Development, Washington, DC.

Pritchard DJ, Hollington PA, Davies WP, Gorham J, Diaz de Leon JL, Mujeeb-Kazi A (2002) K^+/Na^+ discrimination in synthetic hexaploid wheat lines: Transfer of the trait for K^+/Na^+ discrimination from *Aegilops tauschii* to *Triticum turgidum*. Cereal Res Commun 30:261–267.

Qu L-J, Foote TN, Roberts MA, Money TA, Aragon-Alcaide L, Snape JW, Moore G (1998) A simple PCR-based method for scoring the *ph1b* deletion in wheat. Theor Appl Genet 96:371–375.

Rajaram S, Mann, Ch E, Ortiz-Ferrara G, Mujeeb-Kazi A (1983) Adaptation, stability and high yield potential of certain 1B/1R CIMMYT wheats. In: Sakamoto S (ed) Proceedings of 6th. International Wheat Genetic Symposium, Kyoto, Japan, pp. 613–621.

Reynolds M, Trethowan RM (2007) Physiological interventions in breeding for adaptation to abiotic stress. In: Spiertz JHJ, Struik PC, van Laar HH (eds) Scale and complexity in plant systems research: gene-plant-crop relations, Springer, Dordrecht, the Netherlands, pp 129–146.

Reynolds MP, Mujeeb-Kazi A, Sawkins M (2005) Prospects for utilizing plant-adaptive mechanisms to improve wheat and other crops in drought- and salinity-prone environments. Ann Appl Biol 146:239–259.

Reynolds M, Dreccer F, Trethowan RM (2007) Drought-adaptive traits derived from wheat wild relatives and land races. J Exp Bot 58:177–186.

Shah SH, Gorham J, Forster BP, Wyn Jones RG (1987).Salt tolerance in the Triticeae: the contribution of the D genome to cation selectivity in hexaploid wheat. J Exp Bot 38:254–269.

Sharma HC, Gill BS (1983) Current status of wide hybridization in wheat. Euphytica 32:17–31.

Simonite T (2006) Ancient genetic tricks shape up wheat; turning back the evolutionary clock offers better crops for dry regions. Nature (London) On Line: 3 January 2006.

Ter-Kuile N, Jahan Q, Hashmi N, Aslam M, Vahidy AA, Mujeeb-Kazi A (1991) 1B/1R Translocation wheat cultivars detected by A-page electrophoresis and C-banding in the 1990 National Uniform Wheat Yield Trial in Pakistan. Pak J Bot 23.203–212.

Torun A, Gultekin I, Kalayci M, Yilmaz A, Eker S, Cakmak I (2001) Effects of zinc fertilization on grain yield and shoot concentrations of zinc, boron and phosphorus of 25 wheat cultivars grown on a zinc-deficient and boron-toxic soil. J Plant Nutr 24:1817–1829.

Trethowan RM, Mujeeb-Kazi A (2008) Novel germplasm resources for improving environmental stress tolerances of hexaploid wheat. Crop Sci 48:1255–1265.

Trethowan RM, Crossa J, van Ginkel M, Rajaram S (2001a) Relationships among bread wheat international yield testing locations in dry areas. Crop Sci 41:1461–1469.

Trethowan RM, Pena RJ, van Ginkel M (2001b) The effect of indirect tests for quality on grain yield and industrial quality of bread wheat. Plant Breed 120:509–512.

Trethowan RM, van Ginkel M, Rajaram S (2002) Progress in breeding for yield and adaptation in global drought affected environments. Crop Sci 42:1441–1446.

Trethowan RM, Borja J, Mujeeb-Kazi A (2003a) The impact of synthetic wheat on breeding for stress tolerance at CIMMYT. Ann Wheat Newslett 49:67–69.

Trethowan RM, van Ginkel M, Ammar K, Crossa J, Cukadar B, Rajaram S, Hernandez E (2003b) Associations among 20 years of bread wheat yield evaluation environments. Crop Sci 43:1698–1711.

Trethowan RM, Reynolds MP, Sayre KD, Ortiz-Monasterio JI (2005) Adapting wheat cultivars to resource conserving farming practices and human nutritional needs. Ann Appl Biol 146:404–413.

Valkoun J (2001) Wheat pre-breeding using wild progenitors. Euphytica 119:17–23.

Valkoun J, Dostal J, Kucerova D (1990) *Triticum* x *Aegilops* hybrids through embryo culture. In: Bajaj YPS (ed) Biotechnology in agriculture and forestry, 9th ed., Vol. 13,, Springer, Berlin, Heidelberg, pp. 152–166.

Villareal RL, Bañuelos O, Mujeeb-Kazi A (1997) Agronomic performance of related durum wheat (*Triticum turgidum L.*) stocks possessing the chromosome substitution TIBL.IRS. Crop Sci 37:1735–1740.

Villareal RL, Bañuelos O, Mujeeb-Kazi A, Rajaram S (1998) Agronomic performance of chromosome 1B and T1BL.1RS genetic stocks in spring bread wheat (*Triticum aestivum* L.) cultivar Seri M82. Euphytica 103:195–202.

Villareal RL, Sayre K, Banuelos O, Mujeeb-Kazi A (2001) Registration of four synthetic hexaploid wheat (*Triticum turgidum/Aegilops tauschii*) germplasm lines tolerant to waterlogging. Crop Sci 41:274.

Wang RR-C (1989) Intergeneric hybrids involving perennial Triticeae. Genet (Life Sci Adv) 8:57–64.

Wang RR-C, Larson SR, Horton WH, Chatterton NJ (2003a) Registration of W4909 and W4910 bread wheat germplasm lines with high salinity tolerance. Crop Sci 43:746–747.

Wang RR-C, Li XM, Hu ZM, Zhang JY, Larson SR, Zhang XY, Grieve CM, Shannon MC (2003b) Development of salinity-tolerant wheat recombinant lines from a wheat disomic addition line carrying a *Thinopyrum junceum* chromosome. Int J Plant Sci 164:25–33.

Warburton ML, Skovmand B, Mujeeb-Kazi A (2002) The molecular genetic characterization of the 'Bobwhite' bread wheat family using AFLPs and the effect of the T1BL.1RS translocation. Theor Appl Genet 104:868–873.

Yang J, Sears RG, Gill BS, Paulsen GM (2002) Growth and senescence characteristics associated with tolerance of wheat-alien amphiploids to high temperature under controlled conditions. Euphytica 126:185–193.

General Topics

Chapter 17
Survival at Extreme Locations: Life Strategies of Halophytes

H.-W. Koyro, N. Geissler, and S. Hussin

Abstract Seven percent of the land's surface and five percent of cultivated lands are affected by salinity. There are often not sufficient reservoirs of freshwater available and most of the agronomically used irrigation systems are leading to a permanent increase in the soil-salinity and step by step to growth conditions in-acceptable for most of the conventional crops. Significant areas are becoming unusable each year. Although it is a world-wide problem, most acute is in Australasia, the Near East and Africa, North and Latin America and to an increasing degree also in Europe. This large extent of salinity problem reduces crop productivity. In contrast to crop plants, there exist specialists that thrive in the saline environments along the sea shore, in estuaries and saline deserts. These plants, called halophytes, have distinct physiological and anatomical adaptations to counter the dual hazards of water deficit and ion toxicity. The sustainable use of halophytic plants is a promising approach to valorize strongly salinised zones unsuitable for conventional agriculture and mediocre waters. There are already many halophytic species used for economic interests (human food, fodder) or ecological reasons (soil desalinisation, dune fixation, CO_2-sequestration). However, the wide span of halophyte utilisation is not jet explored even to a small degree. For economic utilisation of potential halophytes ecological studies should be complemented with comparative physiological studies about salinity tolerance in halophytes are essential.

Keywords Salt injury • salt tolerance • *Spartina townsendii* • seabeet • *Beta vulgaris* ssp *maritima* • *Inula critmoides* • *Chenopodium quinoa* • single cell measurements

H.-W. Koyro (✉), N. Geissler, and S. Hussin
Institute of Plant Ecology, Justus-Liebig-University Gießen, Heinrich-Buff-Ring 26–32, D-35392 Gießen, Germany
e-mail: Hans-Werner.Koyro@bot2.bio.uni-giessen.de

1 Introduction

Seven percent of the land's surface and five percent of cultivated lands are affected by salinity (Ghassemi et al. 1995; Szabolcs 1994), with salt stress being one of the most serious environmental factors limiting the productivity of crop plants. When soils in arid regions of the world are irrigated, solutes from the irrigation water can accumulate and eventually reach levels that have an adverse affect on plant growth. Of the current 230 million hectares of irrigated land, 45 million hectares are salt-affected (19.5%) and of the 1,500 million hectares under dryland agriculture, 32 million are salt-affected to varying degrees (2.1%). There are often not sufficient reservoirs of freshwater available and most of the agronomically used irrigation systems are leading to a permanent increase in the soil-salinity and step by step to growth conditions inacceptable for most of the conventional crops. Significant areas becoming unusable each year. It is a world-wide problem, but most acute in Australasia (3.1 million hectares), the Near East (1,802 million hectares) and Africa (1,899 million hectares), North and Latin America (3,963 million hectares) and to an increasing degree also in Europe (2,011 million hectares of salt-affected soils; FAO Land and Plant Nutrition Management Service). Despite advances in increasing plant productivity and resistance to a number of pests and diseases, improving salt tolerance in crop plants remains elusive, mainly because salinity simultaneously affects several aspects of plant physiology.

In contrast to crop plants, there exist specialists that thrive in the saline environments along the sea shore, in estuaries and saline deserts. These plants, called halophytes, have distinct physiological and anatomical adaptations to counter the dual hazards of water deficit and ion toxicity. Salinity can affect any

M. Ashraf et al. (eds.), *Salinity and Water Stress*,

process in the plant's life cycle, so that tolerance will involve a complex interplay of characters (Flowers et al. 1977). New insights into the mechanisms by which plants achieve this have emerged from Research projects investigating details of the physiology and biochemistry of salt tolerance. Unfortunately, there are few investigations which combine studies of growth and other measurements on both biophysical and biochemical plant characteristics. Such joint investigations will be particularly important in the discovery of traits which present the ability to maintain high plant productivity in saline environments.

The sustainable use of halophytic plants is a promising approach to valorize strongly salinised zones unsuitable for conventional agriculture and mediocre waters (Boer and Gliddon 1998; Lieth et al. 1999). There are already many halophytic species used for economic interests (human food, fodder) or ecological reasons (soil desalinisation, dune fixation, CO_2-sequestration). However, the wide span of halophyte utilisation is not jet explored even to a small degree.

2 Halophytes – Plants Able to Complete Their Life Cycle on Saline Substrates

Saline conditions reduce the ability of plants to absorb water, causing rapid reductions in growth rate, and induce many metabolic changes similar to those caused by water stress (Epstein 1980).

Halophytes are plants, able to complete their life cycle in a substrate rich in NaCl (Schimper 1891). One of the most important property of halophytes is their salinity tolerance (Koyro and Lieth 1998; Lieth 1999). This substrate offers for obligate halophytes advantages for the competition with salt sensitive plants (glycophytes). There is a wide range of tolerance among the 2,600 known halophytes (Pasternak 1990; Amzallag 1994; Lieth and Menzel 1999). However, information about these halophytes need partially careful checking. A precondition for a sustainable utilisation of suitable halophytes is the precise knowledge about their salinity tolerance and the various mechanisms enabling a plant to grow at (their natural) saline habitats (Greenway and Munns 1980; Koyro et al. 1997; Ashraf, 1994; Marcum 1999; Warne et al. 1999; Weber and Dántonio 1999; Winter et al. 1999). This paper concentrates on eco-physiological mechanisms.

3 The Quick Check System

It is – without doubt – necessary to develop sustainable biological production systems which can tolerate higher water salinity because freshwater resources will become limited in near future (Lieth 1999). A precondition is the identification and/or development of salinity tolerant crops. An interesting system approach lines out that after halophytes are studied in their natural habitat and a determination of all environmental demands has been completed, the selection of potentially useful plants should be started (Lieth 1999). The first step of this identification list contains the characterisation and classification of the soil and climate, under hich potentially useful halophytes grow. Only artificial conditions in sea water irrigation systems in a growth cabinet under photoperiodic conditions offer the possibility to study potentially useful halophytes under reproducible experimental growth and substrate conditions. The supply of different degrees of sea water salinity (0%, 25%, 50%, 75%, 100% [and if necessary higher] sea water salinity) to the roots in separate systems under otherwise identical or/and close to natural conditions gives the necessary preconditions for a comparative study in a quick check system (QCS) for potential cash crop halophytes. The experiments of the QCS started off at steady state conditions in a gravel/hydroponic system imitating the climatic conditions of subtropical dry regions. (Fig. 17.1, Koyro and Huchzermeyer 1999). It is well known that salinity tolerance depends on the stage of development and period of time over which the plants have grown in saline conditions (Munns 2002). Plants were exposed to salinity in the juvenile state of development and were studied until achieving the steady state of adult plants. Variable applicable QCS seems to be valuable for the selection of useful plants and it suggest itself as a first step for the controlled establishment of cash crop halophytes because it provides detailed information about three major goals as there are the threshold of salinity tolerance at idealized growth conditions, how to uncover the individual mechanisms for salt tolerance and about the potential of utilization for the pre-selected halophytic species (cash crop halophytes).

4 Threshold of Salinity Tolerance

In correspondence with the definition for the threshold of salinity tolerance according to Kinzel (1982), the growth reaction and the gas exchange are used

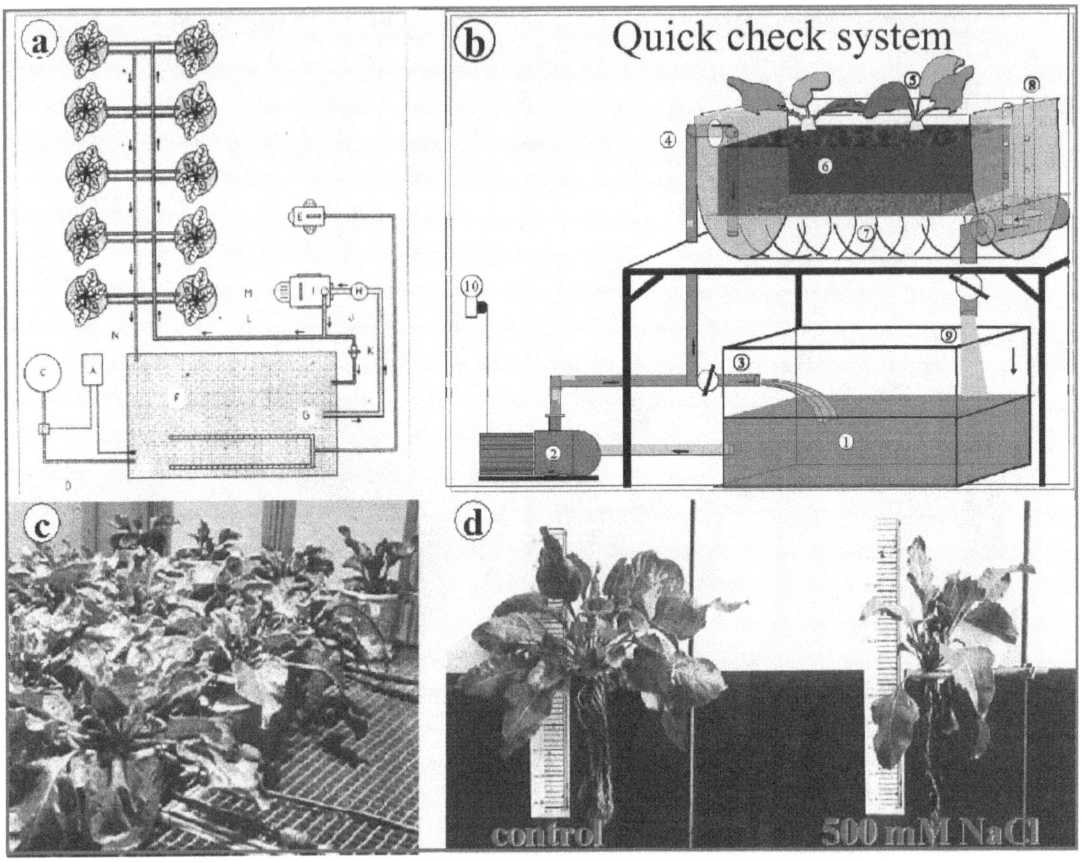

Fig. 17.1 A selection of different culture systems (so-called quick-check-systems) under photoperiodic conditions in a growth cabinet (plant species: *Beta vulgaris* ssp. *Maritima*). (**a**) Gravel/hydroponic quick check system with automatic drip irrigation. (**b**) Interdidal irrigation (alternating water level) quick check system. (**c**) Habitus of seabeet plants grown in artificial nutrient solution (foreground) and in high salinity treatment (background). (**d**) Control plants of the seabeet are visible on the left side, the sea water salinisation treatment on the right side

during the screening of halophytes as objective parameters for the description of the actual condition of a plant (Ashraf and O'Leary 1996). There are now reliable informations available about studies with several halophytic species from different families such as *Chenopodium quinoa* (Fig. 17.2a), *Aster tripolium*, *Plantago* cf. *coronopus*, *Beta vulgaris* ssp. *maritima* (Fig. 17.2b), *Batis maritima*, *Puccinellia maritima*, *Spartina townsendii* (Fig. 17.2c), *Atriplex nummularia*, *Atriplex leucoclada*, *Atriplex halimus*, *Laguncularia racemosa*, *Limoneastrum articulatum*, *Sesuvium portulacastrum* and *Inula critmoides* (Fig. 17.2d) (Pasternak 1990; Koyro and Huchzermeyer 1997, 1999; Koyro et al. 1999; Lieth and Menzel 1999; Koyro 2000; Koyro and Huchzermeyer 2004). The substrate-concentration leading to a growth depression of 50% (refer to freshweight, in comparison to plants without salinty) is easy to calculate with the QCS (by extrapolation of the data) and it leads to a precise specification of a comparative value for the threshold of salinity tolerance (Fig. 17.2a–d). Dramatic differences are found between halophytic plant species. The threshold of salinity tolerance amounts to 350 $mol^{*}m^{-3}$ NaCl in *Chenopodium quinoa* and *Beta vulgaris* ssp. *maritima*, 500 $mol^{*}m^{-3}$ in *Spartina townsendii* and in *Inula critmoides* (Fig. 17.2). These results prove that it is essential to quantify differences in salinity tolerance between halophytic species as one basis for assessment of their potential of utilization.

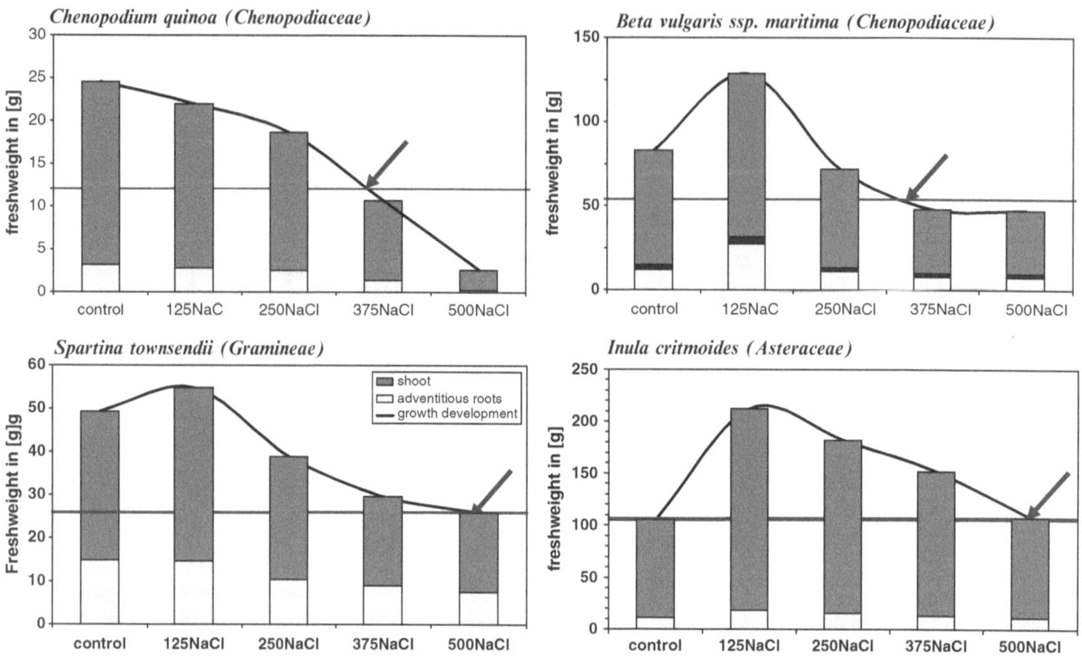

Fig. 17.2 Development of the plant freshweight at treatments with different percentages of sea water salinity. The crossover of the red and the black lines reflects the NaCl-salinity where the growth depression falls down to 50% of the control plant (threshold of NaCl-salinity according to Kinzel 1982). (a) *Chenopodium quinoa*: 75% sea water salinity. (b) *Beta vulgaris ssp. maritima*: 75% sea water salinity. (c) *Spartina townsendii*: 100% sea water salinity. (d) *Inula critmoides*: 100% sea water salinity. 0% sea water salinity = control, 25% = 125NaCl, 50% = 250NaCl, 75% = 375NaCl and 100% = 500NaCl

5 Morphological Structures to Reduce Salt Concentrations

In many cases various mechanisms and special morphological structures are advantageous for halophytes since they help to reduce the salt concentrations especially in photosynthetic or storage tissue and seeds. Salt glands may eliminate large quantities of salt by secretion to the leaf surface. This secretion appears in complex multicellular organs, for example in *Avicennia marina* or by simple two cellular salt glands, for example in *Spartina townsendii* (Fig. 17.3, Sutherland and Eastwood 1916; Walsh 1974; Koyro and Stelzer 1988; Marcum et al. 1998). Several halophytes can reduce the salt concentrations in vital organs by accumulation in bladder hairs (*Atriplex halimus*, *Leptochloa fusca* (L.), *Halimione portulacoides*), enhancing the LMA (leaf mass to area ratio, e.g. by *Suaeda fruticosa*, *Salicornia europaea*, *Salsola kali*, *Sesuvium portulacastrum*), establishing apoplastic barriers (Freitas and Breckle 1992, 1993a, b; Hose et al. 2001), translocating NaCl into special organs (z.B. *Kandelia candel* L.), using of ultrafiltration at the root level to exclude salt. (*Avicennia marina*, *Sonneratia alba*) or shedding of old leaves (*Beta vulgaris* ssp. *maritima*, see literature in Marschner 1995; Schroeder 1998; Glaubrecht 1999; Koyro 2002).

6 Screening Procedure

However, many halophytic species can tolerate high sea water salinity without possessing special morphological structures. To achieve salt tolerance three interconnected aspects of plant activity are important for plants with or without saltglands. Damage must be pre-

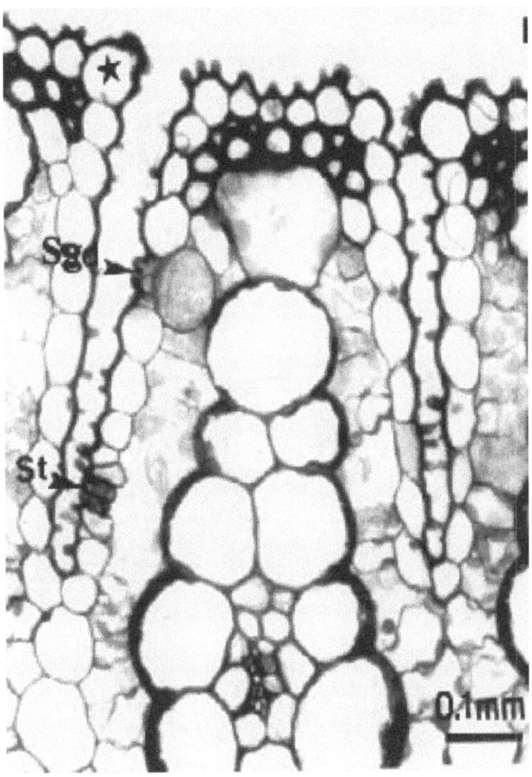

Fig. 17.3 Radial cross section of an adventitious roots of *Spartina townsendii. Sg: saltgland, St: stomate*

vented, homeostatic conditions must be re-established and growth must resume. Growth and survival of vascular plants at high salinity depends on adaptation to both low water potentials and high sodium concentrations, with high salinity in the external solution of plant cells producing a variety of negative consequences (Azaizeh and Steudle 1991; Munns, 1993; Tazuke 1997; Munns et al., 2002). It is the exception, that a single parameter is of major importance for the ability to survive at high NaCl-salinity. A comprehensive study with the analysis of at least a combination of several parameters is a necessity to get a survey about mechanisms constitution leading at the end to the salinity tolerance of individual species. These mechanisms are connected to the four major constraints of plant growth on saline substrates: water deficit, restriction of CO_2 uptake, ion toxicity and nutrient imbalance (Robertson and Wainwright 1987; Munns, 1993, 2002).

Plants growing in saline habitats face the problem of having low water potential in the soil solution and high concentrations of potentially toxic ions such as chloride and sodium. Salt exclusion minimizes ion toxicity but accelerates water deficit and deminishes indirectly the CO_2-uptake. Salt absorption facilitates osmotic adjustment but can lead to toxicity and nutritional imbalance.

The presence of soluble salts can affect growth in several ways (Mengel and Kirkby 2001). In the first place plants may suffer from water stress, secondly high concentrations of specific ions can be toxic and induce physiological disorders and thirdly intracellular imbalances can be caused by high salt concentration.

Terrestric plants at saline habitats are often surrounded by low water potentials in the soil solution and atmosphere. Plant water loss has to be minimized under these circumstances, since biomass production depends mainly on the ability to keep a high net photosynthesis by low water loss rates. Therefore, one crucial aspect of the screening procedure is the study of growth reduction and leaf (plant) water potential especially at the threshold of salinity tolerance (Fig. 17.4). Water deficit is one major constraint at high salinity and can lead to a restriction of CO_2-uptake. The balance between water loss and CO_2-uptake is another basis for assessment of their potential of utilization. Additionally it helps to find weak spot in the mechanisms of adjustment (of photosynthesis) to high salinity.

In principle, salinity tolerance can be achieved by salt exclusion or salt inclusion. Several physiological mechanisms are described in literature which avoid salt injury (and to protect the symplast) are known as major plant responses to high NaCl-salinity (Marschner 1995; Mengel and Kirkby 2001; Munns 2002; Koyro 2003).

Useful parameters for screening halophytes should base on the major plant responses to high NaCl-salinity (Volkmar et al. 1998). It seems to be essential that such a screening system should include salt induced morphological changes such as succulence and LAR (leaf mass to area ratio, Koyro 2002), growth, water relations, gas exchange and composition of minerals (and compatible solutes) at different parts of the root system and in younger and older leaf tissues. The measurement of such general scientific data at plant-, organ- or tissue level reveals general trends – but since these represent a mean behaviour of several cell types, many informations on single cell adjustment are lost. They cannot give sufficient information about the compartmentation inside a cell or along a diffusion

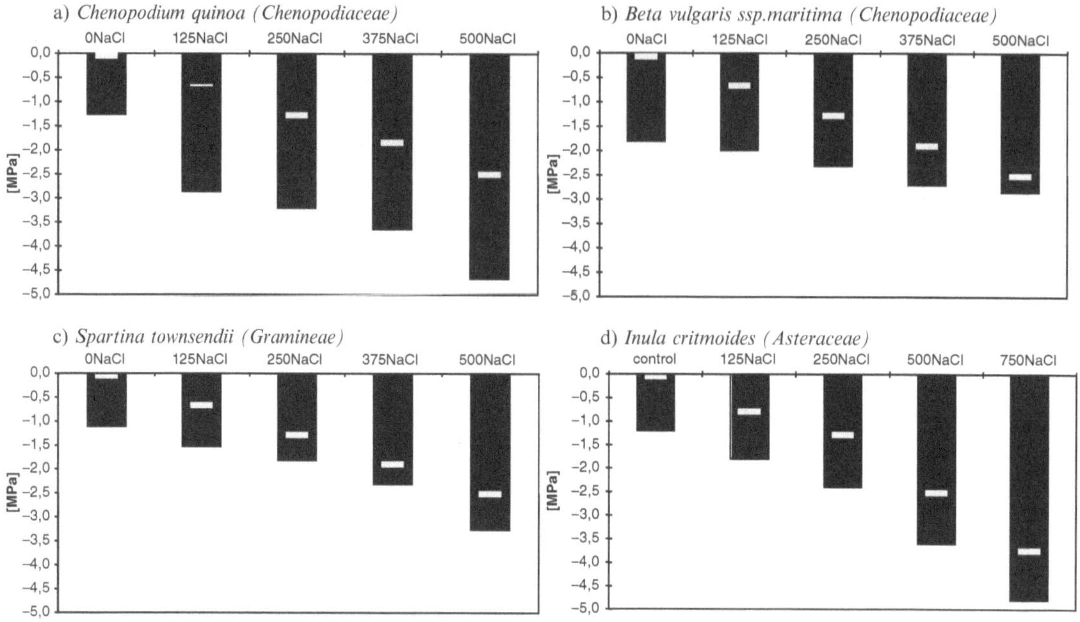

Fig. 17.4 Leaf water potentials (MPa) of (**a**) *Chenopodium quinoa*, (**b**) *Beta vulgaris* ssp. *maritima*, (**c**) *Spartina townsendii* and (**d**) *Inula critmoides*. The red lines in the bars mark the water potentials in the nutrient solutions. Leaf water potentials were always lower than in the assigned nutrient solution potential. The difference between water potentials in the leaves and in the nutrient solutions decreased with increasing NaCl-salinity. 0% sea water salinity = control, 100% sea water salinity = sea water salinity

zone in a root apoplast or about ultrastructural changes such as apoplastic barriers (Hose et al. 2001). The collection of scientific data should be completed if necessary (to uncover the individual mechanisms for salt tolerance) by a special physiological research at single cell level supplemented optionally by methods such as the analysis of the gene-expression and its genetic basis (genomics and proteomics, Winicov and Bastola 1997, 1999; Winicov 1998).

The general scientific data give an impression of various mechanisms of adaptation to high NaCl-salinity. Beside water stress and ion specific toxic physiological disorders on tissue level intracellular ionic imbalances (K^+, Ca^{2+} and Mg^{2+}) can be caused by high salt concentrations (Wolf et al. 1991; Jeschke et al. 1995) Mengel and Kirkby 2001). The capacity of plants to maintain K^+ homeostasis and low Na^+ concentrations in the cytoplasm appears to be one important determinant of plant salt tolerance (Yeo 1998; Läuchli 1999). A possibility to find such limiting factors is the study of the relations inside single cells such as the compartmentation between cytoplasm and vacuole, the distribution of elements in different cell types or along a diffusion zone in a root apoplast and ultrastructural changes.

Beta vulgaris ssp. *maritima* and *Spartina townsendii* keep Na and Cl concentrations low (Fig. 17.5) in their young growing tissues (such as juvenile leaves) and in their storage organs (such as taproot or rhizome). However, *Beta vulgaris* ssp. *maritima* is a typical Cl-includer and *Spartina townsendii* a typical Cl-excluder with high Na-accumulation in the leaves. Both species seem to react similar to salinity with changes of leaf water potential, gas-exchange and nutrients (Koyro and Huchzermeyer 2002). However, this number of partially congruent or complementary results does not allow to conclude analogical intracellular relations. The comparison of their intracellular ionic balance will be used to demonstrate the necessity of special physiological investigations.

In contrast to water stress effects, occurring in the meristematic region of younger leaves, salt toxicity predominantly occurs in adventitious roots and mature leaves (Mengel and Kirkby 2001). Furthermore, most of the Na and Cl are stored mainly in the shoot of halophytes such as *Beta vulgaris* ssp *maritima* and *Spartina townsendii* leading to a growth reduction of the above ground parts much higher than of the root (Koyro 2000;

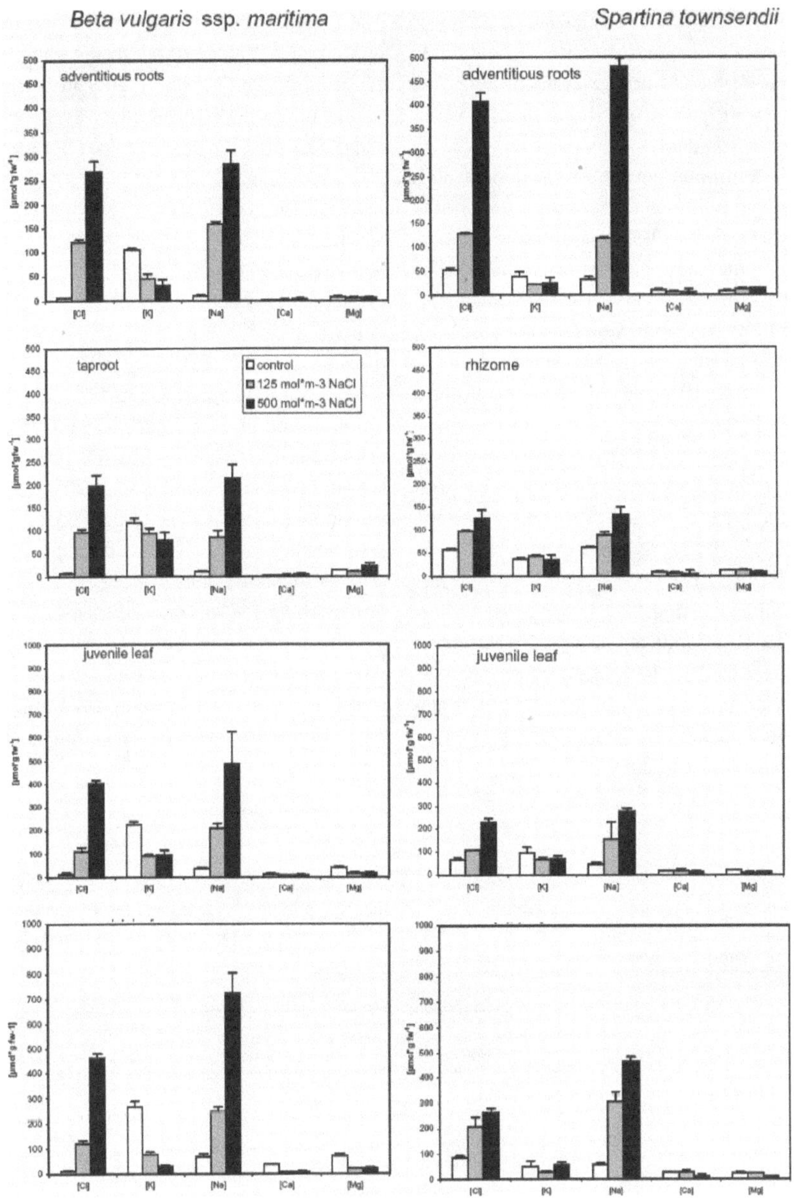

Fig. 17.5 Chlorine-, Potassium-, sodium-, calcium- and magnesium-concentrations in mol*m^{-3} in different tissues of *Beta vulgaris* ssp. *maritima* and *Spartina townsendii*

Koyro and Huchzermeyer 2004). These changes can be interpreted as signs of a critical load. Therefore, to distinguish between the individual mechanisms of salinity tolerance further investigations of the intracellular ionic balance were performed first of all at epidermal leaf cells (the end of the transpiration stream) of these both species.

The single cell data of the vacuolar and cytolasmatic composition in cells of the upper leaf-epidermis are summarized for the controls and the high-salinity treatments (at seawater salinity) in Table 17.1. The intracellular composition of the leaf epidermal cytoplasm and vacuoles of controls of *Beta vulgaris* ssp. *maritima* and *Spartina townsendii* show some more congruities of both species. The epidermal vacuoles of controls of both species contains most of the elements (with the exception of P) in higher concentrations as the cytoplasm indicating the overall picture of a vacuolar buffer. The leaf-vacuoles in its entirety can be described as a voluminous potassium-pool with high storage capacity for sodium and chloride. This pool is needed in case of high NaCl-salinity for the maintenance of the K-homeostasis in the cytoplasm. The dominant elements in the cytoplasm were P and K. The K-concentrations were in the epidermal cytoplasm of control plants in an ideal range for enzymatic reactions (Wyn Jones et al. 1979; Wyn Jones and Pollard 1983; Koyro and Stelzer 1988).

It is obvious that seawater salinity leads to a decrease of P, S, Mg and K in the epidermal vacuoles of both species. The remaining K, S and Mg concentrations were only in *Spartina* two-digit and especially for K much higher as in *Beta*. The vacuolar buffer of the latter one seems to be exhausted.

NaCl salinity led to a significant decrease of the K and P concentrations especially in the cytoplasm of *Beta* and to a breakdown of the homeostasis. This result points at a deficiency for both elements in the cytoplasm. Additionally the concentrations of sodium and chlorine were at high NaCl-salinity below 5 mol*m^{-3} in the cytoplasm of the epidermal cytoplasm and the gradients between cytoplasm and vacuole were higher in comparison with the results of *Spartina* In summary these results support the hypothesis that the sea beet does not sustain ion-toxicity but ion-deficiency! It is hypothesized that such low K^+ levels in the cytoplasm can lead to a reduction of protein synthesis which is of utmost importance in the process of leaf expansion (Mengel and Kirkby 2001). One possible consequence is the supply of sufficient fertilizers (especially K and P) at high NaCl-salinity to reduce the symptoms of K- and P-deficiency in Beta.

The salt induced reductions of the cytoplasmic K and P concentrations were much less pronounced in Spartina as in Beta. The results of Spartina point at a working system to keep ionic homeostasis. However there was one important exception: The sodium concentration increased significantly in the epidermal cytoplasm. Sodium could (try to) substitute potassium in its cytoplasmic functions or it could be the first sign of an intoxication.

The results and interpretations are in agreement with the hypothesis that plant growth is affected by ion imbalance and toxicity and probably leads to the long-term growth differences between the salt-tolerant and sensitive species.

Table 17.1 Chlorine- phosphor-, sulphur-, sodium- magnesium-, potassium- and calcium-concentrations in mol*m^{-3} (measured with EDX-analysis in bulk frozen tissues) in the vacuoles and in the cytoplasm of adaxial epidermis cells of *Beta vulgaris* ssp. *maritima* and *Spartina townsendii*

	Vacuole				Cytoplasm			
Adaxial leafepidermis	Control		480 NaCl		Control		480 NaCl	
Beta vulgaris ssp. *maritima*								
Cl	22.6	±4.7	654.3	±54.8	0.0		<5	
P	22.2	±4.6	6.4	±4.2	58.2	±8.4	28.1	±6.7
S	40.4	±7.9	2.7	±1.9	10.5	±4.0	<5	
Na	12.2	±3.1	724.9	±65.1	<5		<5	
Mg	24.0	±1.2	4.3	±1.7	<5		<5	
K	282.5	±4.7	9.3	±6.1	88.9	±9.5	66.7	±7.2
Ca	<5		<5		<5		<5	
Spartina townsendii								
Cl	21.2	±3.2	324.3	±64.8	<5		<5	
P	11.1	±1.1	5.3	±2.1	81.5	±6.8	71.6	±10.8
S	24.4	±6.9	20.8	±2.6	8.8	±3.0	<5	
Na	16.4	±3.1	521.0	±54.9	<5		15.3	±3.2
Mg	29.6	±2.1	18.8	±1.6	<5		<5	
K	212.5	±34.8	71.5	±6.9	92.7	±12.7	78.4	±6.9
Ca	<5		<5	±2.06	<5		<5	

However, Beta and Spartina are also two excellent examples how important it can be to validate intracellular ionic imbalances (K^+, Ca^{2+} and Mg^{2+}) at high salt concentrations to uncover the individual mechanisms for salt tolerance and to understand the threshold levels of individual species.

The results presented in this paper contain a lot of information about the essential eco-physiological needs of several halophytes at high salinity. The very variable screening of individual species enables to study the characteristic combination of mechanisms against salt injury and the threshold of salinity tolerance. The so-called QCS can be modified to the special characteristics and needs of other species and is therefore useful to study a wide range of suitable halophytes. This screening procedure is a practical first step on the selection of economically important cash crop halophytes.

For future studies on utilisation potentials of halophytes precise data about the ecological demands of halophtic species are required. Comparative physiological studies about salinity tolerance are essential. A precondition for this demand is a precise specification of a comparative value for halophytic species as shown in this paper. The literature has to be screened prior to the selection of priority species (potentially useful species) in order to get first order information about their natural occurrence in dry or saline habitats, existing utilisation (because of their structure, chemical content or other useful properties), natural climatic and substrate conditions, water requirement and salinity tolerance. Soon after the selection of a priority species, the threshold of salinity should be determined according to Kinzel (1982) and Munns (2002). The characteristic major plant response has to be evaluated for precise informations of ecophysiological demands. The data can build up a well-founded basis for the improvement of the utilisation potential. Additionally, research about the genetic composition of chromosomes mastering saline environment is also needed and bases on quantitative precise determination (Winicov and Bastola 1997, 1999; Winicov 1998).

7 Development of Cash Crop Halophytes

The physiological studies with the sea water irrigation system have the potential to provide highly valuable means of detecting individual mechanisms of species against NaCl stress, and may also provide opportunities for the comparison and screening of different varieties for their adaptation to salinity (QCS for cash crop halophytes). After the selection of halophytic species suited for a particular climate and for a particular utilisation green house experiments at the local substrates (and climatic conditions) to select and propagate promising sites (Isla et al. 1997) have to be started. This must be followed by studies with Lysimeters on field site to study the water consumption and ion movements. Last not least a design for a sustainable production system in plantations at coastal areas or at inland sites (for example for economical use) have to be developed.

References

Amzallag GN (1994) Influence of parental NaCl treatment on salinity tolerance of offspring in *Sorghum bicolor* (L) *Moench*. New Phytol 128: 715–723.

Ashraf M (1994) Breeding for salinity tolerance in plants Crit Rev Plant Sci 13: 17–42.

Ashraf M, O'Leary JW (1996) Effect of drought stress on growth, water relations, and gas exchange of two lines of sunflower differing in degree of salt tolerance. Int J Plant Sci 157: 729–732.

Azaizeh H, Steudle E (1991) Effects of salinity on water transport of excised maize (Zea mays L.) roots Plant Physio 97: 1136–1145.

Boer B, Gliddon D (1998) Mapping of coastal ecosystems and halophytes (case study of Abu Dhabi, United Arab Emirates). Mar Freshwater Res 49: 297–301.

Epstein E (1980) Responses of plants to saline environments. In: Rains DW, Valentine RC, Hollaender A (eds) Genetic Engineering of Osmoregulation.Plenum, New York, pp. 7–21. darter.ocps.

Flowers TJ, Troke PF, Yeo AR (1977) The mechanisms of salt tolerance in halophytes. Annu Rev Plant Physiol 28: 89–121.

Freitas H, Breckle SW (1992) Importance of bladder hairs for salt tolerance of field-grown Atriplex-species from a Portuguese salt marsh. Flora 187: 283–297.

Freitas H, Breckle SW (1993a) Progressive cutinization in Atriplex bladder stalk cells. Flora 188: 287–290.

Freitas H, Breckle SW (1993b) Accumulaton of nitrate in bladder hairs of Atriplex species. Plant Physiol Biochem 31(6): 887–892.

Ghassemi F, Jakeman AJ, Nix HA (1995) Salinisation of land and water resources: Human causes, extent, management and case studies. USNW Press, Sydney, Australia.

Glaubrecht M (1999) Mangrove der tropische Gezeitenwälder. Naturw Rdsch 52: 264–271.

Greenway H, Munns R (1980) Mechanisms of salt tolerance in nonhalophytes. Annu Rev Plant Physiol 31: 149–190.

Hose E, Clarkson DT, Steudle E, Schreiber I, Hartung W (2001) The exodermis: A variable apoplastic barrier. J Exp Bot 52: 2245–2264.

Isla R, Royo A, Aragues R (1997) Field screening of barley cultivars to soil salinity using a sprinkler and a drip irrigation. Plant Soil 197: 105–117.

Jeschke WD, Klagges S, Hilpert A, Bhatti AS, Sarwar G (1995) Partitioning and fl ows of ions and nutrients in salt-treated plants of Leptochloa fusca L Kunth.1. Cations and chloride. New Phytol 130: 23–35.

Kinzel H (1982) Pflanzenökologie und Mineralstoffwechsel. Eugen Ulmer, Stuttgart.

Koyro HW (2000) Untersuchungen zur Anpassung der Wildrübe (*Beta vulgaris* ssp. *maritima*) an Trockenstreß oder NaCl-Salinität. Habilitation, Justus-Liebig-University Giessen, Germany.

Koyro HW (2002) Ultrastructural effects of salinity in higher plants. In: Läuchli A, Lüttge U (eds) Salinity: Environment – Plants – Molecules. Kluwer, The Netherlands, pp. 139–158.

Koyro HW, Huchzermeyer B (1997) The physiological response of *Beta vulgaris* ssp. *maritima* to sea water irrigation. In: Lieth H, Hamdy A, Koyro HW (eds) Water Management, Salinity and Pollution Control Towards Sustainable Irrigation in the Mediterranean Region. Salinity Problems and Halophyte Use. Tecnomack, Bari, Italy, pp. 29–50.

Koyro HW, Huchzermeyer B (1999) Infl uence of high NaCl salinity on growth, water and osmotic relations of the halophyte Beta vulgaris ssp. maritima. Development of a quick check In: Lieth H, Moschenko M, Lohmann M, Koyro HW, Hamdy A (eds) Progress in Biometeorology. Volume 13. Backhuys, Leiden, NL, pp. 87–101.

Koyro HW, Huchzermeyer B (2004) Ecophysiological needs of the potential biomass crop *Spartina townsendii* GROV. J Tropical Ecol 45(1):123–139 (in press).

Koyro HW, Stelzer R (1988) Ion concentrations in the cytoplasm and vacuoles of rhizodermal cells from NaCl treated *Sorghum*, *Spartina* and *Puccinellia* plants. J Plant Physiol 133: 441–446.

Koyro HW, Wegmann L, Lehmann H, Lieth H (1997) Physiological mechanisms and morphological adaptation of Laguncularia racemosa to high salinity. In: Lieth H, Hamdy A, Koyro HW (eds) Water Management, Salinity and Pollution Control Towards Sustainable Irrigation in the Mediterranean Region: Salinity Problems and Halophyte Use. Tecnomack, Bari, pp. 51–78. See also http://www.usf.uos.de/ hlieth/publications.html

Koyro HW, Wegmann L, Lehmann H, Lieth H (1999) Adaptation of the mangrove *Laguncularia racemosa* to high NaCl salinity. In: Lieth H, Moschenko M, Lohmann M, Koyro HW, Hamdy A (eds) Progress in Biometeorology. Volume 13. Backhuys, Leiden, pp. 41–62.

Läuchli A (1999) Potassium interactions in crop plants. In: Oosterhuis DM, Berkowitz GA (eds) Frontiers in Potassium Nutrition. New Perspectives on the Effects of Potassium on Physiology of Plants. Marcel Dekker, New York, pp. 71–76.

Lieth H (1999) Development of crops and other useful plants from halophytes. In: Lieth H, Moschenko M, Lohmann M, Koyro KW, Hamdy A (eds) Halophytes Uses in Different Climates, Ecological and Ecophysiological Studies. Backhuys, Leiden, pp. 1–18.

Lieth H, Moschenko M, Lohmann M, Koyro HW, Hamdy A (1999) Halophyte uses in different climates I. Ecological and ecophysiological studies. In: Progress in Biometeorology. Volume 13. Backhuys, Leiden, pp. 258.

Lieth U, Menzel U (1999) Halophyte Database Version 2. In: Lieth H, Moschenko M, Lohmann M, Koyro KW, Hamdy A (eds) Halophytes Uses in Different Climates, Ecological and Ecophysiological Studies. Backhuys, Leiden, pp. 159–258.

Marcum KB (1999) Salinity tolerance mechanisms of grasses in the subfamily *Chloridoideae*. Crop Sci 39: 1153–1160.

Marcum KB, Anderson SJ, Engelke MC (1998) Salt gland ion secretion: A salinity tolerance mechanism among five zoysiagrass species. Crop Sci 38: 806–810.

Marschner H (1995) Mineral nutrition of higher plants. Academic, London/New York/San Diego, CA/Boston, MA/ Sydney/Tokyo/Toronto, pp. 889.

Mengel K, Kirkby EA (2001) Principles of plant nutrition. Kluwer, Dordrecht, Boston, MA/London, p. 849.

Munns R (1993) Physiological processes limiting plant growth in saline soils: Some dogmas and hypotheses. Plant Cell Environ 16: 15–24.

Munns R (2002) Comparative physiology of salt and water stress. Plant Cell Environ 25: 239–250.

Munns R, Husain S, Rivelli AR, James RA, Condon AG, Lindsay MP, Lagudah ES, Schachtman DP, Hare RA (2002) Avenues for increasing salt tolerance of crops, and the role of physiologically based selection traits. Plant Soil 247: 93–105.

Pasternak D (1990) Fodder production with saline water. The Institute for Applied Research, Ben Gurion University of the Negev. Project report BGUN-ARI-35-90. Beer-Sheva/Israel, p. 173.

Robertson KP, Wainwright JJ (1987) Photosynthetic responses to salinity in two clones of Agrostis stolonifera. Plant Cell Environ 10: 45–52.

Schimper AFW (1891) Pflanzengeographie auf physiologischer Grundlage. Fischer Publisher, Jena.

Schroeder FG (1998) Lehrbuch der Pflanzengeographie. Quelle and Meyer, Wiesbaden.

Sutherland GK, Eastwood A (1916) The physiological anatomy of *Spartina townsendii*. Ann Bot 30: 333–351.

Szabolcs I (1994) Soils and salinisation. In: Pessarakli M (ed) Handbook of Plant and Crop Stress. Marcel Dekker, New York, pp. 3–11.

Tazuke A (1997) Growth of cucumber fruit as affected by the addition of NaCl to nutrient solution. J Jpn Soc Hortic Sci 66: 519–526.

Volkmar KM, Hu Y, Steppuhn H (1998) Physiological responses of plants to salinity: A review. Can J Plant Sci 78: 19–27.

Walsh GE (1974) Mangroves. A review. In: Reimold RJ, Queen WH (eds) Ecology of Halophytes. Academic, New York, London, pp. 51–174.

Warne TR, Hickok LG, Sams CE, Vogelien DL (1999) Sodium/potassium selectivity and pleiotropy in stl2, a highly salt-tolerant mutation of *Ceratopteris richardii*. Plant Cell Environ 22: 1027–1034.

Weber E, D'Antonio CM (1999) Germination and growth responses of hybridizing *Carpobrotus* species (*Aizoaceae*) from coastal California to soil salinity. Am J Bot 86: 1257–1263.

Winicov I (1998) New molecular approaches to improving salt tolerance in crop plants. Ann Bot 82: 703–710.

Winicov I, Bastola DR (1997) Salt tolerance in crop plants: New approaches through tissue culture and gene regulation. Acta Physiol Plant 19: 435–449.

Winicov I, Bastola DR (1999) Transgenic overexpression of the transcription factor Alfin1 enhances expression of the endogenous MsPRP2 gene in alfalfa and improves salinity tolerance of the plants. Plant Physiol 120: 473–480.

Winter U, Kirst GO, Grabowski V, Heinemann U, Plettner I, Wiese S (1999) Salinity tolerance in *Nitellopsis obtusa*. Aust J Bot 47: 337–346.

Wolf O, Munns R, Tonnet ML, Jeschke WD (1991) The role of the stem in the partitioning of Na+ and K+ in salt treated barley. J Exp Bot 42: 697–704.

Wyn Jones RG, Pollard A (1983) Proteins, enzymes and inorganic ions. In: Läuchli A, Bieleski RL (eds) Inorganic Plant Nutrition. Encyclopedia of Plant Physiology. Volume 15b, Springer, pp. 528–555.

Wyn Jones RG, Brady CJ, Speirs J (1979) Ionic and osmotic relations in plant cells. In: Laidman DL, Wyn Jones RG (eds) Recent Advances in the Biochemistry of Cereals. Academic, New York.

Yeo A (1998) Molecular biology of salt tolerance in the context of whole-plant physiology. J Exp Bot 49: 915–929.

Chapter 18
Adaptive Mechanisms of Halophytes in Desert Regions

D.J. Weber

Abstract Plants growing in desert regions have to face a number of environmental adversaries such as high temperature, soil salinity and water stress due to low precipitation. Halophytes are among the successful plants that grow in desert saline regions. Halophytes use many different strategies to survive under these conditions. Some halophytes seeds can germinate in the presence of high salinity. Seeds of other halophytes are kept dormant due to the high salinity, but germinate when the rains come and reduce the salinity on the seeds. Persistent seed banks can be a source for new halophyte seedlings which permits seeds to germinate over different time periods when conditions are more favorable. Some root systems of halophytes can exclude salts. Other halophytes accumulate NaCl or synthesis osmotically compatible solutions such as proline, glycine betaine in their shoots to increase their ability to absorb water. Secreting salt from salt glands can reduce the salt level in certain halophytes. In other cases, the salt is compartmentalized in certain tissues of halophytes, which act as salt storage away from growing cells. Generally ion pumps prevent the salt from concentrated in cells that photosynthesize. Often halophytes develop succulence, which dilute the level of salt in the plant and stores water for use during dry periods.

Keywords Halophytes • adaptive mechanisms • desert regions • strategies • salinity • drought

D.J. Weber
Department of Plant and Wildlife Sciences, Brigham Young University, Provo, Utah 84602

1 Introduction

A number of environmental adversaries such as shortage of water, high temperatures, and high saline soils are some saline features in desert regions. In addition, rainfall occurs at irregular intervals, which makes the desert regions a very challenging environment for plants (Khan and Weber 2006). Over a long period of time, the plants that have adapted to the dry saline conditions of desert regions have been halophytes. Halophytes use many different strategies to survive under these conditions (Breckle 1983).

2 Strategies Relating to Halophyte Seed Germination

2.1 Seeds of Halophytes Can Germinate in the Presence of High Salinity and a Range of Temperatures

Seed germination in glycophytes is inhibited by an increase in salinity. Whereas halophytes seeds can germinate in very high salinity. Seeds of *Kochia scoparia* can germinate in high salinity levels and are an example of adaptation of halophytes to high saline soils (Al-Ahmadi and Kafi 2007). Al-Ahmadi and Kafi (2006) found that seeds of *Kochia scoparia* could germinate over a wide range of temperatures (8–40°C). *Salicornia rubra* is one of the most salt tolerant species in the western half of the United States. The seeds of *S. rubra* can germinate at very high salinity levels (Khan et al. 2000). Because glycophytes seeds are inhibited by high salinity whereas halophytes can germinate in the

M. Ashraf et al. (eds.), *Salinity and Water Stress*,

present of high salinity, the halophytes are the plants that become established in the high saline areas.

2.2 Seeds of Halophytes Are Kept Dormant Due to the High Salinity of the Soil. Then, When the Rains Come and Reduce the Salinity in the Seeds, the Seeds Germinate

When the rains come and addition of the moisture reduces the salinity on the seeds, the additional moisture also provides soil moisture for the development of the seedling after seed germination.

Seeds of *Chenopodium glaucum* were inhibited from germinating as salinity increased but the seeds germinated readily when the salinity was removed (Duan et al. 2004). Seeds of *Desmostachya bipinnata*, a perennial grass of near–coastal and inland deserts could germinate in low salinity soils but at high salinity the seeds were inhibited. When the seeds were transferred to low salinity, they germinated rapidly (Gulzar et al. 2007). Song et al. (2005) found that seeds of *Suaeda physophora* and *Haloxylon ammodendron* had a higher concentration of Na^+ in the seed coats as compared to a xerophyte. The higher Na^+ concentration inhibited seed germination.

There are many different interactions of plant growth regulators in halophytes. There is evidence that growth regulators are a factor in the saline dormancy of halophyte seeds. The addition of growth regulators broke the saline dormancy of *Allenrolfea occidentalis* seeds (Gul and Weber 1998; Gul et al. 2000). Even though halophytes take up considerable amount of Na^+ to their plant body, their seeds do not have a high salt content (Weber et al. 2007).

2.3 Some Halophytes Have Two Types of Seeds, One Seed That Germinates Quickly and Another Seed That Is More Resistant to Harsh Conditions and Germinates After a Longer Period of Time

The halophyte, *Suaeda salsa*, produces dimorphic seeds. The soft brown seeds absorb water more quickly and have a higher germination rate than the hard black seeds. The brown seeds can germinate earlier than the black seeds (Li et al. 2005). Dimorphic seeds were also found in *Atriplex rosea* (Khan et al. 2004). The brown seeds germinate in the early part of the growing season, whereas the black seeds are capable of surviving harsher conditions and then germinate at later time periods. Characteristics of the dimorphic seeds increase chances for survival in the harsh saline desert environment (Khan et al. 2004).

2.4 Seed Banks and Plant Runners Are Sources for New Halophyte Plants

Production of new halophytes plants can occur by runners from the mother plant or by the germination of seeds. A seed bank is represented by the number of seeds that are in the soil near the adult plant. The seeds in a seed bank increase the chance of new plants because the seeds germinate over a longer period of time. Inland saline areas normally have large seed banks. But the significance of the seed banks in producing new halophytic plants may not be as significant to inland saline areas because the plants tend to be perennials and often reproduce by roots and runners (Ungar 1995).

3 Strategies for Growth and Development of Halophytes in Desert Regions

3.1 Promotion of Root Development by Salinity and Salt Exclusion

Large root development is important in survival of plants in desert regions. The root mass may represent 75% of the plant mass in desert regions. The root morphology and development was promoted for *Suaeda physophoira*, *Suaeda nitratria* and *Haloxylon ammendendron* at certain salt concentrations (Yi et al. 2007). Large root development is inhibited in glycophytes by high salinity. In monocot halophytes, salt exclusion can be a factor in tolerating saline soils but osmotic compounds need to be produced to pull water into the plant (Weber 1995). Tester and Davenport (2003) found a number of halophytes that selectively excluded toxic ions from the roots.

3.2 *Halophytes in Desert Regions Accumulate NaCl to Increase Their Ability to Absorb Water into the Plant*

The absorption of water from saline soils is critical for plants to survive in desert environments. Halophytes from an inland saline lake were found to accumulate Na^+ and glycine betaine as their major response to increased salinity (Tipirdamaz et al. 2006). Xi et al. (2004) found that most halophytes take up more Na^+ from the soil and hence the Na^+/K^+ ratio in the rhizosphere tends to decrease. A true halophyte needs saline water to develop whereas pure water can often inhibit growth.

Song et al. (2006) found that in succulent halophytes the accumulation of nitrate as well as cation and anions in the vacuole lowered their osmotic potential to a degree that uptake of water from saline soils was accomplished.

Aster tripolium and *Sesuvium portulacastrum* used different strategies to survive saline habitats (Balasuban et al. 2006). In *A. tripolium* more K^+ was accumulated in comparison to Na^+ as the salinity of the soil increased. In contrast in *S. portulacastrum* more Na^+ was accumulated in comparison to K^+. The non-photochemical quenching (NPQ) increased in both plants as the concentration of NaCl increased, but the NPQ for *S. portulacastrum* was almost half the NPQ value for *A. tripolium* (Balasuban et al. 2006). In the halophyte, *Cynara cardunculus*, high concentrations of KCl were more deleterious than high concentrations of NaCl (Benlloch-Gonzalez et al. 2005).

3.3 *Many Halophytes Synthesize Osmotically Compatible Solutions Such as Glycine Betaine, Proline, Sugars and Polyols to Increase the Water Absorption*

The production of osmotically compatible solutions result in the ability of the halophyte to absorb water from the saline soil. In two halophytes, *Carex paleacea* and *Scirpus americanus*, the concentration of glycine betaine increased three fold and in *Puccinellia phryganodes* the proline concentration was increased five times (Ewing et al. 1989).

Monosaccharides and disaccharides play an important role in osmotic adjustment in plants subjected to salinity stress (Gorham et al. 1981). Ashraf et al. (2006) observed mono and disaccharides played a role in water absorption in forage grasses in the Cholistan Desert.

4 Strategies to Maintain Status and Metabolism of the Mature Plant

4.1 *Secreting Salt from Leaves to Reduce Salt Level in the Plant Body*

In certain cases, halophytic plants excrete the salts through salt glands or by concentrating the salts in certain cells of the plant. *Limonium* and *Distichlis* are two examples of halophytes that have salt secreting glands (Weber 1995). In *Aeluroopus littoralis*, a salt secreting halophyte, the main ions secreted from the salt glands were Na^+ and Cl^- (Barhoumi et al. 2006). In *Reaumuria hirtella*, six ions were excreted with Na^+ and Cl^- being the predominate ions and represented 86% of the excreted ions. Of the ions absorbed from the soil, 67% were excreted from the plant (Ramadan 1998).

4.2 *In Some Halophytes, Salt Is Concentrated in Certain Plant Regions. After High Concentrations of Salt Accumulate, the Tissue Dies but Water Conducting Cells Still Function for Moving Water up to the Living Plant Tissue*

In *Salicornia pacifica var utahensis*, salt ions move up into the shoot and accumulate in the cortex region, After a period of time the salt concentration becomes too high and the lower segments of the shoot die but the vascular tissue in the center of the shoot continues to function. The water continues to move up to the upper living segments of the shoot. The dead segments act as a region of deposit for the excess salts (Weber et al. 1977). *Diplachne fusca* has the ability to sequester high levels of Na^+ and Cl^- in the sheath away from the leaf blades as well as maintaining a high selectivity for K^+ over Na^+ (Warwick and Halloran 1992).

4.3 Compartmentalization of the Ions Within Tissues and Cells

Energy Dispersive X-Ray Microanalysis of *Salicornia pacifica var utahensis* indicated a higher concentration of Na^+ in the cortex of the shoot as compared to the palisade regions (Weber et al. 1977). Zhoa et al. (2003) found that *Suaeda salsa* compartmentalized toxic ions into vacuoles, which permitted the halophyte to grow in high saline soils. The distribution of chloride ions in cells of *Salicornia pacifica var utahensis* was determined by silver precipitation in thin sections. The chloride ion was low in organelles such as chloroplasts but present in the cytoplasm and the vacuole of palisade and cortex cells (Hess et al. 1975). The low chloride ion concentration in the chloroplasts would indicate that the cell membranes were able to keep the ion concentrations low so that photosynthesis could continue to function.

4.4 Effect of Ion Content in Cells of Halophytes

A comparison of the uptake of Na^+ by six inland halophytes indicated no toxicity symptoms up to 6000 mg l^{-1}. *Suaeda nudiflora* had the highest Na^+ accumulation. (Shekhawat et al. 2006). Drenovsky and Richards (2003) suggested that the ability of *Sarcobatus vermiculatus* to attain high leaf nitrogen rather than an interaction between Na^+ and N^- enhances it performance at saline sites. In addition, the ability of *Sarcobatus* to maintain high macronutrients cation selectivity despite high salinity allows it to grow on saline and alkaline soils.

4.5 Increase in Sulfolipid Components of Membranes

Membranes are important components in regulating movement of ions into and out of cells. The sulfolipid component is an important component of membranes. An increase in sulfolipid content occurred when salt stress increased in the halophytes, *Aster tripolium* and *Sesuvium portulacastrum* as compared to a glycophytes (Ramani et al. 2004). When *Aster tripolium* was watered with NaCl solutions the sulfolipid contents were enhanced. The enzymatic formation of L-cystine and beta cyanoalanine were drastically reduced in Aster leaves and roots (Balasubramanian et al. 2004). Mansour and Salama (2004) suggested that cell membrane permeability and cytoplasmic viscosity were key factors in salt tolerance in halophytes. They indicated that plasma membrane integrity under salt stress was higher in halophytes than in glycophytes. The ATPase in *Salicornia pacifica var utahensis* was salt tolerant and the enzyme could function in the presence of 3 M NaCl (Weber et al. 1980). Malic enzyme and F-type ATPase from halophytes and glycophytes were identical in their amino acid composition (Huchzermeyer et al. 2004). They suggested that salt tolerance in halophytes was due to a faster and more effective response of the F-type ATPase activity at the plasma lemma, which bring about ion homeostasis by adaptation of transmembrane transport to saline conditions and coordination of metabolism (Huchzermeyer et al. 2004). Jithesh et al. (2006) investigated the antioxidative response mechanisms in halophytes. They concluded that antioxidant enzymes protect halophytes from deleterious reactive oxygen species within the cellular compartments of the plant cell.

4.6 Avoiding High Salinity in Relation to Photosynthesis in Halophytes

The cells that photosynthesize are protected from the entry of salts by ion pumps. Analysis of the ion content in cells of *Salicornia pacifica var utahensis* indicated that ion pumps prevented the accumulation of salinity in the palisade cells. (Weber et al. 1977). An increase in salinity in *Atriplex portulacoides* did not have an adverse effect on photosystem II (Redondo-Gomez et al. 2007). The decrease in growth response appears to depend on the changes in its photosynthetic gas exchange mainly through stomatal conductance. One way to decrease water loss is through the closing of the stomata in the leaves of halophytes. Perera et al. (1994) found that when the NaCl concentration increased in the leaf cells of *Aster tripolium*, there was a suppression of the stomatal opening. He suggested that the accumulation in the cell vacuoles caused the Na^+ ions to accumulate in the apoplast around the guard cells which caused a partial closure of the stomata opening.

4.7 Halophytes That Develop Succulence Can Dilute the Level of Salt in the Plant and Stores Water for Use During Dry Periods

Succulence is another strategy of halophytic plants to maintain moisture under drought conditions. *Haloxylon ammodendron* and *Zygophyllum xanthoxylum* are succulent xerophytes that absorb large quantities of Na^+ for osmotic adjustment. The Na^+ was absorbed by the roots and then transported to the succulent leaves, which resulted in salt dilution (Wang et al. 2004). During the summer, the desert is hot and the protection of water loss is related to the surface wax of the halophyte. In *Allenrolfea occidentalis*, a glabrous succulent bush, Hess and Weber (1995) found the surface wax had hydrocarbons from C_{22} to C_{38}. These large hydrocarbons were stable in the summer heat. The hydrocarbons chain length was longer than the chain length of the surface hydrocarbons of glycophytes.

4.8 Potential of Characterizing the Genes Responsible for Salt Tolerance in Halophytes

In theory the genetic characteristics that adapt halophytes to arid saline environments have great potential if they can to transferred to an economical plant. Chen et al. (2007) have determined 1,082 expressed sequence tags in a halophyte of which they feel that a number of the sequence tags are specific to halophytes.

4.9 Certain Proteins Are Associated with Salt Tolerance of Halophytes

The bases for salt tolerance are genetic and proteins are a reflection of the genetic information. Proteins of *Suaeda aegyptiaca* were analyzed by proteomics in relation to increased salt concentration (Askari et al. 2006). Out of 700 protein spots 102 spots increased in response to increased salinity. Some of the proteins were determined to be glycine betaine synthesis, cytoskeleton remodeling, photosynthesis ATP production, protein degradation, cyanide detoxification and chaperone activities. Kant et al. (2006) suggested that salt-sensitive glycophytes and salt-tolerant halophytes employ common mechanism to cope with salinity, and that the differences in salt tolerance arise because of changes in the regulation of a basic set of salt tolerant genes. The regulation of proline accumulation and catabolism and the tight control of Na^+ uptake in *T. halophila* appeared to be the key factors.

5 Genetic Bases for Salt Tolerant in Halophytes

The different adaptations that are present in halophytes have a genetic basis (Table 18.1). While natural selection is the process for selecting the characteristics that have permitted the halophytes to survive under the saline

Table 18.1 Summary of adaptive strategies of halophytes at different growth stages

Seed stage
1. Seeds germinate in high salinity
2. Seeds are dormant with salt, but after rains, the seeds germinate
3. Seed banks permit seed germination over longer period of time
Root stage
1. Root development promoted by salinity
2. Roots exclude salts
Shoot stage
1. Shoots accumulate salts in their shoots to increase water absorption
2. Shoots synthesize osmotically compatible solutions to increase water uptake
3. Shoots compartmentalize ions within tissues and cells to increase water uptake
4. Shoots regulate salt movement (ion pumps) to keep concentration low for cells that photosynthesis
Mature plant
1. Secreting salts from salt glands to reduce salt level in the plant
2. Development of succulence to dilute the level of salt and to store water

desert conditions. A lot of research is needed before the specific genetic traits can be identified. Flowers and Flowers (2005) indicated that high salt tolerance is a multigenic trait. The adaptive mechanisms reviewed here does represent considerable success in halophytes adapting to the environmental adversaries of saline deserts. Flowers and Flowers (2005) suggested that domestication of halophytes appears to be the best alternative to producing economical plants that could grow in the saline deserts.

References

Al-Ahmadi MJ, Kafi M (2006) Salinity effects on germination properties of *Kochia scoplaria*. Asian J Plant Sci 5: 71–75

Al-Ahmadi MJ, Kafi M (2007) Salinity effects and cardinal temperatures on germination properties of *Kochi scoparia* L. J Arid Environ 68: 308–314

Askari H, Edqvist J, Hajheidari M, Kafi M, Salekdeh GH (2006) Effects of salinity levels on proteome of *Suaeda aegyptiaca* leaves. Proteomics 6: 2542–2554

Ashraf M, Hameed M, Arshad M, Ashraf Y, Akhtar K (2006) Salt tolerance of some potential forage grasses from Cholistan desert of Pakistan. In: Khan MA Weber DJ (eds) Ecophysiology of high salinity tolerant plants. Springer, Dordrecht, The Netherlands, pp. 31–54.

Balasuban R, Reeck T, Debez A, Stelzer R, Huchzermeyer B, Papenbrock J (2006) *Aster tripolium* L. and *Sesuvium protulacastrum* L.: two halophytes, two strategies to survive in saline habitats. Plant Physiol Biochem (Paris) 44: 395–408.

Balasubramanian F, Papenbrock J, Schmidt A (2004) Connecting sulfur metabolism and salt tolerance mechanism in the halophytes *Aster tripolium* and *Sesuvium portulacastrum*. Tropical Ecol 45: 173–182.

Barhoumi Z, Djebali W, Smaoui A, Chaibi W, Abdelly C (2006) Contribution of NaCl excretion to salt resistance of *Aeluropus littorallis* (Willd) Parl. J Plant Physiol 164: 842–850.

Benlloch-Gonzalez M, Fournier JM, Ramos J, Benlloch M (2005) Strategies underying salt tolerance in halophytes are present in *Cynara cardunculus*. Plant Sci (Oxford) 168: 653–659.

Breckle SW (1983) Temperate deserts and semi-deserts of Afghanistan and Iran. In: Goodall EW, West N (eds) Ecosystems of the World. pp 271–319. Elsevier, Amsterdam, The Netherlands.

Chen S, Guo SL, Wang ZL, Zhao JQ, Zhao YX, Zhang H (2007) Experessed sequence tags from the halophyte *Limonium sinense*. DNA Sequence 18: 61–67.

Drenovsky RE, Richards JH (2003) High nitrogen availability does not improve salinity tolerance in *Sarcobatus vermiculatus*. Western N Am Naturalist 63: 472–478.

Duan DY, Liu XJ, Khan MA, Gul B (2004) Effects of salt and water stress on the germination of *Chenopodium glaucum* L. seed. Pak J Bot 36: 793–800.

Ewing K, Earle JC, Piccinin B, Kershaw KA (1989) Vegetation patterns in James Bay costal marshes II Physiological adaptation to salt-induced water stress in three halophytic graminoids. Can J Bot 67: 521–528.

Flowers TJ, Flowers SA (2005) Why does salinity pose such a difficult problem for plant breeders. Agr Water Manage 78: 15–24.

Gorham J, Hughes I, Wyn-Jones RG (1981) Low molecular weight carbohydrates in some salt stressed plants. Physiol Plant 31: 149–190.

Gul B, Weber DJ (1998) Role of dormancy relieving compounds and salinity on the seed germination of *Allenrolfea occidentalis*. Ann Bot 82: 555–562.

Gul B, Khan MA, Weber DJ (2000) Alleviation salinity and darkness enforced dormancy in *Allenrolfea occidentalis* seeds under various thermoperiods. Aust J Bot 48: 745–752.

Gulzar S, Khan MA, Liu X (2007) Seed germination strategies of *Desmostachya bipinnata*: a fodder crop for saline soils. Range Ecol Manage 60: 401–407.

Hess WM, Weber DJ (1995) Morphology of epicuticular wax and comparison of lipids of *Allenrolfea occidentalis*. In: Khan MA, Ungar IA (eds) Biology of Salt Tolerant Plants. Dec. 12–16, 1994. pp. 107–117. BookCrafters, Department of Botany, University of Karachi, Karachi, Pakistan.

Hess WM, Hansen DJ, Weber DJ (1975) Light and electron microscopy localization of chloride ions in cells of *Salicornia pacifica* var. *utahensis*. Can J Bot 53: 1176–1187.

Huchzermeyer B, Hausmann N, Paquet-Durant F, Koyro HW (2004) Biochemical and physiological mechanisms leading to salt tolerance. Tropical Ecol 45: 141–150.

Khan MA, Gul B, Weber DJ (2000) Germination responses of *Salicornia rubra* to temperature and salinity. J Arid Environ 45: 207–214.

Khan MA, Gul B, Weber DJ (2004) Temperature and high salinity effects in germinating dimorphic seeds of *Atriplex rosea*. Western N Am Naturalist 64: 193–201.

Jithesh MN, Prashanth SR, Sivaprakash KR, Parida AK (2006) Antioxidative response mechanisms in halophytes: their role in stress defence. J Genet 85: 237–254.

Kant S, Kant P, Raveh E, Barak S (2006) Evidence that differential gene expression between the halophyte, *Thellungiella halophila* and *Arabidopsis thaliana* is responsible for higher levels of the compatible osmolyte proline and tight control of Na^+ uptake in *T. halophila*. Plant Cell Environ 29: 1200–1234.

Khan MA, Weber DJ (2006) Ecophysiology of High Saline Tolerant Plants. Tasks for Vegetation Science, vol 40. 399 pp. Springer, Dordrecht, The Netherlands.

Khan MA, Bilquees Gul and Darrell J Weber (2004) Temperature and high salinity effects in germinating dimorphic seeds of *Atriplex rosea*. Western N Am Naturalist 64:193–201

Li W, Liu X, Khan MA, Yamaguchi S (2005) The effect of plant growth regulators, nitric oxide, nitrate, nitrite and light on the germination of dimorphic seeds of *Suaeda salsa* under saline conditions. J Plant Res 118: 207–214.

Mansour MM, Salama K (2004) Cellular basis of salinity tolerance in plants. Environ Exp Bot 52: 113–122.

Perera LK, Mansfield TA, Malloch JC (1994) Stomatal responses to sodium ions in *Aster tripolium*: a new hypothesis to explain regulation in above ground tissues. Plant Cell Environ 17: 335–340.

Ramadan BT (1998) Ecophysiology of salt excretion in the xero-halophyte, *Reaumuria hirtella*. New Phytol 139: 273–281.

Ramani B, Zorn H, Papenbrock J (2004) Quantification and fatty acid profiles of sulfolipids in two halophytes and a glycophyte grown under different salt concentrations. Zeitschrift fur Naturforschung 59: 835–842.

Redondo-Gomez S, Mateos-Naranjo E, Davy AJ, Fernandes-Munozn F, Castellance EM, Luque T, Figueroa ME (2007) Growth and photosynthetic responses to salinity of the salt-marsh shrub *Atriplex portulacoides*. Ann Bot 100: 555–563.

Shekhawat VPS, Kumar A, Neumann KH (2006) Effect of sodium chloride salinity on growth and ion accumulation in some halophytes. Comm Soil Sci Plant Anal 37: 1933–1946.

Song J, Feng G, Tian C, Zhang F (2005) Strategies for adaptation of *Suaeda physophora, Haloxylon ammodendron* and *Haloxylon persicum* to a saline environment during seed germination. Ann Bot (London) 96: 399–405.

Song J, Ding XD, Feng G, Zhang F (2006) Nutritional and osmotic roles of nitrate in a euhalophyte and xerophyte in saline conditions. New Phytol 171: 357–366.

Tester M, Davenport R (2003) Na^+ tolerance and Na^+ transport in higher plants. Ann Bot 91: 503–527.

Tipirdamaz R, Gagneul D, Duhaze C, Ainouche A, Monnier C, Ozkum D, Larher F (2006) Clustering of halophytes from an inland salt marsh in Turkey according to their ability to accumulate sodium and nitrogenous osmolytes. Environ Exp Bot 57: 139–153.

Ungar IA (1995) Seed bank ecology of halophytes. In: M. A. Khan MA, Ungar IA (eds). Biology of Salt Tolerant Plants. Dec. 12–16,1994. Pp. 65–82. BookCrafters, Department of Botany, University of Karachi, Karachi, Pakistan.

Wang SM, Wan CG, Wang YR, Chen H, Zhou ZY, Fu H, Sosebee RE (2004) The characteristics of Na^+, K^+, and free proline distribution in several drought-resistant plants of the Alxa Desert, China. J Arid Environ 56: 525–539.

Warwick NWM, Halloran GM (1992) Accumulations and excretion of sodium, potassium and chloride from leaves of two accessions of *Diplachne fusca* (L) Beauv. New Phytol 121: 53–61.

Weber DJ (1995) Mechanisms and reactions of halophytes to water and salt stress. In: Khan MA, Ungar IA (eds) Biology of Salt Tolerant Plants. pp. 170–180. BookCrafter, Department of Botany, University of Karachi, Karachi, Pakistan.

Weber DJ, Rasmussen HP, Hess WM (1977) Electron microprobe analysis of salt distribution in the halophyte *Salicornia pacifica var utahensis*. Can J Bot 55: 1516–1523.

Weber DJ, Hess WM, Kim CK (1980) Distribution of ATPase in cells of *Salicornia pacifica* var. *utahensis* as determined by lead phosphate precipitation and x-ray microanalysis. New Phytol 84: 285–291.

Weber DJ, Ansari R, Gul B, Khan MA (2007) Potential of halophytes as source of edible oil. J Arid Environ 68: 315–321.

Xi J, Zhang F, Chen Y, Mao D, Yin C, Tian C (2004) A preliminary study on salt contents of soil in root-canopy area of halophytes. Yingyong Shengtai Xuebao 15: 53–58.

Yi L, Ma J, Li Y (2007) The comparisons of root systems and root hair morphological characteristics among three desert halophytes. Bull Bot Res 27: 204–211.

Zhoa KF, Fan H, Zhou S, Song JN (2003) Study on the salt and drought tolerance of *Suaeda salsa* and *Kalanchoe daigremonticana* under iso-osmotic salt and water stress. Plant Sci 165: 837–844.

Chapter 19
Is Sustainable Agriculture with Seawater Irrigation Realistic?

S.-W. Breckle

Abstract Since 1966 the use of seawater for agriculture was often studied. Despite intensive research and projects, only few organisms have been found, which can be grown with seawater: some mangrove trees and shrimps. Even today there is still no considerable use of seawater irrigation. Some halophytic vascular plants, however, can fulfil their whole lifecycle with seawater. But they also grow better on half seawater concentration. In many thousands (!) of other projects (with many cash crops) the use of only 10–20% seawater concentration has been tried. But even this concentration is often too high and spoils the soils in their structure, especially if not an efficient leaching is applied. A sustainable agriculture based on irrigation with seawater on a large scale seems to be still an utopic illusion. For special cases certainly a small scale seawater irrigation on anyhow saline coastal areas may be in fact very advisable and even economic, e.g. for production of secondary compounds, for producing fiber material, for horticultural purposes and especially for phytoreclamation of sometimes large areas of salt- and sand-deserts of desiccated seafloors (e.g. Aral Sea) etc. For saline and alkaline degraded lands only real Eu-Halophytes and Recreto-Halophytes can be used for phytomelioration. For their propagation it needs special techniques. And it needs special techniques for planting seedlings and saplings depending on site conditions. There are many applications but very few for food production. Under an arid climate sustainable agriculture with high production of crops per surface area is always only achievable with nonsaline conditions. On the long run it pays more to spend additional costs to maintain sustainable irrigation and leaching systems to keep salinity of soil low. The takehome message is: "No irrigation without drainage!" This also means it pays more to invest in good desalinization technology systems (inverse osmosis, energy sources from high radiation in deserts, photovoltaic devices etc.), to keep soils low in salt, since fresh water is always indispensable for human welfare. Basic facts and ecological principles on climate, aridity and salinity, on ecophysiological behaviour of plants to salinity and definition of halophyte-types, on salt balance in ecosystems, in soils and fields, on saline agriculture and crop yield as well as on sustainable agriculture with seawater at specific sites are discussed with the help of often-heard statements, relevant answers and take-home messages are supplied.

Keywords Birjand-Declaration • drainage • drylands • greening • deserts • halophytes • phytomelioration, salinity

1 Introduction

From the world's land surface area at least 6% are salt-affected lands. This amounts to about 9.5×10^6 km^2 according to UNEP-figures (Flowers and Yeo 1995) or 8.0 10^6 km^2 according to FAO-data (Munns 2005). Large areas in drylands and along coasts are naturally salt-affected, but by irrigation under arid climatic conditions secondary salinisation (Breckle 1989) is a big threat (Table 19.1).

In future, the growing world population needs more food, more agricultural areas are needed and/or an intensification of harvest yields.

S.-W. Breckle
Department of Ecology, Wasserfuhr 24-26, D-33619
Bielefeld, Germany
e-mail: sbreckle@gmx.de

M. Ashraf et al. (eds.), *Salinity and Water Stress*,

As sea-water is practically limitless a few investigators have claimed success using the highly saline water for irrigation of certain crops (Boyko 1966, 1968; Epstein and Norlyn, 1977; Epstein et al. 1980). Apparently, others have not been able to verify their findings.

Some halophytes, like *Sesuvium*, *Batis*, *Salicornia*, *Arthrocnemum*, *Halocnemum*, *Halostachys* and a few others can fulfil their whole lifecycle with seawater (Waisel 1972). But they also grow better on half seawater concentration. In many thousands (!) of other projects (with almost all known cash crops) the use of only 10–15% seawater concentration, rarely 20%, has been tested, it is still not economically feasible. But even this concentration of brackish water is often too high and spoils the soils in their structure, especially if not an efficient leaching is applied. This depends on water- and salt-balance of the relevant ecosystems resp. fields. And water- and salt-balance is also always dependant on climatic conditions.

Table 19.1 Secondary salinization of the world's irrigated lands

Country	Cultivated (Mha)	Irrigated (%)	Sec. salinized (%)
Argentina	35.8	4.8	33.7
Australia	47.1	3.9	8.7
China	97.0	46.2	15.0
Egypt	2.7	100	33.0
India	169	24.9	16.6
Iran	14.8	38.7	30
Pakistan	20.8	77.5	26.2
USA	190	9.5	23.0
World	1,473	15.4	20.0

Source: After Ghassemi et al. 1995

2 Salinity and Aridity

It is appropriate to point out some basic facts on water- and saltbalance in ecosystems and landscapes. Under an arid climate regime, most of the water brought in by rain, snow, dew, irrigation is lost by evaporation and transpiration (Fig. 19.1). A more or less small percentage only, depending on surface properties, may reach the scarce lower groundwater in soil. Thus, the predominant flow of capillary water is upwards. As a consequence the carried soluble salts are deposited after evaporation in the upper soil horizons or on top as a salt crust. On a larger scale the scarce run-off feeds inland lakes which in drylands are often endorrheic salt lakes with no outlet to the ocean. These are the regions where primary salinity in salt deserts developed (Chapman 1960; Reimold and Queen 1974; Breckle 2002a, b). The input of water by precipitation in humid areas is higher than the loss by evaporation. As a consequence water-budget is positive and a river-system develops with wells, creeks, streams, rivers which discharge to the ocean.

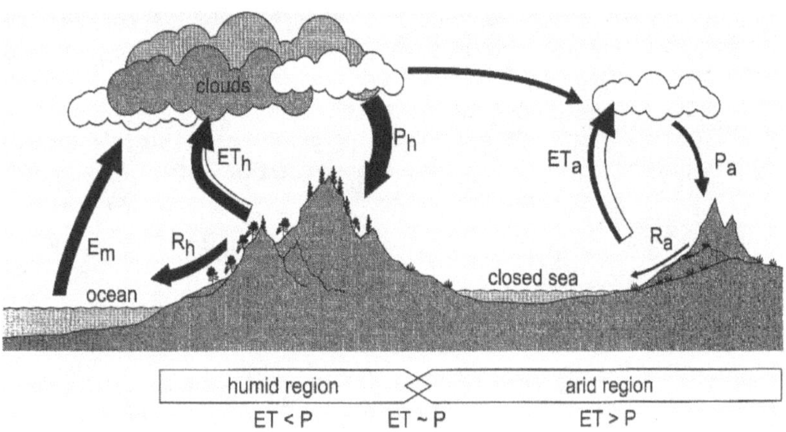

Fig. 19.1 Scheme of the hydrological cycle of the earth in humid and arid regions. A = arid, E = evaporation, ET = evapotranspiration, h = humid, m = marine, P = precipitation, R = surface and subsurface runoff. The hydrological cycle is coupled with the salt cycle! The closed area ("inland sea") as well as the ocean contain considerable amounts of salt

3 Halophyte-Types and Ecophysiological Behaviour of Plants to Salinity

There are not only two types of halophytes, thus, the general statement: "There are two types of halophytes: includers and excluders" is not sufficient. In the last 100 years dozens of schemes to define a halophyte and to define different types of halophytes were proposed and published (Waisel 1972; Albert 1982; Breckle 1990) There are excluders, the glykophytic grasses and sedges, and most of the non-halophytes. They do not accumulate NaCl. There are includers, like the halosucculent Chenopodiaceae, which accumulate NaCl (and other solutes, Na_2Ox, CaOx, comp. solutes etc.) in their tissues and become halo-succulent in leaves and/or stems. These are the eu-halophytes, which many of them can conclude their whole life-cycle with sea-water conditions. They have evolved for survival, not for maximal productivity. But there are others, like *Atriplex* or *Halimione* species, which again excrete NaCl by bladder. Excretion of salts can also be seen by many different types of salt glands, like in many grasses, as e.g. *Spartina*, or in dicots like *Glaux*, *Frankenia*, *Tamarix*, *Cressa*, *Limonium* etc. The halophytic scene is very complex (Reimann and Breckle 1993; Breckle 1995, 2000b). The various strategies even can often be active jointly in many species (Breckle 2000a, b). In fact, the salt factor is not only an ecological factor, active on the level of whole plants or in fields or in a salt desert vegetation plot, but also on higher and lower levels (Table 19.2) and exhibits specific interactions and effects which normally cannot extrapolated adequately between levels. On the level of cells, tissues and whole plants the osmotic adaptation plays a major role (Fig. 19.2). Here, halophytes have evolved various adaptive mechanisms by becoming halo-succulent, by accumulating inorganic salts and by synthesis of osmolytes, which are often taxon-specific compatible solutes. All this needs energy. These are tools for survival, not for high productivity.

4 Salt Balance in Soils, Fields and Ecosystems

Since all irrigation water contain salts to some extent (see Table 19.3) consecutive irrigation through many years or few decades leads inevitably to an increase of the salt load in the fields under dryland conditions. Often this salt load is fluctuating between soil horizons according to irrigation treatments. The concentration of soluble salts in such irrigated soils increases with water application and evapotranspiration rates, because the salt is left behind as most of the applied water is removed by evaporation and transpiration. Thus, salinity problems can develop over rather short time from use of saline water for irrigation without proper management. This can only be prevented by leaching with additional water, if a drainage system exists. The more arid the climate is, the more salts will accumulate and the more one needs drainage water for leaching. Table 19.3 demonstrates the water quality of some rivers and their

Table 19.2 The various levels of interaction of NaCl

Complexicity level of effect or interaction	Examples for reactions or effects of salt (NaCl)
Interactions in the biosphere, in biomes	Salt cycle, energy turnover (evaporation), weathering, erosion, sedimentation, salt dust storms
Interactions and influences in ecosystems	Salt cycle, transport, accumulation, species selection and composition, lifeforms
Influence on populations	Growth, reproduction, dispersal, distribution, age classes, competitive force, selection, lifeforms, ecotypes
Interactions or effects on intact, whole plants	Water balance and budget, (stomates: transpiration/ photosynthesis), growth, mineral metabolism, adaptations, modifications, mutations and selection
Effects, interactions with plant tissues	Formative effects, selective transport, adaptation and differentiation, storage and recreting tissues
Effects on cells, interactions with cells	Formative effects; defects, hormonal changes, ion balance, differentiation
Effects on cell-organelles	Respiration, photosynthesis, organell-synthesis, ion balance, sec. compounds, compatible solutes, oxidative stress, cell wall structure, cuticle
Effects on membranes	Permeability, water pores, ion channels, ion selectivity, electr. potential
Biochemical effects	Gen-regulation, mutations, gen- and enzyme-activities, DNA-synthesis

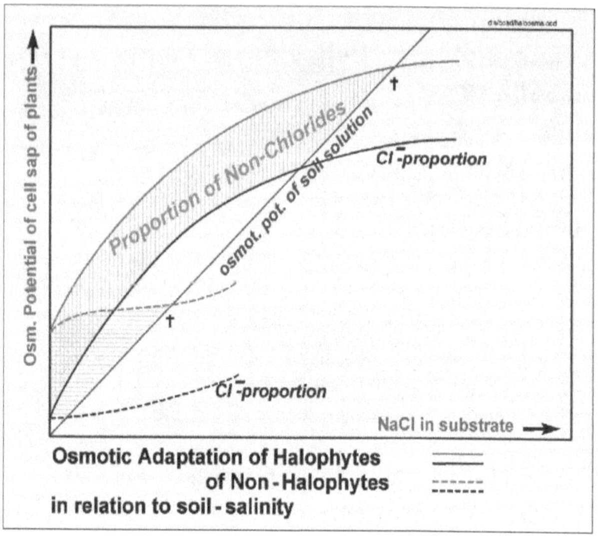

Fig. 19.2 Osmotic adaptation of halophytes and non-halophytes with increasing salinity. Non-halophytes only have a small proportion of inorganic salts in their tissues and rather high proportion of organic osmolytes, halophytes have about the same concentration of osmolytes, but often a very high salt content. Osmotic adaptation can only be maintained to a distinct concentration (marked with †)

Table 19.3 Percentage of salt in river water, used for irrigation and some water quality parameters (ionic concentration of major ions, TDS: total dissolved solids, EC: electric conductivity, SAR: sodium adsorption ratio)

Water source	Na (mmol/l)	Ca (mmol/l)	Mg (mmol/l)	Cl (mmol/l)	SO_4 (mmol/l)	TDS (g/l)	EC (dS/m)	SAR (mmolc/l)½
Arkansas River/USA	25.2	18.8	15.7	4.1	50.6	4.2	4.3	6.1
Medjerda River, Tunisia	34.0	12.8	8.8	19.6	21.2	–	5.3	10.0
Colorado River/USA	0.1	4.6	2.9	4.3	9.2	0.6	1.5	6.1
Tozeur/Tunisia (wells)	17.6	9.0	6.7	17.6	13.0	2.1	3.1	6.3
Zarzis/Tunisia (wells)	81.3	14.8	6.2	70.2	32.6	6.5	9.2	24.8

Source: FAO-data

ionic content. For comparison also the total dissolved solids (TDS) is shown, which for the Nile (in Cairo) is 0.2–0.5, for the Lower Euphrat (Iraq) TDS is 0.9, for the upper Euphrat 0.3, the Tigris near Bagdad has 0.33, whereas the Rhine or the Danube in climatically humid Germany exhibit only TDS values of about 0.17–0.2.

A simple model-calculation can demonstrate the accumulation of salt and thus salinisation in an irrigated field (Table 19.4). Nile-water formerly was used during the seasonal Nile-floods, which always meant also drainage of the fields. Modern irrigation from the Nile water reservoirs sometimes neglect drainage to save water. Thus, only after 30–50 years the salinity can be so high that agriculture has to cease. This has been often the detrimental case also in huge developmental projects of constructing water-reservoirs and cultivation of new irrigation farming land, but without drainage-systems, in USA, in Pakistan, in Australia.

Table 19.4 Model-calculation to demonstrate increasing salinity by irrigation without drainage in drylands

Checked or calculated parameter	Irrigation with Nile-water without leaching
Salinity of the Nile	400 mg/l (= 400 ppm)
Annual water use for irrigation	1,200 mm (= 1,200 l/m² soil surface)
Annual salt amount input	480 g/m²
Salt amount after 30 years	14 kg/m²
Weight of 1 m soil column	1,400 kg
Salinity percentage in soil after 45 years	1%

Source: Modified from Kreeb 1964

5 Crop Yield and Saline Agriculture

Despite research and projects since decades to use seawater for agriculture (seawater irrigation was probably first proposed by the Israeli Boyko already in 1966), only few organisms have been found, which can be grown with seawater: some mangrove trees and shrimps. And even after 40 years of Boyko's studies there is still no use of seawater irrigation in Israel or other countries. Some halophytes can at least fulfil their lifecycle with seawater. But they also grow better on half seawater concentration. Halophytic crops are not yet common. "Who wants to eat every day *Salicornia* – salad or – vegetable?" (Fig. 19.3). Their productivity is low, by biochemical, physiological (Fig. 19.2) and ecophysiological reasons as well as anatomical and morphological adaptations their survival on seawater salinity is maintained but this inhibits high crop yields (Munns 1993). Thus, the statement: "Agriculture can also be done with seawater" is not yet true.

The use of callus-cultures and germ-plasms is often used as a means of selecting salt-tolerant traits, but again it turned out long time ago, that specific tissues of halophytes and non-halophytes may be rather salt-tolerant, but the intact individual plant behaves differently (Hedenström and Breckle 1974).

In many thousand other projects (with cash crops) the use of only 10–20% seawater concentration has been somehow economically feasable. Often the use of brackish water of 20% sea water has been declared a big success, but 20% instead of 10% or 15% is still no sea-water agriculture, and many crops still suffer from ionic and/or osmotic damages under saline conditions (Castillo et al. 2007). But according to Ayars (1996), even water of greater than 3 dS/m in EC are severely restricted in their use for irrigation. However, water of many different compositions ranging in salinity up to at least 8 dS/m (6,000 mg/l TDS) are currently being used more or less productively for irrigation in numerous places throughout the world under widely varying conditions of soil, climate, irrigation and cropping. To remember: 100% seawater has an EC of 45 dS/cm.

Flowers (2007) has checked 195 publications since 1993 related to transformations to enhance salt tolerance in plants. This involved 21 species. The transformations were dealing with about 120 genes. The major categories have been: Compatible solutes (46 papers), Transport-related proteins (32 papers), signalling molecules (26 papers), ROS-related (16 papers), transcription factors (15 papers). Flowers: "90% of these papers imply an enhancement of salt tolerance in the title or abstract by the transfer of one or two genes. This is really surprising, given the multigenic nature of salt tolerance" (see also Ma et al. 2006). In many papers the treatments or the material were missing or the control incomplete, and almost all were lacking any field-trials. Most of the papers again were dealing only with brackish water, not seawater. Some examples are given in Tables 19.5–19.7.

In all these examples it is obvious that yield is decreasing in all crops to a varying extent even with relatively low saline water. This is a problem in all larger irrigation regions like the Karakalpakstan area in Uzbekistan or the Nile Valley in Egypt, where irrigation water has to be reused several times (Tables 19.7 and 19.8).

There are several good attempts for sustainable agriculture in drylands, as e.g. in Egypt. One example is the Siwa Oasis. The Siwa Oasis in Egypt has the

Fig. 19.3 *Salicornia europaea*, an extreme eu-halophyte tolerating sea-water salinity, at the current coast-line of the retreating Aral Sea (Photo: SWBr, 23-05-2003). "Who wants to eat every day salty *Salicornia* salad"?

largest naturally flowing springs in the New Valley. Siwa once contained a thousand springs, of salinity ranging from EC 2 to 4 dS/m, which were used successfully to irrigate olive and date-palm orchards, with some scattered forage areas. Due to over-irrigation without appropriate drainage facilities, seepage as well as run off to low lying land, salinity and waterlogging have developed in some lands of the oasis. To reduce drainage water volumes, minimize water pollution and safely dispose of the ultimate unusable final drainage water, new strategies are being developed and experimented by the Government authorities in Siwa Oasis (FAO, Rady 1990).

These include:

- Use of natural flowing springs to irrigate winter crops such as cereals and forage
- Use of saline water over 5 dS/m to irrigate salt tolerant crops like barley, vetches, Rhodes grass, sugarbeet, etc.
- Use of biologically-active drainage water for the production of windbreak and growing wood trees
- Use of drainage water for stabilization of sand dunes
- Reuse of drainage water (average salinity is EC 6.0 dS/m with SAR values of 10 to 15) after blending with good quality water (recently drilled deep well of salinity EC 0.4 dS/m with SAR of 5) or by alternating the drainage water with good water

The use of saline drainage water in Egypt was reported by Abu-Zeid (1988). About 2.3×10^{9} m^{3} of drainage wastewater are discharged annually to the Mediterranean Sea via return to the Nile River in Upper Egypt; 12×10^{9} m^{3} are discharged directly into the sea and northern lakes; $2-3\times10^{9}$ m^{3} are used for irrigating about 405,000 ha of land. About 75% of the drainage water discharged into the sea has a salinity of less than 3,000 mg/l. People normally use drainage water directly for irrigation if its salinity is less than 700 mg/l; to mix it 1:1 with good Nile water (180–250 mg/l) if the concentration is 700 to 1,500; or 1:2 or 1:3 with Nile water if its concentration is 1,500–3,000 mg/l; and to avoid reuse if the salinity of the drainage water exceeds 3,000 mg/l. This has strong potential disadvantages of increasing salinization of the fields.

Table 19.5 Representative yields (in %) by crop and irrigation water salinity in survey of Hissar area of Haryana, India

	Tubewell salinity (EC in dS/m)		
Crop	2–4	4–6	6–8
Cotton	100	70	55
Millet	100	79	52
Wheat	100	89	60
Mustard	100	86	67
Average	100	81	59

Source: After Boumans et al. 1988

Table 19.6 Relative salt tolerance (50% EC values) of various crops during growth to maturity

Crop		EC saturated soil extract
Common name	Botanical name	dS/m – 50% yield
Barley	*Hordeum vulgare*	18
Cotton	*Gossypium hirsutum*	17
Sugarbeet	*Beta vulgaris*	15
Sorghum	*Sorghum bicolour*	15
Wheat	*Triticum aestivum*	13
Cowpea	*Vigna unguiculata*	9.1
Alfalfa	*Medicago sativa*	8.9
Tomato	*Lycopersico n lycopersicum*	7.6
Cabbage	*Brassica oleracea capitata*	7.0
Maize	*Zea mays*	5.9
Onion	*Allium cepa*	4.3
Rice	*Oryza sativa*	3.6
Bean	*Phaseolus vulgaris*	3.6

Source: After Maas 1986

Table 19.7 Effect of irrigation with different salinity levels on principal crop yield (t/ha) and percentage reduction (in brackets) by low salinity, grown in the Nile delta area (Fayoum Gov.)

Source of irrigation water (EC in dS/m)	Wheat grain	Onion	Maize	Summer tomato	Winter tomato	Pepper
Drainage water (2.8 dS/m with SAR 22)	5.0 (100%)	6.5 (67%)	1.8 (72%)	2.5 (33%)	8.0 (64%)	12.5 (50%)
Fresh Nile water for seedling establishment, then drainage water	3.0 (60%)	6.5 (67%)	2.0 (80%)	4.0 (53%)	8.7 (69%)	20.0 (80%)
Fresh Nile water (0.5 dS/m with SAR 4)	5.0 (100%)	9.7 (100%)	2.5 (100%)	7.5 (100%)	12.5 (100%)	25.0 (100%)

Source: Adapted by Mashali, based on data reported by Rady 1990

Table 19.8 Quantity of drainage water, salinity levels and estimated reuse in years 1988 and 1992 in the Nile delta

	Quantity of drainage water in MCM							
	Salinity levels (EC in dS/m)						Estimated reuse	
Regions	<1	1–2	2–3	3–4	>4	Total	Year 1988	Year 1992
Eastern Delta	949	1,565	1,055	310	433	4,312	1,130	2,000
Middle Delta	330	1,421	1,832	273	1,191	5,047	686	1,400
Western Delta	473	412	1,291	901	1,914	4,991	554	1,050
Total	1,752	3,398	4,178	1,484	3,538	14,350	2,370	4,450

Source: Adapted by Mashali based on data reported by Amer and Ridder 1988; Rady 1990

A sustainable agriculture on a large scale based on seawater irrigation is an utopic illusion. For special cases certainly a small scale seawater irrigation on anyhow saline coastal areas may be in fact very advisable and economic, e.g. for production of Salsolin by *Salsola*, or other special secondary compounds, for producing fiber material for paper industry (from *Juncus*), for horticultural purposes etc. Producing halophytic fodder has the big disadvantage of being very rich in sodium. The production of oil-seed crops may have a greater potential due to the low ash content and the widely available technology for harvesting and processing (Flowers 2007). Certainly, on anyhow degraded land saline agriculture even with low crop yields may be advisable as an alternative to bare wasteland.

But, an intensive agriculture with really high production per surface area is always only achievable with nonsaline conditions. Additional costs to maintain sustainable irrigation and leaching systems to keep salinity of soil low, pays more on the long run. Takehome message: "No irrigation without drainage!" This also means it pays more to invest in desalinization systems (inverse osmosis, energy source from high radiation in deserts, photovoltaic systems, wind generators etc.) on the long run, to keep soils low in salt, since fresh water is always indispensable for human livelihood and welfare. And it might be also more economic even to invest in intensive greenhouse agriculture with recycled water (e.g. solar glasshouses in Bolivia 4,000 m asl). There are still many ways to improve land use techniques, as e.g. by use of halophytic companian plants (Colla et al. 2006) or by specific mixed two- or three storey cultures, as it has been practised in Tunisian oasis since centuries.

However, there are indeed many other additional and very useful applications for halophyte use. As already mentioned, specific cash crops for secondary compounds, but an even larger application is the phytomelioration of salt-deserts, as e.g. in the Aralkum. There the desiccated seafloor exhibits 55,000 km^2 of a new land surface, covered by sand- and salt-deserts with various types of solonchak soils. The salt crust is puffy or even powdery, and is the source of several severe salt dust storms per year (Breckle et al. 2001; Breckle 2003; Breckle and Wucherer 2006, 2007). Mainly in spring, this threatens the whole adjacent agricultural lands of the Syrdarya and Amudarya river basins, as well as human health of these and neighbouring countries. Large scale plantings by saplings, mainly of saxaul (*Haloxylon aphyllum*) can substantially reduce wind-speeds and thus dust storm sources (http://earthobservatory.nasa.gov/NaturalHazards/natural_hazards_v2.php3?img_id=14253). After a few years a spontaneous vegetation cover has established (Breckle et al. 2001; Wucherer et al. 2005).

There is the statement of Sardo (2006): "Fertility is a complex concept which cannot be limited to physical or chemical aspects, though important they are. One should not overlook, for instance, the microbiological aspects, nor the climate; in a desert one can find a potentially good soil, but pedologic and climatic conditions must be jointly considered. And then: good for what? The same soil can be perfect for growing, red clover and unusable for citrus. Therefore why not to use a good desert soil with appropriate climate conditions for growing halophytes? Better halophytes than nothing.

According to our experience full strength seawater can be used to sustainably irrigate *Sesuvium*, *Aster* and *Spartina* along the sandy seashores, without impairing soil quality provided that a sufficient drainage exists.

I don't believe that anybody thinks of practicing an "intensive agriculture with high production per surface area", but using saline or brackish water to irrigate marginal lands or wastelands in order to increase plant production and CO_2 sequestration, to fix erodible soils, or for greenification and biodiversity enhancement is perfectly reasonable (again, provided that a sufficient drainage exists)."

It is agreed on that and it is long known, that saline soils (solonchaks) only can be used after intensive and thorough reclamation procedures (USDA 1954; Breckle and Wucherer 2007) and it might be advisable to use deep rooting plants for phytomelioration as e.g. *Glycyrrhiza glabra* under moderate salinities (Khushiev et al. 2005) or *Haloxylon aphyllum* (Breckle and Wucherer 2002, 2006).

6 Seawater for Green Deserts

Deserts on earth, in fact, cover a huge surface area. But is the following statement right?: "Deserts on the earth cover a huge surface area. It is urgently necessary to make them green by seawater irrigation to reduce CO_2 by producing biomass, and thus to reduce global warming". It will not work to make the deserts green by seawater irrigation. Sodium would damage the clay minerals, and a good desert soil would be totally and irreversibly spoiled. On the other hand, it is not at all economic to pump seawater over long distances (with corrosion problems on pumps and pipes), to accumulate salt in concrete basins in inland deserts and to spoil soils in oasis or lower aquifers of good water. Fresh water aquifers are known from many deserts, which is demonstrated by good Artesian springwater. How greenification should occur? Planting longliving trees is the only choice for binding CO_2 for a longer period. But which trees thrive on seawater? only mangroves, and from that almost only *Avicennia*, all other species need from time to time diluted seawater, tidal activity or riverwater flushes for nutrient supply. *Avicennia* in the deserts? Only in some coastal deserts or especially at all stands of the former mangrove coasts where mangroves were cut and replaced by shrimp farms (in the Indopacific etc., where mangroves could have had stopped at least partly the deadly tsunami) it is strongly advised to increase again the area of mangroves and thus to recover a protective mangrove belt.

The arguments by Sardo (2006): **"**Of course sodium damages soil colloids and structure, but desert soils are frequently sandy and a-structural and in all cases securing sustainability is assumed as a prerequisite (e.g. provided that a sufficient drainage exists, once again!).

"How greenification should occur? Planting longliving trees is the only choice for binding CO_2 for a longer period.**"** Sorry, I find here some confusion between two different scopes: greenification, which is basically an embellishment of the landscape and maybe also an encouragement to biodiversity, and CO_2 sequestration; however I fail to understand why greenification with, say, perennial grasses or *Atriplex* bushes is not an apt tool to also capture and sequester carbon dioxide.

"It would be much wiser to stop further deforestation**"**, and who could disagree on the need to stop or at least slow down deforestation? But nothing prevents us from giving at the same time a hand by increasing plantations: there is no contrast between the two approaches.

"Degraded and desertified lands should be used for greenification**"**: all right, why not to plant halophytes whenever soil and/or climate conditions bar the plantation of other plants?"

But it should be considered, that in contrast to trees, perennial grasses or *Atriplex* are not accumulating much biomass, their turnover by mineralization is fast.

Greenification of deserts, even by fresh water, would mean to accelerate the carbon-cycling but not to increase considerably the biomass on landsurface. It would be much wiser to stop further deforestation in all the humid climatic zones (ZB I, II, IV, V, VI, VIII, Breckle 2002a), especially in the tropics, in the Siberian and Canadian taiga, and to let grow all secondary forests to a greater extent, instead of transforming them from highly biodiverse primary forests to poor MacDonald beef sources. Degraded and desertified land should be used for greenification – and it is important to be aware that deserts have not much to do with the desertification process caused by man. Even some desert areas are known to be desertified by humans.

The global carbon balance is not to be equalled by the present known flux rates. There is a rather big sink, which is not yet exactly known. The ocean itself may be a great reservoir for CO_2 by formation of $CaCO_3$ sediments and other processes.

Additionally there are other much more effective heat-trapping gases, like CH_4 and N_2O, the reduction of their emission is an even bigger problem. Paddy-fields and cows are one of the main sources for CH_4, but are in future strongly needed to feed more people. *Avicennia* from the desert by no means can replace it.

By the way, during the last million years the CO_2-content of the earth's atmosphere has fluctuated tremendously. The glacial and interglacial periods were caused by various reasons, not by human activity. Anyhow the 380 ppm from today are alarming, since the speed of increase is sharp and no substantial reduction is visible. Taking this into account, the takehome message can only be: "Let us plant trees and stop deforestation". This is the best indirect way for greenification of the desert.

7 Conclusions

Salinity (and alkalinity) of soils in drylands is the biggest threat to productive agriculture. Saline agriculture will play an increasing role in future. However, the progress in breeding salt tolerant genotypes of conventional crops is slow. Salt tolerance in plants is a complex phenomenon and has evolved with many aspects, demonstrated by rather differing halophyte types. Future focus could be on developing halophytes as crops, on a better understanding of the physiology and biochemistry of halophytes. On the other side, land-use practice should try to avoid salt accumulation in the agricultural system by precise and water saving irrigation techniques and by intact drainage systems.

Thus a modified version of the Birjand-Declaration should be taken into account:

1. Salinity is a global problem. Salinization of soil and water resources and their management and rehabilitation is of great importance. Therefore governments should consider it as a serious challenge.
2. The conventional water resources and crops do not meet all the requirements of human society in dry and saline areas. Seawater and/or brackish water and crops of salt tolerant plants should be considered for research.
3. Comprehensive plans for sustainable use of saline fields, saline waters and halophytes as useful plants and crops are needed for each region.
4. Priority, however, should be given to any means to prevent further salinization of agricultural fields, since non-saline agriculture is always more productive than saline agriculture. Non-saline soils never should be spoiled with saline water.
5. Holding workshops and conferences for exchange of experiences and ideas is highly recommended.
6. It is suggested that the different scientific disciplines should cooperate in the solution of sustainable and economically feasible management of saline systems.
7. Salinity affected countries need a strong human base of the research on saline agriculture. However, coordination among various government agencies and academia is not sufficient. Therefore an aggressive plan may be made to increase collaboration and planning for phytomelioration of saline areas to make them available for agriculture.
8. Salinity is a global problem. Therefore there is a need to create national authorities and international funds and collaboration for the management of saline lands and to enlarge the mandate of UNCCD to encourage international collaboration and financing at the government's levels to make indispensable contributions both locally and globally.

Acknowledgements The author is indebted to DFG (German Research Foundation), to the DAAD (German Academic Exchange Service), to BMBF (German Ministry of Science and Technolgy) as well as to the Schimper-Foundation, who provided partial funding support. For technical help with field experiments and laboratory analysis we have to thank I. Meier and A. Scheffer, and several students. Many local people in various countries were very helpful, their hospitality is greatly acknowledged.

References

Abu-Zeid M (1988) Egyptian policies for using low quality water for irrigation. In: Bouchet R (ed) Proceedings of the Cairo/Aswan seminar "Reuse of low quality water for irrigation in Mediterranean Countries", Cairo

Albert R (1982) Halophyten. In: Kinzel H (ed) Pflanzenökologie und Mineralstoffwechsel. Ulmer, Stuttgart, pp 33–215

Amer MH, Ridder NA (1988) Land drainage in Egypt. Drainage Research Institute, Water Research Center, Cairo, 376 pp

Ayars JE (1996) Managing irrigation and drainage systems in arid areas in the presence of shallow groundwater: case studies. Irrig Drain Syst 10: 227–244

Boumans JH, Hoorn JWV, Krusermann GP, Tenwar BS (1988) Water table control, reuse and disposal of drainage water in Haryana. Agr Water Manage 14: 537–545

Boyko H (1966) Basic ecological principles of plant growing by irrigation with highly saline or seawater. In: Boyko H (ed) Salinity and aridity. Dr. W. Junk Publishers, The Hague

Boyko H (1968) Saline irrigation for agriculture and forestry. UNESCO Symposium. Dr. W. Junk Publishers, The Hague
Breckle S-W (1989) Role of salinity and alkalinity in the pollution of developed and developing countries. In: Öztürk MA (ed) International Symposium on the effect of pollutants to plants in developed and developing countries in Izmir, Turkey 22–28.8.1988, pp 389–409
Breckle S-W (1990) Salinity tolerance of different halophyte types. In: Bassam N et al. (eds) Genetic aspects of plant mineral nutrition. Plant Soil 148: 167–175 (Kluwer, Dordrecht, The Netherlands)
Breckle S-W (1995) How do plants cope with salinity? In: Khan MA, Ungar IA (eds) Biology of salt tolerant plants (Proceedings of the International Symposium). Department of Botany, University of Karachi, Pakistan, pp 199–221
Breckle S-W (2000a) Salinity, halophytes and salt affected natural ecosystems. In: Läuchli A, Lüttge U (eds) Salinity: environment - plants – molecules. Kluwer Dordrecht, pp 53–77
Breckle S-W (2000b) Wann ist eine Pflanze ein Halophyt? Untersuchungen an Salzpflanzen in Zentralasien und anderen Salzwüsten. In: Breckle S-W, Schweizer B, Arndt U (Hrsg) Ergebnisse weltweiter ökologischer Forschungen (Proceedings of the 1st Symposium of the A.F.W. Schimper-Foundation, established by H & E Walter, Hohenheim). Verlag Günter Heimbach, Stuttgart, pp 91–106
Breckle S-W (2002a) Walter's vegetation of the earth. The ecological systems of the geo-biosphere. 4th edition. Springer, Berlin, 527 pp
Breckle S-W (2002b) Salt deserts in Iran and Afghanistan. In: Böer B, Barth, H-J (eds) Sabkha ecosystems. Kluwer, Dordrecht, The Netherlands, pp 109–122
Breckle S-W (2003) Rehabilitation of the Aral Sea environment, Kazakhstan. Proceedings of the International Workshop (Aleppo, May 2002): "Combating desertification – rehabilitation of degraded drylands and biosphere reserves". UNESCO-MAB dryland series no.2, pp 47–57
Breckle S-W, Wucherer W (2002) The Aral Sea Crisis (in Arabic). Environment & Development, Beirut, pp 36–40
Breckle S-W, Wucherer W (2006) Combatting desertification in the Northern Aral Sea region. In: Gao J et al. (eds) Restoration and stability of ecosystems in arid and semi-arid areas. Science Press, Beijing, pp 304–316 (Peking-Symposium China Herbst 2004)
Breckle S-W, Wucherer W (2007) What will be the future of the Aral Sea? In: Lozan JL et al. (eds) Global change: enough water for all? Wissenschaftliche Auswertungen. GEO, Hamburg, pp 142–146
Breckle S-W, Veste M, Wucherer W (eds) (2001) Sustainable land-use in deserts (Proceedings of the International Königswinter Workshop). Springer, Berlin, 465 pp
Castillo EG, Tuong TP, Ismail AM, Inubushi K (2007) Response to salinity in rice: comparative effects of osmotic and ionic stresses. Plant Prod Sci 10: 159–170
Chapman VJ (1960) Salt marshes and salt deserts of the world. In: Polunin N (ed) Plant Science Monographs. Hill, London, 392 pp
Colla G, Rouphael Y, Fallovo C, Cardarelli M, Graifenberg A (2006) Use of *Salsola soda* as a companion plant to improve greenhouse pepper (*Capsicum annuum*) performance under saline conditions. New Zeal J Crop Hortic Sci 34: 283–290
Epstein E, Norlyn JD (1977) Seawater-based crop production: a feasibility study. Science 197 (4300): 249–251
Epstein E, Norlyn JD, Rush DW, Kingsbury RW, Kelley DB, Cunningham GA, Wrona AF (1980) Saline culture of crops: a genetic approach. Science 210: 399–404
Flowers TJ (2007) Halophytes: Plants for the Future. Plenary Lecture at the Biennial Meeting of the German Botanical Society; Symposium 1: Halophytes (by Breckle S-W, Veste M), August 2007, Hamburg
Flowers TJ, Yeo AR (1995) Breeding for salinity resistance in crop plants: where next? Aust J Plant Physiol 22: 875–884
Ghassemi F, Jakeman AJ, Nix HA (1995) Salinisation of land and water resources: human causes, extent, management and case studies. UNSW Press, Sydney, Australia
Hedenström Hv, Breckle S-W (1974) Obligate halophytes? A test with tissue culture methods. Z Pflanzenphysiol 74: 183–185
Khushiev H, Noble A, Abdullaev I, Toshbekov U (2005) Remediation of abandoned saline soils using *Glycyrrhiza glabra*: a study from the hungry steppe of central Asia. Int J Agr Sustain 3: 102–113
Kreeb KH (1964) Ökologische Grundlagen der Bewässerungskulturen in den Subtropen. Fischer, Stuttgart, 149 pp
Ma S, Gong Q, Bohnert HJ (2006) Dissecting salt stress pathways. J Exp Bot 57: 1097–1107
Maas EV (1986) Salt tolerance of plants. Appl Agr Res 1: 12–26
Munns R (1993) Physiological processes limiting plant growth in saline soils: some dogmas and hypotheses. Plant Cell Environ 16: 15–24
Munns R (2005) Genes and salt tolerance: bringing them together. New Phytol 167: 645–663
Rady AHM (1990) Water, soil and crop management relating to the use of saline water. In: Proceedings of Expert Conservation on Water, Soil and Crop Management relating to the Use of Saline Water, October 1989, AGL/MISC/16/90. FAO, Rome
Reimann C, Breckle S-W (1993) Sodium relations in Chenopodiaceae, a comparative approach. Plant Cell Environ 16: 323–328
Reimold RJ, Queen WH (1974) Ecology of halophytes. Academic, New York/ London, 605 pp
Sardo V (2006) E-mail-exchange on "Biosaline agriculture"
USDA - US Salinity Laboratory Staff (1954) In: Richards LA (ed) Diagnosis and improvement of saline and alkali soil. US Department of Agriculture Handbook No 60
Waisel Y (1972) Biology of halophytes. Academic, New York/London
Wucherer W, Veste M, Herrera Bonilla O, Breckle S-W (2005) Halophytes as useful tools for rehabilitation of degraded lands and soil protection. Proceedings of the First International Forum on Ecological Construction of the Western Beijing, Beijing, pp 87–94 (English); 169–175 (Chinese)

Chapter 20
Enhanced Tolerance of Transgenic Crops Expressing Both Superoxide Dismutase and Ascorbate Peroxidase in Chloroplasts to Multiple Environmental Stress

S.-S. Kwak, S. Lim, L. Tang, S.-Y. Kwon, and H.-S. Lee

Abstract Oxidative stress is one of the major damaging factors reducing crop growth and productivity. In order to develop transgenic crops such as sweetpotato (*Ipomoea batatas* L. Lam) and potato (*Solanum tuberosum* L.) plants with enhanced tolerance to multiple stress, the genes of both CuZn superoxide dismutase (CuZnSOD) and ascorbate peroxidase (APX) were expressed in chloroplasts under the control of an oxidative stress-inducible peroxidase (*SWPA2*) promoter (referred to SSA plants). SSA sweetpotato and potato plants showed enhanced tolerance to oxidative stress caused by the application of methyl viologen (MV, paraquat), a ROS-generating non-selective herbicide. The expression of introduced CuZnSOD and APX genes was significantly induced in leaves of SSA plants after MV treatment. The APX activity in chloroplast fractions from SSA sweetpotato was approximately 15 times higher than that from non-transgenic (NT) plants. SSA sweetpotato plants showed higher tolerance to chilling stress than NT plants, whereas SSA potato plants showed higher tolerance to high temperature. In addition, SSA sweetpotato plants showed a strong tolerance to the application of sulfur dioxide and dehydration. Our results strongly suggested that the rational manipulation of antioxidative mechanism in chloroplasts will be applicable to the development of all plant species with enhanced tolerance to multiple environmental stresses.

Keywords Antioxidative mechanism • metabolic engineering • multiple environmental stress • potato • sweetpotato

1 Introduction

Cells of all organisms including plants produce reactive oxygen species (ROS) such as the superoxide anion radical ($\bullet O_2^-$), hydrogen peroxide (H_2O_2), and hydroxyl radicals (•OH) as by-products of normal metabolic processes (Foyer et al. 1994; Asada 1999). Excessive production of ROS can also result from various environmental stresses such as drought, salinity, extreme temperature, or strong light. Thus oxidative stress caused by ROS is one of the major damaging factors reducing plant productivity. ROS-scavenging antioxidative mechanisms include antioxidant enzymes such as superoxide dismutase (SOD), ascorbate peroxidase (APX), peroxidase (POD), catalase, and low molecular weight antioxidants such as ascorbic acid, glutathione and phenolic compounds (Noctor and Foyer 1998; Asada 1999).

The chloroplast, which is the cellular compartment with a high energetic photosynthetic electron transport system and a generous supply of oxygen, is a rich source of ROS (Asada 1999). Two key ROS detoxification enzymes in the chloroplast are SOD and APX. SOD catalyzes the dismutation of two molecules of superoxide anion radical into oxygen and hydrogen peroxide. APX reduces hydrogen peroxide to water by utilizing ascorbate as an electron donor. The formation of most toxic hydroxyl radicals by superoxide anion

S.-S. Kwak (✉), S. Lim, S.-Y. Kwon, and H.-S. Lee
Environmental Biotechnology Research Center, Korea Research Institute of Bioscience and Biotechnology (KRIBB), Oun-dong 52, Yuseong-gu, Daejeon 305–806, Korea
e-mail: sskwak@kribb.re.kr

L. Tang
Biotechnology and Food Science College, Tianjin University of Commerce, East Entrance of Jinba Road Beichen District, Tianjin 300134, China

radical and hydrogen peroxide can be controlled by the combined reactions of SOD and APX. To maintain the productivity of plants under stress conditions, it is possible to fortify the antioxidative mechanism of the chloroplasts by manipulating the antioxidant enzymes present in chloroplasts. As we have previously reported, transgenic tobacco plants expressing both CuZnSOD and APX in chloroplasts under the control of CaMV 35S promoter provided strong protection to methyl viologen (MV)-induced oxidative stress (Kwon et al. 2002).

A strong oxidative stress-inducible POD (*SWPA2*) from cultured cells of sweetpotato was isolated and its function was characterized in transgenic tobacco plants and cultured cells during environmental stress (Kwak et al. 1995; Kim et al. 1999, 2003). Results from this study suggested that this promoter will be biotechnologically useful for the development of transgenic plants with enhanced tolerance to environmental stress (Wang et al. 2005).

Sweetpotato [*Ipomoea batatas* (L.) Lam.] and potato (*Solanum tuberosum* L.) are major food crops in many parts of the world and rank seventh and fourth in annual production in the world. They are a good source of energy, supplying sugars and carbohydrates. Specially, sweetpotato is used for food processing as well as for starch and alcohol production. Therefore, the sweetpotato is one of the most important crops for securing a stable food supply and has the potential to alleviate some global environmental problems in the 21st century. In this report, we developed transgenic sweetpotato (cv. Yulmi) and potato (cv. Atlantic) plants expressing both genes of CuZnSOD and APX in chloroplasts under the control of an oxidative stress-inducible *SWPA2* promoter (referred to as SSA plants) and evaluated their phenotypes in terms of oxidative stress and temperature stress responses.

2 Results and Discussion

2.1 Development of Transgenic Sweetpotato and Potato Plants

Transgenic sweetpotato and potato plants were successfully generated expressing both CuZnSOD and APX in a chloroplast expression manner using chloroplast-targeted signal peptide under the control of an oxidative stress-inducible *SWPA2* promoter (Fig. 20.1). Presence and integration of the introduced SOD, APX and NPTII in genomic DNA of transgenic plants were confirmed using PCR and Southern blot analysis (data not shown).

2.2 Enhanced Tolerance of SSA Sweetpotato Plants to Oxidative Stress

To investigate oxidative stress tolerance on the whole plants level, SSA sweetpotato and non-transgenic (NT) plants were evaluated for photosynthetic activity (Fv/Fm) after treatment with solutions containing 0, 100, 150, or 200 μM MV. Figure 2A shows large differences in leaf damage between NT and SSA plants at 5 days after treatment (DAT). The leaf damage in NT plants was dramatically increased in proportion with the MV concentration, whereas damage in SSA plants was considerably less. SSA sweetpotato plants retained high photosynthetic activity compared to NT plants, even though there were no visible damage in both NT and SSA plants at 2 DAT. The SSA4 plant at 200 μM MV treatment was slowly decreased to about 41% compared to non-treatment of SSA plants (Fig. 20.2B). In the fourth leaf after MV treatment, the Fv/Fm value was significantly reduced in NT plants whereas it was

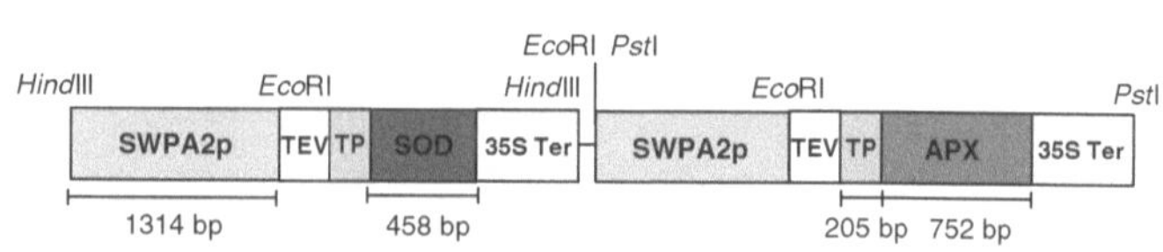

Fig. 20.1 Plant transformation vector pSSA-K. SWPA2pro, sweetpotato peroxidase (SWPA2) promoter; SOD, cassava CuZnSOD (mSOD1); APX, pea ascorbate peroxidase; TEV, tobacco etch virus 5′-UTR; 35S Ter, CaMV 35S terminator; TP, chloroplast-targeted transit peptide

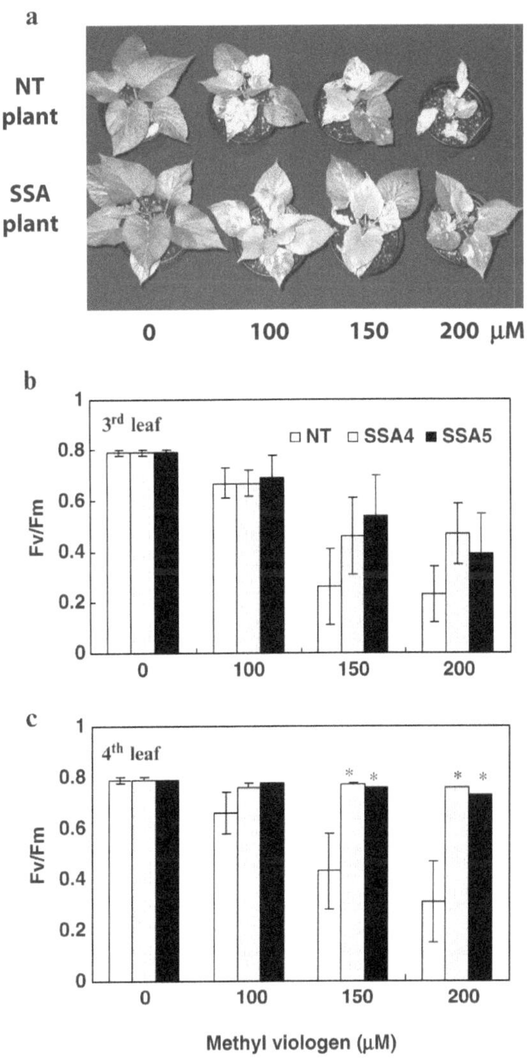

Fig. 20.2 Effect of MV on NT and SSA sweetpotato plants after spraying with different MV dosage (0, 100, 150 and 200 μM). (a) Visible damages in the leaves of NT and SSA plants at 5 days after spraying with different MV dosage. (b and c) Photosynthetic activity (Fv/Fm) in the leaves of SSA and NT plants. Data are means of three replicates. *Significant differences in mean activity between NT and SSA plants determined by *t*-test ($P = 0.05$)

not significantly changed in SSA plants compared to untreatment (Fig. 20.2C).

To understand the tolerance mechanism in SSA sweetpotato plants to MV-mediated oxidative stress, we investigated changes in gene expression and enzyme activities of introduced CuZnSOD and APX after 100 μM MV treatment with time course. Gene expression of CuZnSOD was induced at low level after 1 DAT and showed highest level at 2 DAT, whereas that of APX was gradually increased from 1 DAT (data not shown). The levels of SOD and APX in total protein fractions of both NT and SSA plants were increased by MV treatment. The SOD and APX activity in total proteins of SSA plants was increased from 2 DAT, which was significantly higher than that of NT plants (Fig. 20.3A, B). The SOD activity in chloroplasts from SSA plants was five times higher than that from NT plants from 2 DAT (Fig. 20.3C). Surprisingly, The APX activity in chloroplast fractions from SSA plants was approximately 15 times higher than that from NT plants (Fig. 20.3D).

The expression of combinations of antioxidant enzymes in transgenic plants might have synergistic effects on stress tolerance. However, only a few reports have been conducted to address scavenging systems expressing more than one enzyme in the chloroplasts of higher plants (Aono et al. 1995; Allen et al. 1997; Kwon et al. 2002). In the SSA plants of this study, the transcription levels of introduced CuZnSOD and APX genes in leaves of SSA plants after MV treatment were parallel to that of the tolerance to MV-mediated oxidative stress. As expected, the levels of SOD and APX in chloroplast fractions showed a dramatic increase reflecting the chloroplast-targeted expression and stress-inducible promoter of the expression system in this study. These results indicate that CuZnSOD and APX proteins in SSA plants were correctly targeted into chloroplasts. It is likely that the sequential reactions of CuZnSOD and APX in chloroplasts of SSA plants may play an important role in enhanced tolerance to MV-mediated oxidative stress by synergistic effects of both enzymes.

2.3 Enhanced Tolerance of SSA Potato Plants to Oxidative Stress

To investigate oxidative stress tolerance, SSA potato and NT plants were evaluated for visible damage 5 days after spraying with solutions containing 0, 150, 200, or 250 μM (Fig. 20.4A). Visible leaf damage on NT plants

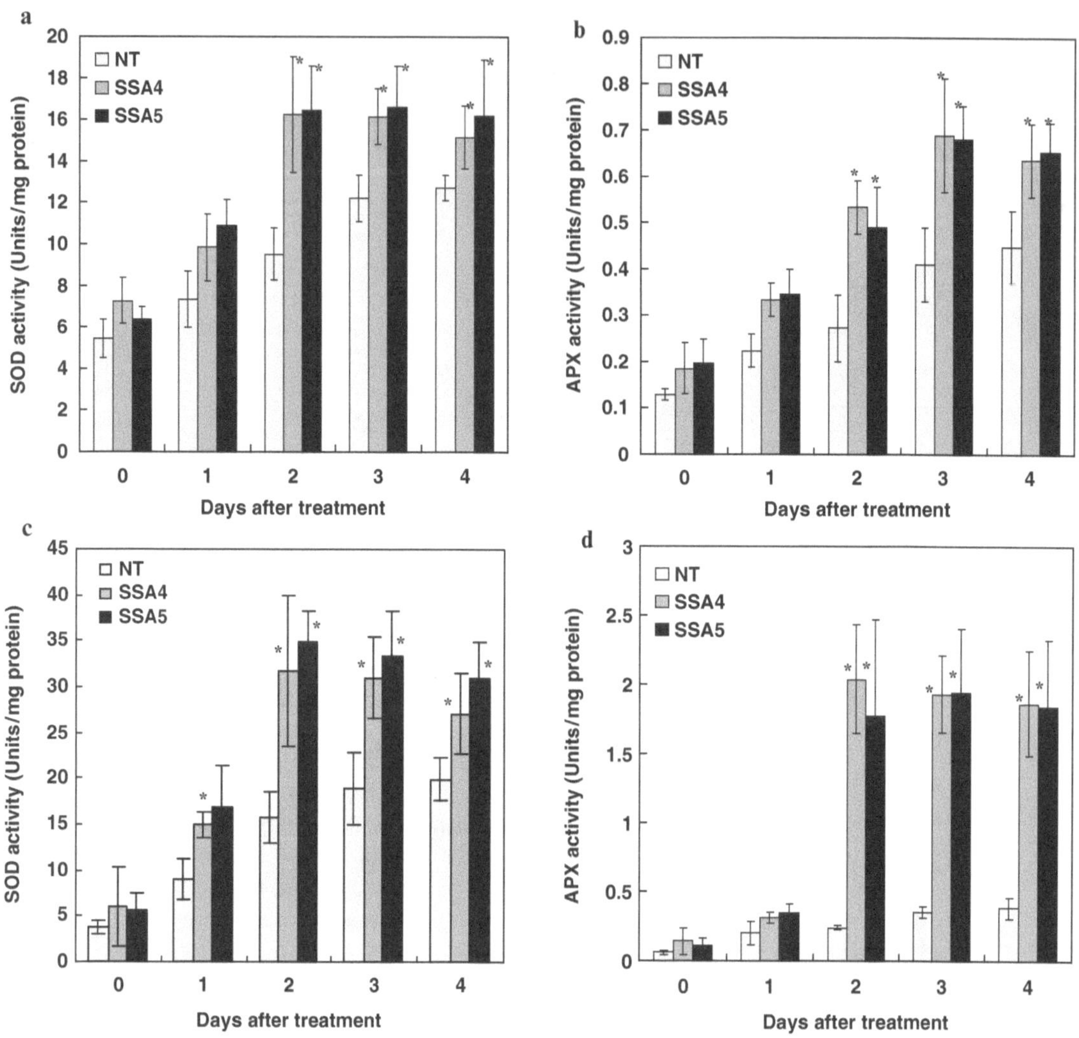

Fig. 20.3 Enzyme activity (units/mg protein) of SOD and APX in total proteins and soluble proteins from isolated chloroplast fraction in leaves from NT and SSA sweetpotato plants subjected to 100 μM MV spray. (a) SOD activity in total proteins; (b) APX in total proteins; (c) SOD in soluble proteins from isolated chloroplast fraction; (d) APX activity in soluble proteins from isolated chloroplast fraction. Data are means of three replicates. *Significant differences in mean activity between NT and SSA plants determined by *t*-test ($P = 0.05$)

became more severe with increased concentration. When 250 μM MV treatment was applied, NT plants showed 60% damage, whereas SSA11 plants exhibited only 17% leaf damage (Fig. 20.4B). In addition to visible damage, the effect of MV spray was also investigated at 5 days post-MV treatment by determining the dry weight of non-damaged leaf tissues from the first to fourth leaves. (Fig. 20.4C). At 200 μM MV treatment, no reduction in the dry weight of SSA11 plants was detected, although NT plants showed 50% reductions in dry weight at 250 μM MV treatment. The chlorophyll contents of SSA plants were not significantly affected by high MV concentration (250 μM), but these were significantly reduced in NT plants (Fig. 20.4D). However, SSA11 plants showed less chlorophyll content loss following MV treatment, showing 96% chlorophyll content at 250 μM MV treatment.

As shown in Fig. 20.5, gene expression of CuZnSOD and APX was induced at low level after 12 h of 150 μM MV treatment and then increased gradually until 72 h,

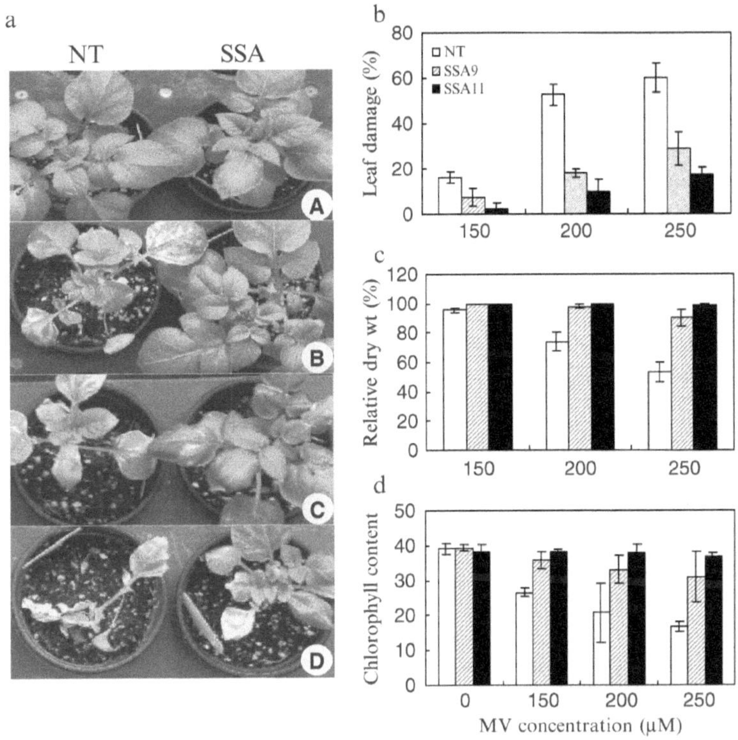

Fig. 20.4 Effect of MV on the NT and SSA potato plants at 5 days after treatment with 0, 150, 200, or 250 μM MV. (a) Visible damages in the leaves of NT and SSA plants at 5 days after treatment. (b) Quantitative estimate of visible damage that appeared on leaves. (c) Relative dry weight of leaves from non-damaged parts in NT and SSA plants. (d) The chlorophyll contents ($mg \cdot cm^{-2}$) in the fifth leaves. Data are means ± SE of three independent measurements

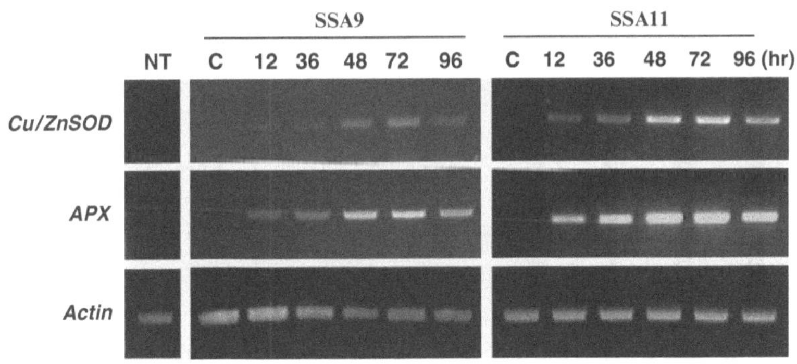

Fig. 20.5 RT-PCR analysis of the expression of CuZnSOD and APX genes in leaves from NT and SSA potato plants subjected to 150 μM MV spray. Total RNA was extracted from leaves 12, 24, 36, 48, 72, and 96 h after treatment with MV. Actin was used to control for equal loading. NT, non-transgenic potato plants; SSA, transgenic potato plants; C, non-treated SSA plants

after which a slight decrease was observed. However, transcripts of CuZnSOD and APX genes were not detected in NT and non-treated SSA plants. Interestingly, expression of those genes in SSA11 plants was higher than SSA9 plants. These results indicate that SOD and APX transcript levels in SSA plants well reflected tolerance to MV and that the synergistic effects of CuZnSOD and APX may play an important role in enhanced tolerance to MV-mediated oxidative stress.

2.4 Enhanced Tolerance of SSA Plants to Chilling and High Temperature Stress

When whole plants of SSA sweetpotato and NT were exposed to chilling at 4°C for 24 h, leaves of NT plants were severely wilted and curled, whereas SSA plants appeared to be only slightly damaged (Fig. 20.6A). Using the fluorescence parameter (Fv/Fm), the photosynthetic activity was determined in the fourth leaf from top of NT, SSA4 and SSA5 plants (Fig. 20.6B). The Fv/Fm of NT plants was decreased by 33% at 24 h after chilling treatment, whereas the activity from SSA4 and SSA5 plants only decreased by 19%. Furthermore, after 12 h of recovery following chilling, the Fv/Fm of SSA5 plants almost fully recovered to the initial levels, whereas that of NT plants further decreased to the lowest level. SSA 4 plants showed an intermediate result between NT and SSA5 plants. These results suggest that SSA plants had a tolerance to the oxidative stress mediated by chilling exposure.

When whole plants of SSA potato and NT were exposed to high temperature at 42°C for 20 h, NT plants were wilted from heat shock after 10 h, whereas SSA plants appeared to remain healthy (Fig. 20.7A).

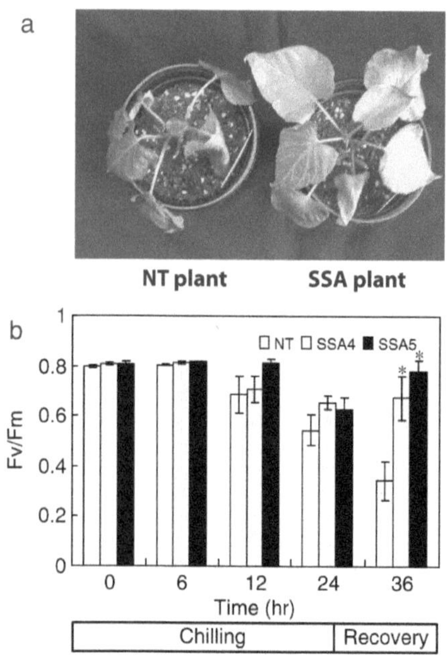

Fig. 20.6 Effect of chilling stress at 4°C on NT and SSA sweetpotato plants. (a) Visible differential damages in the leaves of NT and SSA4 pants at 24 h after chilling treatment and at 12 h of recovery at 25°C following chilling. (b) Photosynthetic activity (Fv/Fm) in the leaves of NT and SSA4 plants at 24 h after chilling treatment and at 12 h of recovery at 25°C following chilling. Data are means of three replicates. *Significant differences in mean activity between NT and SSA plants determined by *t*-test ($P = 0.05$)

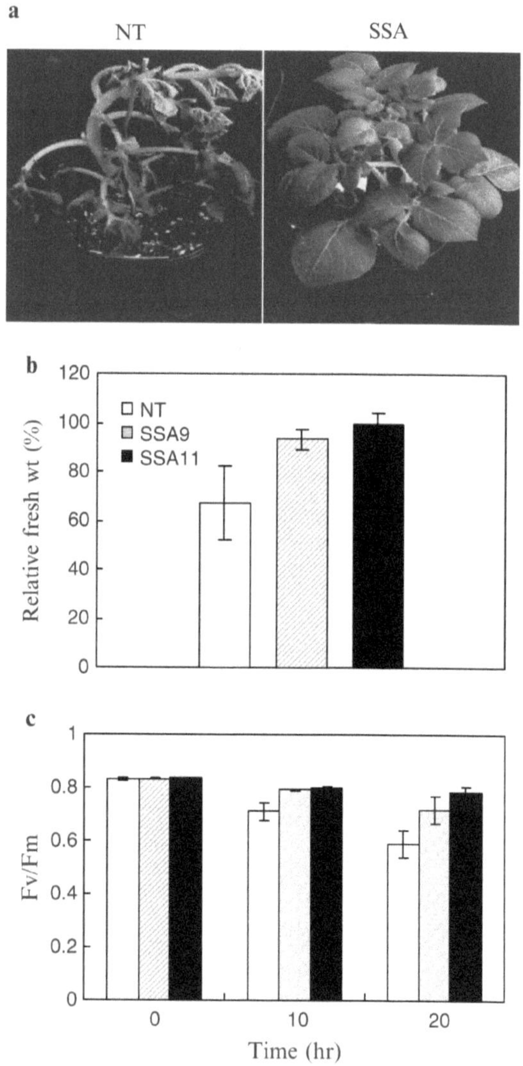

Fig. 20.7 Effects of high temperature (42°C) on NT and SSA potato plants. (a) Visible differential damages in the leaves of NT and SSA (SSA11) plants at 20 h after treatment; (b) Fresh weight of plants at 20 h after treatment; (c) Photosynthetic activity (Fv/Fm) in the leaves of NT and SSA plants for 20 h after treatment. Data are means ± SE of three independent measurements

NT plants showed a 32.7% decrease in fresh weight after treatment. However, SSA plants remained at similar fresh weights compared with the SSA plants grown at 25°C, showing only a decrease of 1.5% (SSA11) (Fig. 20.7B). The Fv/Fm of NT plants was decreased by 29% at 20h after high temperature treatment, respectively, while the activity from SSA plants only decreased 6% (Fig. 7C). Furthermore, after 3h of recovery following heat stress, the Fv/Fm of SSA plants almost fully recovered to the initial levels, while that of NT plants remained low (data not shown).

3 Conclusion

We successfully developed transgenic sweetpotato and potato plants expressing both CuZnSOD and APX in chloroplasts under the control of a stress-inducible promoter. SSA plants showed a strong tolerance to MV-mediated oxidative stress and chilling, high temperature stress. In addition, SSA sweetpotato plants showed enhanced tolerance to the application of drought (Lin et al. 2006) and air pollution stress, such as 500ppb sulfur dioxide (unpublished data). The field application of SSA plants under unfavorable growth conditions such as in dry and cold areas remains to be conducted. Our results in this study suggest that SSA plants might be useful for commercial cultivation in such harsh conditions. Furthermore, we anticipate that SSA plants can be used in breeding programs aimed at creating new varieties with new characteristics such as producing various biomaterials in the tubers. In addition, the manipulation of the antioxidative mechanism in chloroplasts can be applied in the development of increased tolerance to multiple environmental stresses in various other important crops.

Acknowledgements This research was supported by grants from BioGreen21 Program, Rural Development Administration, Korea, from the Environmental Biotechnology National Core Research Center, KOSEF/MOST, Korea, and from the International Collaboration Project, Ministry of Science and Technology (MOST), Korea.

References

Allen RD, Webb RP, Schake SA (1997) Use of transgenic plants to study antioxidant defenses. Free Radical Bio Med 23:473–479

Aono M, Saji H, Sakamoto A et al. (1995) Paraquat tolerance of transgenic *Nicotiana tabacum* with enhanced activities of glutathione reductase and superoxide dismutase. Plant Cell Physiol 36:1687–1691

Asada K (1999) The water-water cycle in chloroplasts: scavenging of active oxygens and dissipation of excess photons. Annu Rev Plant Physiol Plant Mol Biol 50:601–639

Foyer CH, Descourvieres P, Kunert KJ (1994) Protection against oxygen radicals: an important defence mechanism studied in transgenic plants. Plant Cell Environ 17:507–523

Kim KY, Hur KH, Lee HS et al. (1999) Molecular characterization of two anionic peroxidase cDNAs isolated from suspension cultures of sweetpotato. Mol Gen Genet 261:941–947

Kim KY, Kwon SY, Lee HS et al. (2003) A novel oxidative stress-inducible peroxidase promoter from sweetpotato: molecular cloning and characterization in transgenic tobacco plants and cultured cells. Plant Mol Biol 51:831–838

Kwak SS, Kim SK, Lee MS et al. (1995) Three acidic peroxidases from suspension-cultures of sweetpotato. Phytochemistry 39:981–984

Kwon SY, Jeong YZ, Lee HS et al. (2002) Enhanced tolerance of transgenic tobacco plants expressing both superoxide dismutase and ascorbate peroxidase in chloroplasts against methyl viologen-mediated oxidative stress. Plant Cell Environ 25:873–882

Lin Y, Deng XP, Kwak SS et al. (2006) Enhanced drought tolerance in transgenic sweetpotato expressing both Cu/Zn superoxide dismutase and ascorbate peroxidase. J Plant Physiol Mol Biol 32:451–457

Noctor G, Foyer CH (1998) Ascorbate and glutathine: keeping active oxygen under control. Annu Rev Plant Physiol Plant Mol Biol 49:249–279

Wang FZ, Wang QB, Kwon SY et al. (2005) Enhanced drought tolerance of transgenic rice plants expressing a pea manganese superoxide dismutase. J Plant Physiol 162:465–472

Chapter 21

Adaptation to Iron-Deficiency Requires Remodelling of Plant Metabolism: An Insight in Chloroplast Biochemistry and Functionality

A. Castagna, S. Donnini, and A. Ranieri

Abstract Iron-deficiency is a widespread plant disorder particularly frequent in neutral or alkaline soils, where iron availability often falls below the threshold required to meet plant needs. Plants have developed efficient mechanisms to acquire iron: strategy I, adopted by all dicots and non-graminaceous monocots, and strategy II, adopted by graminaceous plants. The activation of co-ordinated mechanisms able to perceive and respond to changes in cell iron levels is of main importance in maintaining iron homeostasis, and, consequently, to prevent biochemical and physiological alterations.

Iron deficiency primarily affects structure and functioning of the chloroplast and induces a marked reduction in chlorophyll and, although less intense, in carotenoid content. Due to the presence of haem groups and Fe-S clusters in many electron carriers, the thylakoid photoelectron transport rate is particularly sensitive to iron starvation, mainly at the PSI level. A decrease in the potential and in the actual quantum yield of PSII is also often observed in iron-deficient plant species, accompanied by an increase in non-photochemical quenching and the activation of the photoprotective xantophyll cycle, pointing to an increased need to dissipate excess excitation energy. Alterations in the physiological parameters result from deep changes in thylakoid organisation and polypeptide composition of photosystems. Quali-quantitative modifications observed in polypeptide composition of chlorotic thylakoids of both herbaceous and tree plants, including LHCII phosphorylation which, by allowing antennae dissociation limits over-excitation of the reaction centers, point to a major sensitivity of PSI to iron starvation as well as to lime-induced chlorosis.

Keywords Chloroplast • Fe(III)-chelate reductase • iron deficiency • photosystem • thylakoids • xanthophyll cycle

1 Iron Availability

Iron-deficiency represents a very frequent plant disorder affecting both spontaneous and cultivated plants in a great proportion of earth lands. Despite iron is present in most soils at high quantities, usually exceeding the plant requirements, its bioavailability in aerobic environments, characterised by neutral or alkaline pH, is limited due to its poor solubility (Guerinot and Yi 1994; Mengel et al. 2001). In aerobic soils, iron is in fact present mainly in the form of Fe (III) as a constituent of scarcely soluble oxyhydroxide polymers. In addition, the high pH typical of calcareous soils, which make up about 30% of the land surface, often reduce iron availability below the threshold sufficient to meet plant needs. The region of minimum Fe solubility occurs between pH 7.5 and 8.5, which is the pH range of most calcareous soils. Moreover, also the presence of bicarbonate is believed to affect both iron uptake and its translocation and availability to the green tissues, although contrasting evidences are reported (Lucena 2000; Mengel et al. 2001; Nikolic and Römheld 2002; Kosegarten et al. 2004). A great number of iron chelates have been developed in order to provide available iron to plants growing under alkaline and calcareous soils and to allow commercial yield (García-Marco et al.

A. Castagna (✉) and A. Ranieri
Dipartimento di Chimica e Biotecnologie Agrarie, Pisa University, Pisa, 56124 Italy

S. Donnini
Dipartimento di Produzione Vegetale, Milan University, Milan, 20133 Italy

M. Ashraf et al. (eds.), *Salinity and Water Stress*,

2006) but the application of supplementary iron chelate implies high management costs and may represent a considerable hazard for soil and water pollution.

2 Mechanisms of Iron Acquisition

Plants respond to limited iron availability by inducing responses aimed at efficiently acquiring the element from the rhizosphere. Depending on the mechanism developed to facilitate iron mobilisation and uptake, plants can be classified into two groups: strategy I plants (all dicots and non-graminaceous monocots) and strategy II plants (graminaceous plants) (Römheld and Marschner 1986) (Fig. 21.1).

In strategy I plants, iron mobilization is achieved by the combined action of both morphological and metabolic changes, mainly at the root level. Increased numbers of root hairs and the formation of transfer cells enlarge absorption area of plant for taking up more iron. Since iron is taken up by the roots only in the ferrous form, the activation of Fe(III)-chelate reductase (FC-R) is fundamental for iron acquisition. At the same time, the plasma membrane-localised H^+-ATPase enhances extrusion of protons in the rhizosphere thus increasing the solubility of the sparingly soluble iron form and generating the driving force for the uptake of the ion (Rabotti and Zocchi 1994). To avoid cytoplasm alkalinization as a consequence of active proton extrusion, the activation of phosphoenolpyruvate carboxylase (PEPC) seems to be an obligatory step to balance the pH change under iron starvation (De Nisi and Zocchi 2000). Ferrous ions are taken up by the roots by means of a Fe^{2+} transporter. IRT1 (iron-regulated transporter 1), belonging to the ZIP family (Eide et al. 1996), is now believed to be the major transporter for high affinity Fe uptake, as indicated by the lethal chlorotic phenotype of *IRT1* knockout mutants of *Arabidopsis* (Connolly et al. 2002; Henriques et al. 2002; Vert et al. 2002). A variety of iron-regulated transporters has been identified during the past few years. IRT2, a protein with close homology to IRT1 at the amino acid level, is also expressed in root epidermal cells under Fe-deficiency (Eckhardt et al. 2001). However, it cannot substitute for loss of IRT1 (Grotz and Guerinot 2002) and seems to play a minor role in iron uptake. NRAMP proteins represent a further family of transporters with possible functions in iron homeostasis. Members of the NRAMP family are

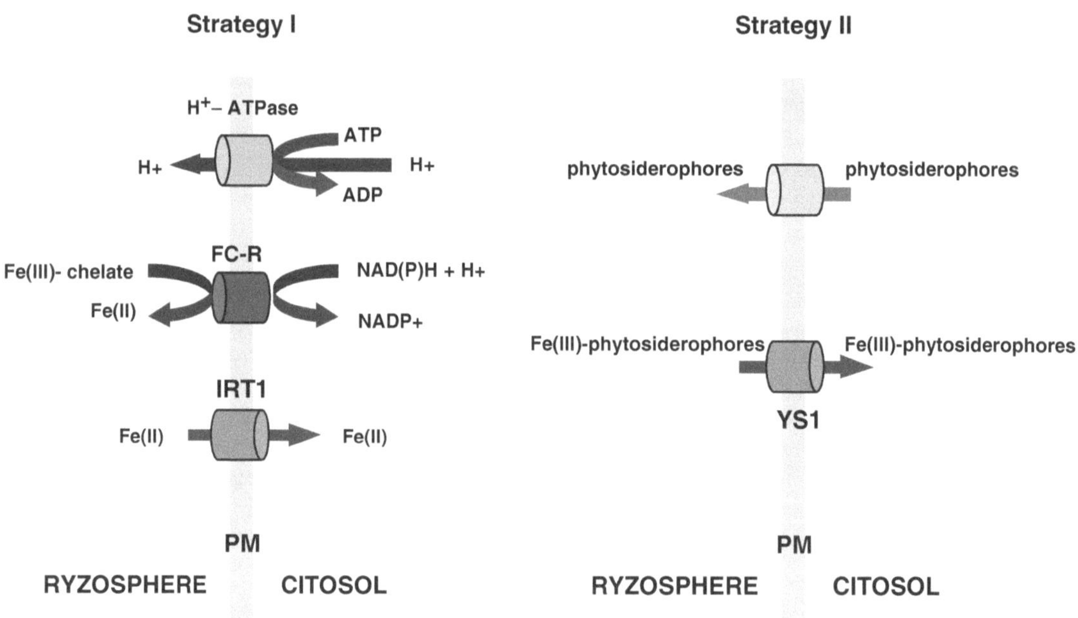

Fig. 21.1 Mechanism of iron uptake by higher plants

up-regulated by iron deficiency in *Arabidopsis*, suggesting a function in the transport of iron (Curie et al. 2000; Thomine et al. 2003).

Plants adopting strategy II respond to iron shortage by releasing high-affinity Fe(III) chelators (phytosiderophores), which, after binding soil iron, are taken up by transporters specific for the Fe(III)-siderophore complex (Römheld and Marschner 1986; Curie et al. 2001). A transporter, classified as a member of the oligopeptide transporter (OPT) family, has been cloned from maize and named yellow stripe1 (YS1) after the phenotype of a maize mutant deficient in phytosiderophore uptake (Curie et al. 2001). In grasses, both the efflux of phytosiderophores and the steady-state level of YS1 are strongly enhanced by iron deficiency (Mori 1999; Curie et al. 2001).

The plant ability to co-ordinately activate the adaptive processes under iron shortage ultimately determines the efficiency of iron acquisition. Plant species, as well as cultivar and clones within species, widely differ in their relative susceptibility to chlorosis. In 1971 a spontaneous tomato (*Lycopersicon esculentum* M.) mutant was isolated from the T3238 cultivar. The mutant, named T3238*fer* (T*fer*), was shown to be iron-deficient when grown in the same conditions as the asymptomatic wild-type, T3238*FER* (T*FER*) (Brown et al. 1971). Differently from the wild type plants, this mutant can only survive when fed with high amounts of easily degradable Fe chelates and is unable to activate neither FC-R nor H^+-ATPase and PEPC (Donnini et al. 2004).

Similarly to herbaceous dicots, some tolerant woody genotypes are able to improve Fe acquisition through enzymatic Fe reduction by FC-R. Under iron starvation the apex roots of pear (*Pyrus communis* L., cv Conference; chlorosis-tolerant genotype) exhibited higher FCR activity, which was found to increase, even if to a lesser extent, also when cuttings were grown in the presence of bicarbonate. Conversely, no increase in FC-R activity was observed in quince rootstocks (*Cydonia oblonga* Mill., MA and BA29; chlorosis-susceptible genotypes) grown under the same conditions of pear. Moreover, differently to the two quince genotypes, cv Conference was able to decrease the pH medium by enhancing proton extrusion by root H^+-ATPase and to control cytoplasm pH by activating PEPC (Donnini et al. 2006), suggesting that the major tolerance of pear to lime-induced chlorosis could be linked to the synergetic and co-ordinated activation of the reducing-acidifying enzymes.

3 Iron Nutrition and Plant Metabolism: The Chloroplast

Iron is an essential nutrient for plants and it is crucial for a variety of cellular functions owing to its tendency to form chelate complexes with various ligands and its ability to undergo valency change, which gives this element a pivotal role in the cell redox systems. The most well known function of iron is linked to its presence in the prosthetic groups haem and haemin of many enzyme systems (catalases, peroxidases, cytochrome oxidases and cythochromes) as well as in many proteins involved in metabolic processes such as photosynthesis, respiration and N_2 fixation, as Fe-S cluster. Another Fe-containing enzyme of universal importance is ribonucleotide reductase which, bringing about the reduction of ribonucleotide diphosphate to deoxy-ribonucleotide, is fundamental for DNA synthesis.

Iron deficiency primarily affects structure and functioning of the chloroplast (Tognini et al. 1996; Soldatini et al. 2000; Morales et al. 2001). It has been in fact calculated that about 80% of the leaf iron is localised in this organelle, mainly in the thylakoid membranes, which contain about 60% of total leaf iron (Terry and Abadía 1986). In particular, iron is an essential element for both chlorophyll synthesis and thylakoid stability, since it is an essential constituent of the cytochrome b559 of the photosystem II (PSII), the two cytochromes and the Rieske protein of the b6-f complex, the Fe-S proteins of photosystem I (PSI) and the ferredoxin (Nishio et al. 1985; Abadía 1992; Tognini et al. 1996). Iron is also present, as ferric ion, at the oxidising site and, as quinone-Fe complex, at the reducing site of PSII. Moreover, thylakoid PSI and PSII pigment-binding apoproteins are unstable and undergo proteolysis, unless associated with chlorophylls (Preiss and Thornber 1995). Iron-deficiency has been also found to reduce leaf photosynthetic capacity by decreasing both RUBISCO gene expression and carboxylation activity (Winder and Nishio 1995; Ferraro et al. 2003). Thus, iron deficiency has an important influence on photosynthesis and a detrimental impact on carbon assimilation and, consequently, on plant growth and yield.

3.1 Iron Deficiency and Chloroplast Functionality

The presence of iron in haem groups and Fe-S clusters of many electron carriers along the photosynthetic transport chain makes thylakoid functionality particularly susceptible to iron shortage (Table 21.1). Photoelectron transport rate of thylakoid membranes, measured by polarographic techniques in the presence of artificial electron donors and acceptors, was found to be significantly affected by iron starvation. In particular, in both Fe-inefficient (T*fer*) and Fe-efficient (T*FER*) tomato genotypes the electron transport activity of the whole chain decreased as iron concentration in the nutrient medium was reduced, with a particular striking effect on the mutant (T*fer*) (Donnini et al. 2003). Similar results are reported in iron deficient field-grown peach plants (*Prunus persica* L., Nedunchezhian et al. 1997) and in iron-deprived sunflower, where the decrease in thylakoid electron transport rate was found to be mainly due to reduced photosystem I (PSI) activ-

Table 21.1 A summary of the main effects induced by iron shortage on chloroplast functionality and on pigment and polypeptide composition of thylakoid membranes

Effect	Treatment	Species	Reference
Decrease in thylakoid electron transport rate	Iron deficiency	*Helianthus annuus*; *Lycopersicon esculentum*	Tognini et al. (1996), Donnini et al. (2003)
	Growth on calcareous soils	*Prunus persica*	Nedunchezhian et al. (1997)
Changes in chlorophyll *a* parameters	Iron deficiency	*Beta vulgaris*; *Lycopersicon esculentum*; *Prunus persica*; *Quercus suber*	Morales et al. (1991), Morales et al. (1998), Gogorcena et al. (2001), Donnini et al. (2003), Molassiotis et al. (2006)
	Bicarbonate supply	*Prunus persica*	Molassiotis et al. (2006)
	Growth on calcareous soils	*Prunus persica*; *Pyrus communis*	Nedunchezhian et al. (1997), Morales et al. (2000)
Decrease in chlorophyll-carotenoid content	Iron deficiency	*Beta vulgaris*; *Cydonia oblonga*; *Helianthus annuus*; *Lycopersicon esculentum*; *Pyrus communis*; *Quercus suber*	Morales et al. (1990), Tognini et al. (1996), Gogorcena et al. (2001), Ranieri et al. (2001), Donnini et al. (2003, 2006)
	Bicarbonate supply	*Cydonia oblonga*; *Pyrus communis*	Donnini et al. (2006)
	Growth on calcareous soils	*Pyrus communis*	Morales et al. (1994)
Activation of the xanthophyll cycle	Iron deficiency	*Beta vulgaris*; *Cydonia oblonga*; *Helianthus annuus*; *Lycopersicon esculentum*; *Pyrus communis*	Morales et al. (1990), Soldatini et al. (2000), Donnini et al. (2003, 2006)
	Bicarbonate supply	*Cydonia oblonga*	Donnini et al. (2006)
	Growth on calcareous soils	*Pyrus communis*	Morales et al. (2000)
Rearrangement of polypeptide composition of photosystems	Iron deficiency	*Chlamydomonas reinhardtii*; *Cydonia oblonga*; *Dunaliella salina*; *Helianthus annuus*; *Pyrus communis*; *Rhodella violacea*	Soldatini et al. (2000), Moseley et al. (2002), Desquilbet et al. (2003), Naumann et al. (2005), Varsano et al. (2006), Donnini et al. (2006)
	Bicarbonate supply	*Cydonia oblonga*	Donnini et al. (2006)

ity (Tognini et al. 1996). However, although at a lesser extent than PSI, also photosystem II (PSII) undergoes functional alteration following iron deprivation. The chlorophyll *a* fluorescence analysis supplies useful information about PSII efficiency. When grown under limited iron supply, the Fe-inefficient tomato mutant displayed a −25% decrease in the Fv/Fm ratio (Donnini et al. 2003), a parameter which indicates the potential quantum yield of PSII under dark-adapted conditions. A decrease in this parameter, which is commonly considered a symptom of photoinhibition, was also observed in iron deficient sugar beet (Morales et al. 1991) and cork oat (Gogorcena et al. 2001) and in peach grown in calcareous soils (Nedunchezhian et al. 1997) or following iron omission or bicarbonate addition to the nutrient solution (Molassiotis et al. 2006). At steady-state photosynthesis, the effects of iron starvation on the actual PSII efficiency (Φ_{PSII}) were even more accentuated and were associated to an increase in the proportion of closed PS II centers (estimated by qp) and also to a decrease in the intrinsic PS II efficiency (Φ_{exc}) (Morales et al. 1998, 2000; Donnini et al. 2003). Moreover, Fe deficiency modified the allocation of the light absorbed by the PS II antenna at steady-state photosynthesis, increasing the proportion of light dissipated thermally within the PS II antenna (as indicated by the enhanced non-photochemical quenching NPQ), pointing to an increased need to dissipate the excess light that cannot be used in photosynthesis (Abadía et al. 1999; Gogorcena et al. 2001).

3.2 Photosynthetic and Accessory Pigments Under Iron Shortage

The first visible symptom of inadequate iron nutrition is the appearance of interveinal chlorosis in young leaves. Paradoxically, chlorotic leaves, often display a higher content of iron than the green ones. This phenomenon, known as chlorosis paradox, is due to immobilisation of iron in the apoplast in an unavoidable form. Chlorotic leaves are characterised by a marked reduction in the chlorophyll content (Table 1; Morales et al. 1990, 1994; Tognini et al. 1996; Gogorcena et al. 2001). The iron-inefficient tomato mutant (T*fer*) and the wild type (T*FER*) plants grown in hydroponic solution with 40 μM Fe-Na EDTA, which was found to be the most suitable iron level to induce phenotypic differences between the two genotypes without being lethal for the mutant, exhibited a significant decrease in total leaf iron content and in the levels of chlorophyll *a* and *b* as compared to plants grown under optimal iron concentration (80 μM Fe-Na EDTA). The reduction was particularly evident in T*fer*, being about twice the decrease measured in the wild type (Donnini et al. 2003). Similarly, iron deprivation induced a decrease down to approximately 15% of the control value in total chlorophyll content of sunflower (*Helianthus annuus* L.) (Ranieri et al. 2001). Also the chlorophyll content of woody plants, as pear (cv Conference) and quince genotypes (MA and BA29), underwent a decrease following both iron starvation (ranging from about −60% to −90%) and the presence of bicarbonate in the medium (ranging from about −15% to −70%), although the cv Conference was less affected than the quince genotypes, in particular under bicarbonate nutrition (Donnini et al. 2006).

The loss of chlorophyll is usually accompanied by a significant decrease in the carotenoid content (Table 1). However, depending on plant species as well as on the kind of carotenoid, the impact of iron-chlorosis can be less intense than on chlorophylls. This is particularly true for the sum of the three carotenoids (violaxanthin, V; antheraxanthin, A and zeaxanthin, Z) involved in the xantophyll cycle, as observed in both Fe-inefficient (T*fer*) and Fe-efficient (T*FER*) tomato genotypes (Donnini et al. 2003), in cork oak (*Quercus suber*, Gogorcena et al. 2001) and in pear (Morales et al. 2000). The activation of the xanthophyll cycle, which involves the de-epoxidation of V to A and Z, represents one of the main processes involved in dissipation of the harmful excess energy in an attempt to avoid thylakoid overenergization and prevent photodamage (Demmig-Adams and Adams 1996). Such an activation, which is well described by the increase in the de-epoxidation index (calculated as percentage of the de-epoxidated forms over the sum of the three xanthophylls), is frequently observed in chlorotic leaves of many species (Table 1), as sugar beet (*Beta vulgaris* L.; Morales et al. 1990), sunflower (Soldatini et al. 2000), tomato (Donnini et al. 2003), pear (Morales et al. 2000; Donnini et al. 2006) and in quince rootstocks grown without iron or with bicarbonate in the nutrient solution (Donnini et al. 2006).

3.3 Biochemical Effects of Iron Deprivation on Photosystem Polypeptides

Alterations in the physiological parameters aforementioned result from deep changes in thylakoid organisation and polypeptide composition of photosystems induced by limited iron availability (Table 21.1). Soldatini et al. (2000) performed a detailed analysis of the pigment-protein complexes of iron-deficient sunflower plants. These authors, by western blot analysis and two dimensional electrophoresis of thylakoid proteins, revealed quali-quantitative modification in polypeptide composition of chlorotic thylakoid membranes. In particular, the content of I_A and I_B proteins of the PSI core complex (CC) was found to decrease by about 35% following iron starvation. Conversely, the two polypeptides of the PSII reaction center (D1 and D2), as well as the two inner PSII antennae (CP43 and CP47) were unaffected by iron shortage, and the major PSII light harvesting complex (LHCII) even increased (+20%) over the control (Soldatini et al. 2000), confirming that PSI is the major target of iron starvation. The cytochrome b6/f complex, which transfers electrons between the reaction center complexes of PSI and PSII, is reported to be the thylakoid complex most sensitive to iron deficiency after PSI (Perez et al. 1995). In chlorotic sunflower plants, the tetramethylbenzidine staining of haem groups of polypeptides resolved by SDS-PAGE revealed indeed a decrease of both cytochrome b-563 and cytochrome f which, together with the decline in PSI CC, may account for the decrease in the electron transport activity of thylakoid membranes (Soldatini et al. 2000). Iron shortage induced a marked alteration in the polypeptide composition of photosystems in pear and quince genotypes as well, these latter being affected also by bicarbonate supply to the nutrient solution. Major decreases in polypeptide content were again observed at the PSI level, although, differently from sunflower, LHCI was more affected than the core complex (Donnini et al. 2006).

Under environmental stresses the incident light harvested by antenna complexes can exceed the energy need to drive photochemistry. As a consequence the photosystems have to regulate the energy distribution to the reaction centers to prevent photoinhibition. In addition to the activation of xanthophyll cycle, plants have evolved another protective mechanism, consisting in the phosphorilation of LCII complexes, which allows their dissociation from the CC, thus limiting their over-excitation. The use of antibodies raised against phosphorilated threonine showed an enhancement of LHCII phosphorilation in the pear cv Conference following bicarbonate feeding (Donnini et al. 2006), confirming the deep rearrangement of the photosystems in chlorotic plants.

Changes in the structure and composition of photosystems under iron deprivation have been reported also in algae. In the green alga *Chlamydomonas reinhardtii* antenna proteins of the two photosystems were found to be differentially affected by iron deficiency, LHCI polypeptides being either increased or suppressed whereas LHCII content and composition were unaffected (Naumann et al. 2005). Moreover, in *C. reinhardtii* as well as in the red alga *Rhodella violacea*, iron deprivation induced uncoupling of LHCI from PSI reaction centers, leading to a decrease in the efficiency of energy transfer (Moseley et al. 2002; Desquilbet et al. 2003). In the halotolerant alga *Dunaliella salina*, iron deficiency induced a deep remodelling of PSI, due to the induction of a new Chl *a/b* binding protein homolog, termed Tidi, most probably localised at the peripheral antenna of PSI (Varsano et al. 2006). Tidi accumulation was correlated with a large decrease in PSI RC polypeptides, and with higher rates of PSI electron transport, leading the Authors to hypothesise that *D. salina* may be able to maintain high growth rate and photosynthetic activity under iron limitation by creating fewer but more efficient PSI units (Varsano et al. 2006).

4 Concluding Remarks

Although iron makes up about 5% by weight of the Earth's crust, its low solubility in oxygenated environments, and the high levels of bicarbonate often present in soils, can decrease iron availability far below that required by plants. Chloroplast, in particular thylakoid membranes, being extremely rich in haem groups and Fe-S clusters, is very sensitive to iron shortage. This reflects on chloroplast metabolism, with deep alteration at both biochemical and physiological level, and, on a larger scale, on carbon assimilation and, consequently, on plant growth and yield. The plant ability to efficiently acquire iron depends on the activation of co-ordinated mechanisms that can sense and respond to changes in iron levels, either within the cell or in its immediate environment. In higher plants, uptake seems to be correlated with the requirement of the shoot rather than with the iron concentration of the root cells, implying the involvement of an iron shoot sensor that responds to overall iron needs and transfer the information to the roots by the mean of a signal communicating network, ultimately leading to control the physiological root responses. The challenge is to unravel the still largely unknown nature of signals involved in the regulation of iron homeostasis and their reciprocal interactions.

References

Abadía J (1992) Leaf responses to Fe deficiency: a review. J Plant Nutr 15: 1669–1713.

Abadía J, Morales F, Abadía A (1999) Photosystem II efficiency in low chlorophyll, iron-deficient leaves. Plant Soil 215: 183–192.

Brown JC, Chaney RL, Ambler JE (1971) A new tomato mutant inefficient in the transport of iron. Physiol Plant 25: 48–53.

Connolly EL, Fett JP, Guerinot ML (2002) Expression of the IRT1 metal transporter is controlled by metals at the levels of transcript and protein accumulation. Plant Cell 14: 1347–1357.

Curie C, Alonso JM, Le Jean M, Ecker JR, Briat J-F (2000) Involvement of NRAMP1 from *Arabidopsis thaliana* in iron transport. Biochem J 347: 749–755.

Curie C, Panaviene Z, Loulergue C, Dellaporta SL, Briat JF, Walker EL (2001) Maize *yellow stripe1* encodes a membrane protein directly involved in Fe(III) uptake. Nature 409: 346–349.

Demmig-Adams B, Adams WW III (1996) The role of the xanthophyll cycle carotenoids in the protection of photosynthesis. Trends Plant Sci 1: 21–26.

De Nisi P, Zocchi G (2000) Phosphoenolpyruvate carboxylase in cucumber (*Cucumis sativus* L.) roots under iron deficiency: activity and kinetic characterization. J Exp Bot 51: 1903–1909.

Desquilbet TE, Duval JC, Robert B, Houmard J, Thomas JC (2003) In the unicellular red alga *Rhodella violacea* iron deficiency induces an accumulation of uncoupled LHC. Plant Cell Physiol 44: 1141–1151.

Donnini S, Castagna A, Guidi L, Zocchi G, Ranieri A (2003) Leaf responses to reduced iron availability in two tomato genotypes: T3238*FER* (iron efficient) and T3238*fer* (iron inefficient). J Plant Nutr 26: 2137–2148.

Donnini S, Zocchi G, Soldatini GF, Castagna A, Ranieri A (2004) Influenza della ridotta disponibilità di ferro su alcune risposte biochimiche di genotipi di pomodoro: T3238*FER* (ferro-efficiente) e T3238*fer* (ferro-inefficiente). *Atti XXI Convegno Società Italiana Chimica Agraria*, pp. 168–175.

Donnini S, Zocchi G, Castagna A, Abdelly C, Ranieri A (2006) Identification of morphological, biochemical and physiological parameters useful to characterize nutritional stress status in arboreous species differently tolerant to chlorosis. In: Abdelley C, Ozturk M, Ashraf M, Grignon C (eds) Biosaline agriculture and high salinity tolerance, Birkhauser Verlag, Basel, pp 53–63

Eckhardt U, Margues AM, Buckhout TJ (2001) Two iron-regulated cation transporters from tomato complement metal uptake-deficient yeast mutants. Plant Mol Biol 45: 437–448.

Eide D, Broderius M, Fett J, Guerinot ML (1996) A novel iron regulated metal transporter from plants identified by functional expression in yeast. Proc Natl Acad Sci USA 93: 5624–5628.

Ferraro F, Castagna A, Soldatini GF, Ranieri A (2003) Tomato (*Lycopersicon esculentum* M.) T3238*FER* and T3238*fer* genotypes. Influence of different iron concentrations on thylakoid pigment and protein composition. Plant Sci 164: 783–792.

García-Marco S, Martínez N, Yunta F, Hernández-Apaolaza L, Lucena J (2006) Effectiveness of ethylenediamine-*N*(*o*-hydroxyphenylacetic)-*N'*(*p*-hydroxyphenylacetic) acid (*o,p*-EDDHA) to supply iron to plants. Plant Soil 279: 31–40.

Gogorcena Y, Molias N, Larbi A, Abadía J, Abadía A (2001) Characterization of the responses of cork oak (*Quercus suber*) to iron deficiency. Tree Physiol 21: 1335–1340.

Grotz N, Guerinot ML (2002) Limiting nutrients: an old problem with new solutions? Curr Opin Plant Biol 5: 158–163.

Guerinot ML, Yi Y (1994) Iron: nutritious, noxious, and not readily available. Plant Physiol 104: 815–820.

Henriques R, Jásik J, Klein M, Martinoia E, Feller U, Schell J, Pais MS, Koncz C (2002) Knock-out of *Arabidopsis* metal transporter gene *IRT1* results in iron deficiency accompanied by cell differentiation defects. Plant Mol Biol 50: 587–597.

Kosegarten H, Hoffmann B, Rroco E, Grolig F, Glüsenkamp KH, Mengel K (2004) Apoplastic pH and FeIII reduction in young sunflower (*Helianthus annuus*) roots. Physiol Plant 122: 95–106.

Lucena JJ (2000) Effect of bicarbonate, nitrate and other environmental factors on iron deficiency chlorosis. A review. J Plant Nutr 23: 1591–1606.

Mengel K, Kirkby EA, Kosegarten H, Appel T (2001) Iron. In *Principles of Plant Nutrition*. Kluwer, Dordrecht, pp. 553–571.

Molassiotis A, Tanou G, Diamantidis G, Patakas A, Therios I (2006) Effects of 4-month Fe deficiency exposure on Fe reduction mechanism, photosynthetic gas exchange, chlorophyll fluorescence and antioxidant defense in two peach rootstocks differing in Fe deficiency tolerance. J Plant Physiol 163: 176–185.

Morales F, Abadía A, Abadía J (1990) Characterization of the xanthophyll cycle and other photosynthetic pigment changes induced by iron deficiency in sugar beet (*Beta vulgaris* L.). Plant Physiol 94: 607–613.

Morales F, Abadía A, Abadía J (1991) Chlorophyll fluorescence and photon yield of oxygen evolution in iron-deficient sugar beet (*Beta vulgaris* L.) leaves. Plant Physiol 97: 886–893.

Morales F, Abadía A, Abadía J (1998) Photosynthesis, quenching of chlorophyll fluorescence and thermal energy dissipation in iron-deficient sugar beet leaves. Aust J Plant Physiol 25: 403–412.

Morales F, Abadía A, Belkhodja R, Abadía J (1994) Iron deficiency-induced changes in the photosynthetic pigment composition of field-grown pear (*Pyrus communis* L.) leaves. Plant Cell Environ 17: 1153–1160.

Morales F, Belkhodja R, Abadía A, Abadía J (2000) Photosystem II efficiency and mechanisms of energy dissipation in iron-deficient, field-grown pear trees (*Pyrus communis* L.). Photosynth Res 63: 9–21.

Morales F, Moise N, Quílez R, Abadía A, Abadía J, Moya I (2001) Iron deficiency interrupts energy transfer from a disconnected part of the antenna to the rest of Photosystem II. Photosynth Res 70: 207–220.

Mori S (1999) Iron acquisition by plants. Curr Opin Plant Biol 2: 250–253.

Moseley JL, Allinger T, Herzog S, Hoerth P, Wehinger E, Merchant S, Hippler M (2002) Adaptation to Fe deficiency requires remodeling of the photosynthetic apparatus. EMBO J 21: 6709–6720.

Naumann B, Stauber EJ, Busch A, Sommer F, Hippler M (2005) N-terminal processing of Lhca3 is a key step in remodeling of the photosystem I-light harvesting complex under iron deficiency in *Chlamydomonas reinhardtii*. J Biol Chem 280: 20431–20441.

Nedunchezhian N, Morales F, Abadía A, Abadía J (1997) Decline in photosynthetic electron transport activity and changes in thylakoid protein pattern in field grown iron deficient peach (*Prunus persica* L.). Plant Sci 129: 29–38.

Nikolic M, Römheld V (2002) Does high bicarbonate supply to roots change availability of iron in the leaf apoplast? Plant Soil 241: 67–74.

Nishio JN, Abadia J, Terry N (1985) Chlorophyll-proteins and electron transport during iron nutrition-mediated chloroplast development. Plant Physiol 78: 296–299.

Perez C, Val J, Monge E (1995) Effects of iron deficiency on photosynthetic structure in peach (*Prunus persica* L. Batsch) leaves. In *Iron Nutrition in Soils and Plants*. Kluwer, Dordrecht, pp. 183–189.

Preiss S, Thornber JP (1995) Stability of the apoproteins of light harvesting complex I and II during biogenesis of thylakoids in the chlorophyll *b*-less mutant chlorina f2. Plant Physiol 107: 709–717.

Rabotti G, Zocchi G (1994) Plasma membrane-bound H^+-ATPase and reductase activities in Fe-deficient cucumber roots. Physiol Plant 90: 779–785.

Ranieri A, Castagna A, Baldan B, Soldatini GF (2001) Iron deficiency differently affects peroxidase isoforms in sunflower. J Exp Bot 354: 25–35.

Römheld V, Marschner H (1986) Evidence for a specific uptake system for iron phytosiderophores in roots of grasses. Plant Physiol 80: 175–180.

Soldatini GF, Tognini M, Baldan B, Castagna A, Ranieri A (2000) Alterations in thylakoid membrane composition induced by iron starvation in sunflower plants. J Plant Nutr 23: 1717–1732.

Terry N, Abadía J (1986) Function of iron in chloroplasts. J Plant Nutr 9: 609–646.

Thomine S, Lelièvre F, Debarbieux E, Schroeder JI, Barbier-Brygoo H (2003) AtNRAMP3, a multispecific vacuolar metal transporter involved in plant responses to iron deficiency. Plant J 34: 685–695.

Tognini M, Castagna A, Ranieri A, Soldatini GF (1996) Effects of iron starvation on photosynthetic electron transport and pigment composition in sunflower plants. Plant Physiol Biochem (Special Issue): pp 102.

Varsano T, Wolf SG, Pick U (2006) A chlorophyll *a/b* binding protein homolog which is induced by iron deficiency is associated with enlarged photosystem I units in the eucaryotic alga *Dunaliella salina*. J Biol Chem 281: 10305–10315.

Vert G, Grotz N, Dedaldechamp F, Gaymard F, Guerinot ML, Briat JF, Curie C (2002) IRT1, an *Arabidopsis* transporter essential for iron uptake from the soil and plant growth. Plant Cell 14: 1223–1233.

Winder TL, Nishio JN (1995) Early iron deficiency stress response in leaves of sugar beet. Plant Physiol 108: 1487–1494.

Chapter 22
Boron Deficiency in Rice in Pakistan: A Serious Constraint to Productivity and Grain Quality

A. Rashid, M. Yasin, M.A. Ali, Z. Ahmad, and R. Ullah

Abstract Recently, boron (B) deficiency has been established to be a serious constraint to crop productivity in Pakistan. Despite being categorized as tolerant to B deficiency, we suspected that rice also suffers with this nutrient disorder. Multi-location, multi-year field experiments were conducted during 2002–2004 in B-deficient (HWE B, 0.21–0.42 mg kg^{-1}), calcareous ($CaCO_3$, 1.5–5.7%) soils in the major rice-growing areas of Punjab and Sindh. Basmati type rice cultivars, i.e., *Super Basmati* and *Basmati–385*, and coarse grain cultivars, i.e., IR–6 and KS–282, were included in the study. Average paddy yield increase with B was 21% each in cv. *Super Basmati* and cv. *Basmati-385*, 30% in cv. IR–6, and 14% in cv. KS–282. The substantial yield increases were primarily the consequence of reduced panicle sterility and increased productive tillers. Better B nutrition of plants also enhanced milling recovery and head rice recovery, and improved kernel quality traits, like stickiness, elongation-upon-cooking, and bursting-upon-cooking. Thus, soil B deficiency not only hampers rice productivity but also impairs its cooking quality. The economics of B use in rice is highly attractive, with value-cost ratios of 45:1 to 26:1 in various cultivars. Cooking quality improvement and residual beneficial effect of B are additional benefits. Consequently, this research has led to B use recommendation (@0.75 kg B ha^{-1}) in rice in Punjab province. As fertilizer B use adoption may remain problematic in the foreseeable future, rice plant improvement for B use efficiency merits attention of biotechnologists.

Keywords Boron deficiency • rice • productivity • cooking quality

1 Introduction

Rice (*Oryza sativa* L.) is an important staple cereal in Pakistan, second only to wheat, and is grown on 2.3 Mha alluvial, calcareous, low organic matter soils, with an average paddy yield of 2.00 t ha^{-1} (GOP 2004). Fertilizer use in rice predominantly pertains to nitrogen (N), and to a lesser extent to phosphorus (P) and zinc (Zn) (Rashid et al. 2000a). Thus, one major cause of low rice productivity, compared with much higher yield potential (i.e., 4.00–4.50 t ha^{-1}) is imbalanced nutrient management.

Despite being considered tolerant to boron (B) deficiency (Shorrocks 1997; Savithri et al. 1999), yield increases with B use have been observed. In Pakistan, first-ever positive responses of rice to B use were observed about three decades ago (Chaudhry et al. 1977). Though the yield increases with B, in two predominant rice cultivars (i.e., *Basmati-370* and IR-6) were appreciable, only recently this micronutrient disorder has received adequate research attention (Rashid et al. 2002b). Whereas Savithri et al. (1999) stated that B deficiency is not a severe problem in rice, they postulated that it may be a problem in calcareous, sodic, and excessively permeable soils in riverine flood plains. In the mean time, Dunn et al. (2005) have reported rice yield increases with B use in Missouri, USA. This paper reports the role of B nutrition on rice productivity and cooking quality, observed in our multi-year extensive field experiments in major growing areas of Pakistan.

A. Rashid (✉)
Pakistan Atomic Energy Commission, P.O. Box 1114, Islamabad, Pakistan
e-mail: abdul.rashid@comsats.net.pk

M. Yasin and R. Ullah
National Agricultural Research Center Islamabad-45500, Pakistan

M.A. Ali
Agricultural Research & Adaptive Research, Department, Lahore, Pakistan

Z. Ahmad
Engro Chemicals Pakistan Ltd., Thatta, Pakistan

M. Ashraf et al. (eds.), *Salinity and Water Stress*,

2 Materials and Methods

During 2002–2004, field experiments were carried out in major rice-growing areas of Punjab and Sindh provinces, of Pakistan using elite fine-grain basmati-type aromatic rice cultivars as well as medium-long grain coarse grain cultivars. There were two sets of field experiments. In Experiment 1, on cvs. *Super Basmati* and *Basmati*-385 in Punjab and IR-6 in Sindh, B rates were 0, 0.5, 1.0, 1.5/2.0 kg ha^{-1}, and Experiment 2 (conducted on cvs. *Super Basmati*, *Basmati*-385 and KS-282 in Punjab) had only two B rates, i.e., control (no B applied) and 1.0 kg B ha^{-1}. Experiment 1 was laid out in a randomized complete block design, with four replications, and Experiment 2 was completely randomized design, treating locations as replications. Soil B was determined by 0.1 M HCl extraction (Rashid et al. 1994) and colorimetry using Azomethine–H (Bingham 1982). Soil properties of field sites are presented in Table 22.1.

In all field experiments, B was broadcast applied as borax (10.5% B), prior to transplanting, along with basal fertilizers. Blanket fertilization consisted of 120 kg N ha^{-1} as urea, 44 kg P ha^{-1} as DAP, and 10 kg Zn ha^{-1} as zinc sulfate. Full dose of all nutrients, except for N, were applied prior to transplanting. Nitrogen was applied in three splits: one-third each prior to transplanting, 21 days after transplanting and 45 days after transplanting. Average plot size was about 1,000 m^2. Experimental data included plant height, productive tillers, panicle sterility, and paddy and straw yields. Flag leaves, sampled at heading, and mature paddy were analyzed for B content by dry ashing (Gaines and Mitchell 1979) and colorimetry using Azomethine–H (Bingham 1982). Rice grain were analyzed for physico-chemical characteristics, organoleptic parameters, and eating quality.

3 Results and Discussion

3.1 Yield Increases with Boron Fertilization

In Experiment 1, carried out during 2003 and 2004 at 14 locations employing graded levels of B, paddy yield increases with B use were substantial in both rice varieties (i.e., cvs. *Super Basmati* and IR-6) at all field sites ($P \leq 0.05$; Figs. 22.1, 22.2). Almost in all cases, maximum paddy yield was mostly obtained with 1.0 kg B ha^{-1} (Figs. 22.1, 22.2). Paddy yield increases with B were 11–31% during 2003 and 7–30% during 2004 in cv. *Super Basmati*, and 18–34% in cv. IR-6. Mean paddy yield increases with B use were 26% in cv. *Super Basmati* and 27% in cv. IR-6. However, near-maximum (95% of maximum) paddy yield was associated with 0.75 kg B ha^{-1} in cv: *Super Basmati* and 0.85 kg B ha^{-1} in cv. IR-6 (Figs. 22.1, 22.2). In Experiment 2, having 0 and 1.0 kg ha^{-1} B rates, paddy yield increases with B use were 17–20% in cv. *Super Basmati*, 20–23% in cv. *Basmati*-385, and 14% in KS-cv. 282 ($P \leq 0.05$; Table 22.2). Yield increases were consistent in all rice cultivars at all field sites during both experimental years.

In Experiment 1, straw yield increases with graded levels of B were 15–36% during 2003 and 6–23% during 2004 in cv. *Super Basmati* and 18–34% in cv. IR-6

Table 22.1 Soil properties of field experimental sites in Punjab and Sindh, Pakistan

	2002		2003		2004			
	Punjab, 5 sites		Punjab, 6 sites		Sindh, 3 sites		Punjab, 8 sites	
Parameter	Range	Mean	Range	Mean	Range	Mean	Range	Mean
pH (1:1)	7.9–8.4	8.1	7.9–8.8	8.2	8.1–8.8	8.5	7.9–8.3	8.1
Organic matter (%)	0.4–1.1	0.6	0.8–1.8	1.1	0.7–0.9	0.8	0.9–1.5	1.0
$CaCO_3$ (%)	1.9–4.5	2.9	1.5–5.7	2.5	11.5–13.5	12.5		
Electrical conductivity (1:1) (dS m^{-1})	0.4–0.6	0.5	0.3–1.5	0.8	2.4–2.5	2.4	0.3–1.2	0.6
AB-DTPA extractable (mg kg^{-1})								
P	1.8–10.4	6.3	5.3–14.0	10.2	3.4–7.3	5.4	3.4–6.8	5.4
K	86–186	131	88–200	146	58–180	130	58–180	109
Zn	0.8–2.4	1.8	0.7–2.5	1.5	0.9–3.1	1.8	0.9–3.1	1.0
Dilute HCl extr. B (mg kg^{-1})	0.26–0.44	0.35	0.15–0.42	0.34	0.43–0.51	0.45	0.21–0.44	0.29

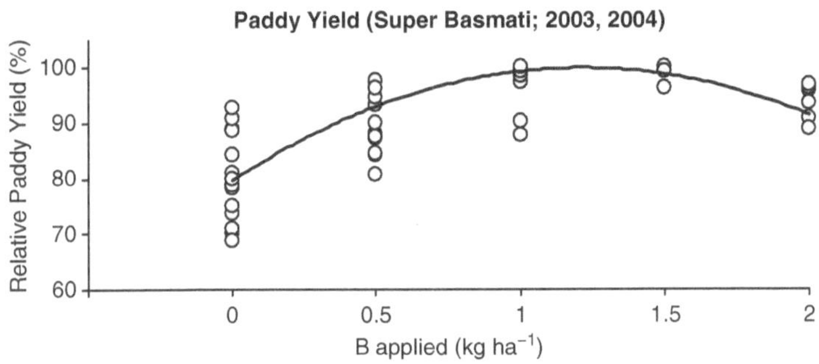

Fig. 22.1 Relationship between fertilizer B rate and paddy yield of cv. *Super Basmati* (maximum yield: 2003, 3.51 t ha^{-1}; 2004, 3.61 t ha^{-1})

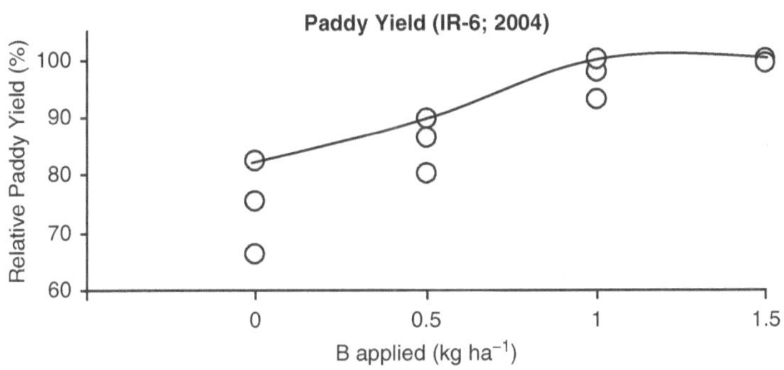

Fig. 22.2 Relationship between fertilizer B rate and paddy yield of cv. IR-6 (maximum yield: 5.81 t ha^{-1})

during 2004 (data not shown). Mean increase in straw yield with B use was 19% in cv. *Super Basmati* and 23% in cv. IR-6. In Experiment 2, straw yield increases with B application ($P \leq 0.05$; Table 22.2) were only slightly lesser than paddy yields. Overall, straw yield increases were not much different than paddy yield increases. Thus, contrary to general perception that B deficiency in cereals hampers grain production more than vegetative growth (Rerkasem and Jamjod 1997), our results affirm that B deficiency affects vegetative growth and paddy yield almost equally. This is in conformity to our earlier findings with rice (Rashid et al. 2002b; 2002c), wheat (Rashid et al. 2002a), sorghum (Rashid et al. 2002b) and rapeseed (Rashid et al. 2002c).

The generally suggested critical level of B in youngest mature leaves of rice is 6 mg B kg^{-1} (Jones et al. 1991; Reuter et al. 1997). Whereas B concentration in flag leaves of *Basmati* cultivars was lesser than 6 mg kg^{-1} during 2002, it was 7.2 mg kg^{-1} in leaves of both cultivars and 6.9 in leaves of cv. KS-282. Thus, internal B requirement in rice leaves as affected by environmental conditions warrants investigation. Boron concentration in mature paddy is hardly reported in the literature. Moreover, nutrient concentration in mature grain was believed to remain unaffected by nutrient status of the soil. However, in this investigation B concentration in rice paddy increased appreciably with B application ($P \leq 0.05$; Table 22.2). This is in conformity to our earlier finding with Zn nutrition of a number of field crops (Rashid and Fox 1992; Rafique et al. 2006).

Thus, this multi-location, multi-year field research revealed that B deficiency is a serious nutrient constrain in fine-grain rice cvs. *Super Basmati* and *Basmati*-385 as well as in coarse-grain rice cvs. KS-282 and IR-6.

Though Dunn et al. (2005) observed maximum rice yields in acid soils of Missouri having hot water

Table 22.2 Paddy yield, agronomic traits and plant B content as affected by B application in calcareous soils of Pakistan

Cultivar	B applied (kg ha^{-1})	Yield (t ha^{-1}) Paddy	Yield (t ha^{-1}) Straw	Panicle sterility (%)	Plant height (cm)	Productive tillers hill^{-1}	1000-grain weight (g)	B concentration (mg kg^{-1}) Leaves	B concentration (mg kg^{-1}) Paddy	B uptake (g ha^{-1})	B use efficiency (%)
					2002 (5 Sites)						
Super Basmati	0	3.23	5.15a	23a	116a	18.4	19.0a	5.5a	1.73a	17.2a	
	1	3.89	5.88b	14b	122b	20.1	20.2b	8.5a	2.51b	34.0b	1.68
	LSD (0.05)	0.68	0.41	4	2	NS	0.8	0.6	0.29	1.4	
Basmati-385	0	3.77a	5.43a	28a	134a	14.3	19.4a	5.3a	1.33a	17.4a	
	1	4.72b	6.63b	16b	140b	16.1	20.1b	8.2b	2.49b	35.2b	1.78
	LSD (0.05)	0.45	0.29	8	5	NS	0.6	0.5	0.18	1.5	
					2003 (6 Sites)						
Super Basmati	0	3.78b	9.74b	18a	131b	19b	21.2	7.2b	1.60b	31.1b	3.1
	1	4.39a	11.30a	12b	135a	22a	21.9	9.3a	2.74a	61.7a	
	LSD (0.05)	0.60	0.36	6	1	3	NS	1.3	0.24	12.9	
Basmati-385	0	3.69b	9.46b	24a	158b	16b	19.4	7.2b	1.63b	29.2b	-
	1	4.54a	10.86a	19b	160a	19a	20.3	8.8a	2.60a	57.1a	2.8
	LSD (0.05)	0.79	0.87	5	2	2	NS	1.0	0.20	13.7	
KS-282	0	4.82b	12.17b	15a	109b	12b	24.5	6.9b	1.69b	35.4b	
	1	5.48a	13.30a	12b	113a	15a	25.2	8.3a	2.58a	69.6a	3.4
	LSD (0.05)	0.64	0.94	5	3	2	NS	1.2	0.08	8.9	

extractable (HWE) B levels as low as 0.25–0.35 mg B kg^{-1} soil, the generally suggested threshold level of HWE soil B for adequate B nutrition of most field crops is >0.5 mg B ha^{-1} (Rashid et al. 1994). And this has been true in this extensive field investigation as well, because appreciable yield increases with B use were observed in calcareous soils having B levels as high as 0.35–0.51 mg kg^{-1} soil (Tables 22.1, 22.2; Figs. 22.1, 22.2). Thus, all the field soils included in this study were deficient in B to varying degrees. However, the literature generally suggests a possibility of B toxicity in salt-prone calcareous soils (Sillanpaa 1982; Yau 1997). Contrarily, in our experience, rice crop grown in high pH sodic soils (pH up to 8.8; Table 22.1) suffered with B deficiency. In fact, incidence B deficiency has also been observed in rice grown in salt-affected (i.e., saline as well as saline-sodic) calcareous soils of Pakistan (Aslam et al. 2002). Previously, we had observed that cotton grown in saline calcareous soils (EC > 4 dS m^{-1}) also suffered with B deficiency rather than toxicity (Yasin et al. 2002). Therefore, crops grown in alluvial, calcareous soils of Pakistan are prone to B deficiency problem rather than toxicity (Rashid et al. 2002b).

Boron-bearing soil mineral, tourmaline, is believed to be the ultimate source of B in soil; however, soil organic matter (SOM) is the immediate source of B supply to plant roots (Berger and Truog 1944). Because of peculiar agro-climatic conditions in rice-wheat belt of Pakistan, the soils are inherently low in organic matter (Table 22.1). Moreover, SOM has got reduced badly, post Green Revolution, from an average of 1.02% during 1967–1974 to 0.59% during 1985–1994 (Byerlee et al. 2003). This sharp decline in SOM appears to be a major cause of soil B stock decline and, hence, increased incidence of B deficiency in rice crop. Additionally, we postulate that enhanced mining of soil B with more intensive cropping and greater biomass production per unit field area, during the post-Green Revolution era, coupled with the suspected leaching of B beyond the root zone, are responsible for more widespread and severe B deficiency in rice.

3.2 Boron Deficiency and Rice Panicle Sterility

Though B application enhanced plant height and rice grain rice ($P \leq 0.05$; Table 22.2), paddy yield increases accrued primarily because of appreciable reduction in panicle sterility (Table 22.2; Fig. 22.3) and increases in productive tillers per hill (Fig. 22.4). Boron deficiency is already known to induce panicle sterility in cereal crops including wheat (Rerkasem and Jamjod 2004). However, there is hardly any report of B-induced panicle sterility in rice. Though sterility in rice may be caused by many biotic and abiotic stresses, this research adequate establishes that B deficiency induced panicle sterility is a major cause of low rice productivity in calcareous soils.

3.3 Improvement in Milling Return and Grain Quality with Boron Use

This research revealed that B deficiency in calcareous soils not only hampers paddy yield but also impairs grain quality. In our experience, fertilizer B application to rice improved milling return as well as head rice recovery ($P < 0.05$; Table 22.3). An improvement in B nutrition of rice plants also improved desirable cooking quality traits like quality index, kernel elongation ratio upon cooking, bursting-upon-cooking, and alkaline spreading value ($P \leq 0.05$; Table 22.3). Improvement in grain quality traits is attributed to better grain filling, facilitated with adequate B nutrition of rice plants. Rice is not only a staple cereal food for the local population and a delicacy at local festivals, it is also a major foreign exchange earner for the country. Its improved milling and cooking quality with adequate

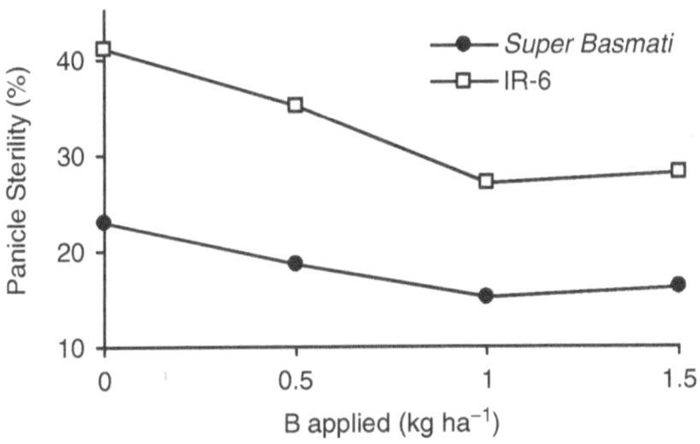

Fig. 22.3 Panicle sterility in rice as affected by boron fertilization

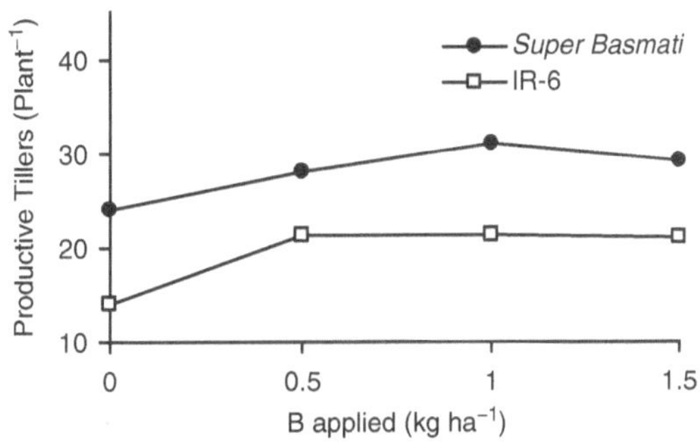

Fig. 22.4 Productive tillers in rice as affected by boron fertilization

Table 22.3 Impact of boron application on rice (cv. *Super Basmati*) milling recovery and cooking quality during 2004

	Boron applied (kg ha^{-1})				
Grain characteristic	0.0	0.5	1.0	1.5	LSD (0.05)
Total milled rice (%)	70.4c	71.6b	72.3a	71.7b	0.5
Head rice (%)	51.6c	54.6b	57.0a	56.4a	0.6
Quality index (L/ BT)	2.96b	2.96b	3.0a	2.97b	0.03
Elongation ratio upon cooking	1.88d	1.94c	2.00a	1.98b	0.02
Bursting-upon-cooking (%)	9.9a	9.0b	8.0c	8.4c	0.5
Alkali spreading value (score 1–7)[1]	4.5c	4.8b	4.9a	4.8b	0.1

[1]Alkali spreading value: 4–5 score = intermediate G.T. type rice
Means followed by different letters are statistically different at LSD (0.05)

Table 22.4 Efficiency and economics of fertilizer boron use in rice (cv. *Super Basmati*) in Punjab

B applied (kg ha^{-1})	Yield (t ha^{-1})		B uptake (g ha^{-1})	B use efficiency (%)	Value: cost ratio
	Paddy	Straw			
			2003 (6 Sites)		
0	2.80c	6.84c	17.6	–	–
0.5	3.20b	7.98b	31.2	2.73	46:1
1.0	3.51a	8.97a	45.1	2.75	42:1
2.0	3.28	8.29b	42.4	2.48	14:
LSD (0.05)	0.05	0. 55			
			2004 (8 Sites)		
0	2.88c	5.0	13.4	–	–
0.5	3.25b	5.3	25.3	2.38	38:1
1.0	3.51ab	5.6	31.6	1.82	32:1
1.5	3.61a	5.8	37.2	1.59	25:1
LSD (0.05)	0.27	N.S.			

B nutrition – particularly of fine grain Basmati type rices – is of paramount significance both for local as well as international markets. Thus, adequate B fertility of calcareous soils, or B fertilizer use in deficient situations, is a prerequisite for harvesting optimum yields of good quality rice.

3.4 Economics of Boron Use in Rice

Considering paddy yield increases alone, B application in rice proved highly cost-effective, with value–cost ratio (VCR) of 32–42:1 in cv. Super Basmati (Table 22.4), 55:1 in cv. Basmati-385, and 36:1 in cv. IR-6. Improvements in milling return, head rice recovery, and cooking quality traits were added advantages of great economic significance. Also, fertilizer B leaves an appreciable residual effect as well as improves efficiency of other farm inputs including major nutrient fertilizers. As the recommended B dose in rice is 0.75 kg ha^{-1}, farmer-level VCRs are expected to be better than the ones observed in our research investigations. Thus, B fertilizer use in rice crop will increase farm productivity, farmer profitability, national economy and soil resource sustainability.

4 Conclusions

Soil B deficiency causes low rice productivity, by inducing panicle sterility and less productive tillering. Grain cooking quality is also impaired. Thus, rice, a major staple cereal, appears to be suffering yield and quality losses due to soil B deficiency. Luckily, its remedy is simple and effective as B use enhances crop productivity as well as improves milling return and cooking quality. Use of B is highly cost-effective. Thus, rice growers must include B in their fertilizer use program. As plant B deficiency is caused by low availability of soil B rather than less total B content, biotechnologists may strive for developing B efficient rice genotypes.

Acknowledgements We thank Chaudhry Abdul Ghaffar, Director General, Agricultural Extension and Adaptive Research, Punjab and M/S Engro Chemical Pakistan Ltd. for support in field experimentation, Mr. M. Asif Ghumman for statistical analysis, and Mr. M. Irfan for word processing, graphics and composing of the manuscript.

References

Aslam M, Mahmood IH, Qureshi RH, Nawaz S, Akhtar J (2002) Salinity tolerance of soil as affected by boron nutrition. Pak J Soil Sci 21: 110–118.

Berger KC, Truog E (1944) Boron tests and determination for soils and plants. Soil Sci 57: 25–36.

Bingham FT (1982) Boron. In: Page AL (ed) Methods of Soil Analysis, Part 2: Chemical and Microbiological Properties. American Society of Agronomy, Madison, WI, pp. 431–448.

Byerlee D, Ali M, Siddiq A (2003) Sustainability of the rice-wheat system in Pakistan's Punjab: how large is the problem? In: Ladha JK, Hill JE, Duxbury JM, Gupta RK, Buresh RJ (eds) Improving the Productivity and Sustainability of Rice-Wheat Systems: Issues and Impacts. ASA Special Publication 65. American Society of Agronomy, Madison, WI, pp. 77–96.

Chaudhry FM, Latif A, Rashid A, Alam SM (1977) Response of rice varieties to field application of micronutrient fertilizer. Pak J Sci Indus Res 19:134–139.

Dunn D, Stevens G, Kending A (2005) Boron fertilization of rice with soil and foliar applications. Online Crop Management. doi: 10.1094/CM–2005–0210–01–RS.

Gaines TP, Mitchell GA (1979) Boron determination in plant tissue by the azomethine-H method. Commun Soil Sci Plant Anal 10: 1099–1108.

GOP, 2004. Pakistan Economic Survey 2004–2005. Ministry of Food, Agriculture and Livestock, Government of Pakistan, Islamabad.

Jones, Jr JB, Wolf B, Mills HA (1991) Plant Analysis Handbook: A Practical Sampling, Preparation, Analysis, and Interpretation Guide. Micro–Macro Publishing, Athens, GA, 312 pp.

Rafique E, Rashid A, Ryan J, Bhatti AU (2006) Zinc deficiency in rainfed wheat in Pakistan: Magnitude, spatial variability, management, and plant analysis diagnostic norms. Commun Soil Sci Plant Anal 37: 181–197.

Rashid A, Fox RL (1992) Evaluating the internal zinc requirements of grain crops by seed analysis. Agron J 84: 469–474.

Rashid A, Rafique E, Bughio N (1994) Diagnosing boron deficiency in rapeseed and mustard by plant analysis and soil testing. Commun Soil Sci Plant Anal 25: 2883–2897.

Rashid A, Kausar MA Hussain F, Tahir M (2000a) Managing zinc deficiency in transplanted flooded rice grown in alkaline soils by nursery enrichment. Tropical Agr (Trinidad) 77: 156–162.

Rashid A, Muhammad S, Rafique E (2000b) Genotypic variation in rice susceptibility to boron deficiency. Intel Rice Res Notes (IRRI, Philippines) 25(3): 29–30.

Rashid A, Rafique E, Bughio N (2002a) Boron deficiency in rainfed calcareous soils of Pakistan. II. Incidence and boron requirement of wheat In: Goldbach HE, Brown PH, Rerkasem B, Thellier T, Wimmer MA, Bell RW (eds) Boron in Plant and Animal Nutrition. Kluwer, New York, pp. 339–348.

Rashid A, Rafique E, Ryan J (2002b) Establishment and management of boron deficiency in field crops in Pakistan. A country report. In: Goldbach HE, Boron PH, Rerkasem B, Thellier T, Wimmer MA, Bell RW (eds) Boron in Plant and Animal Nutrition. Kluwer, New York, pp. 339–348.

Rashid A, Rafique E, Muhammad S, Bughio N (2002c) Boron deficiency in rainfed alkaline soils of Pakistan: incidence and genotypic variation in rapeseed-mustard. In: Goldbach HE et al. (ed) Boron in Plant and Animal Nutrition. Kluwer/Plenum, New York, pp. 363–370.

Rerkasem B, Jamjod S (2004) Boron deficiency in wheat: a review. Field Crop Res 89: 173–186.

Rerkasem B, Jamjod S (1997) Boron deficiency induced male sterility in wheat (*Triticum aestivum* L.) and implications for plant breeding. Euphytica 96: 257–262.

Reuter DJ, Edwards DG, Wilhelm NS (1997) Temperate and tropical crops. In: Reuter DJ, Robinson JB (eds) Plant Analysis: An Interpretation Manual, 2nd ed. CSIRO Publishing. Collingwood, Victoria, Australia, pp. 81–284.

Savithri P, Perumal R, Nagarajan R (1999) Soil and crop management technologies for enhancing rice production under micronutrient constraints. In: Balasubramanium V et al. (eds) Resource Management in Rice Systems: Nutrients. Kluwer, Dordrecht, pp. 121–135.

Shorrocks VM (1997) The occurrence and correction of boron deficiency. Plant Soil 193: 121–148.

Sillanpaa M (1982) Micronutrients and the Nutrient Status of Soils: A Global Study. Food and Agriculture Organization of the United Nations, Rome.

Yasin M, Rashid A, Rafique E (2002) Boron status of cotton soils: Relationship with salinity. In: Abstracts, 9th Congress of Soil Science, 18–20 March 2002, Faisalabad. Soil Science Society of Pakistan, Islamabad, p. 25.

Yau SK (1997) Differential responses of barley, durum and bread wheat to high levels of soil boron. In: Ryan J (ed), Accomplishments and Future Challenges in Dryland Soil Fertility Research in the Mediterranean Area. ICARDA, Aleppo, Syria, pp. 209–218.

Chapter 23
Potential Role of Sabkhas in Egypt: An Overview

H.M. El Shaer

Abstract Salinization and desertification are the major constraints to agriculture development and consequently food production in Egypt. The salt affected soils and salt marshes are located mainly in the Mediterranean and Red Sea coasts and also spotted in some areas in middle, western and eastern parts of the Nile delta. Salt marshes are integral components of the Egyptian coastal and inland ecosystems, serving as important areas of primary production for food, feed, wood, paper, fiber, etc. They represent an important habitat for production of grazing animals, waterfowl and fish. Reeds (i.e. *Phragmites* and *Typha*) and rushes (*Juncus actus* and *Juncus rigidus*) form an extremely important component of the saline lands in Egypt as they provide habitat for a wide range of living organisms (invertebrates, fish, birds, animals, etc). Some other halophytes in the salt marsh could provide great potential resources for agriculture and environment in some parts of saline in Egypt. The diversity of halophytes and other natural recourses in the salt marshes are facing severe threats due to uncontrolled human interference and other environmental factors. In this review, magnitude of salt marsh plants, its economic benefits with special reference to their nutritive values, utilization by ruminants and some constraints limit their utilization are discussed.

Keywords Salinity • soil • halophytes • animal • feeds • aridity • biodiversity

H.M. El Shaer
Desert Research Center, 1 Mathaf El Mataria St., P.O. Box 11753, Mataria, Cairo, Egypt
e -mail: drc_elshaer@hotmail.com

1 Introduction

Approximately 97.5% of earth's water is salt water; the remaining 2.5% is fresh water. Low precipitation associated with inefficient recharge of aquifers, especially in arid and semi-arid areas, has been one of the main reason for lower availability of good quality water (ICARDA 1997). Salinity has been one of the important environmental stresses that have significantly impaired agricultural production all over the world, and constituting a major proportion of the total world salt-affected areas (Szabolcs 1994). Saline soils of various nature and degree occupy over 80 million hectares in the Mediterranean basin (ICARDA 1997); in these salt affected areas, halophytes are abundantly growing there, some of which dominate well recognized plant communities of widespread occurrence.

Salt marshes occupy a large part of the territorial lands of Egypt. The main salt marshes in Egypt are located at the Red Sea coastal belt in South Sinai and east of the eastern desert; at all lakes of the Mediterranean Sea such as: Bardawil Lake, Manzalah Lake, Mariout Lake, and Buroulus Lake (Zahran and Willis 1992). The Western Desert of Egypt, which covers about two-thirds of Egyptian area, has large depressions and oases such as Siwa, Moghra, Wadi EL-Natroun and Bahariya, Kharga, Dakhla and Dungul oases. All these oases and depressions are rich of salt marshes which are generally considered as being among the most productive ecosystems in Egypt. It has been demonstrated that re-vegetation of saline habitats with halophytic species is profitable; therefore, finding proper approaches to make marginal land and saline wasteland productive (El Shaer and Zahran 2002). The objective of this article is to focus on the characteristics and habitats of most common salt marshes in

M. Ashraf et al. (eds.), *Salinity and Water Stress*,

Egypt; its potentialities and usages will be also presented and discussed.

2 Salt Marshes in Egypt

2.1 Red Sea Salt Marshes

In Egypt, the Red Sea coastal land extends from Suez (Lat. 300 N) to Mersa Halaib (Lat. 220 N) at the Sudanese borders. The land adjacent to the sea is generally mountainous, flanked on the western side by the range of coastal mountains. Along the Egyptian Red Sea coast (El-Khouly and Zahran 2002), the mean annual rainfall ranges from 25 mm in Suez (Lat. 300 N), 4 mm in Hurghada (Lat. 270 N) to 3.4 mm in Qusseir (Lat. 260 N). The main bulk of rain occurs in winter, i.e. Mediterranean affinity, and summer, in general, is rainless.

The soil features of the littoral vegetation are significantly affected by the climatic aridity of the Egyptian Red Sea coasts. As precipitation is low, there is insufficient leaching and salts accumulate as crusts on soil surface (Kassas and Zahran 1967; Younes et al. 1983). The soil features are apparently one of the main factors influencing the plant growth, the plant cover, the zonation pattern as well as the geographical distribution of the littoral halophytes. Three main littoral halophytic types within the Red Sea coastal belts are recognized, namely: 1 – mangrove swamps, 2 – reed swamps and 3 – salt marshes (Kassas and Zahran 1967; Zahran 1993). The mangrove swamps are dominated by a widespread species (*Avicennia marina*) which may be co-dominated with *Rhizophora mucronata* in the southern parts of the shorelines. *Phragmites australlis* and *Typha domingensis* are the dominant or co-dominant species of the reed swamp habitat. The salt marsh habitat, however, is inhabited by more than 30 dominant and characteristic halophytes. These include: *Arthrocnenuim macrostuchyum*, *Halocnemum strobilaceum*, *Halopeplis perfoliata*, *Zygophylium album*, *Z. coccineum*, *Limoniun axillare*, *L.pruinosum*, *Aeluropus spp.*, *Sporobulus spicatus*, *Halopyrum mucronaum*, *Nitraria retusa*, *Suaeda monoica*, *S. vermiculata*, *S. pruinosa*, *Atriplex farinosa*, *Juncus rigidus*, *Imperata cylindrica*, *Anabasis setifera*, *Alhagi marorun*, *Cressa crcilea*, *Tamarix nilotica, and T.passerinoides.*

2.2 Mediterranean Salt Marshes

There are many salt marshes existed at the lakes of the Mediterranean Sea costal belt such as: Bardawil Lake, Manzalah lake, Mariout lake, Buroulos lake. Due to the magnitude and broad biodiversity of Bardawill Lake, the salt marshes ecosystem of this lake will be, only, stressed herein.

Bardawil Lake is an oligotrophic, shallow and hypersaline coastal lake. It is situated in Northern Sinai, and its area is about 60,400 ha. The environment of Bardawil Lake differs from that of the other Egyptian lakes in terms of climatic factors, geomorphology and salinity. It is the only oligotrophic hyper saline lake along the Mediterranean coast of Egypt (Krumglaz et al. 1980). There is no fresh water supply into the lake and the only non-marine water source is the scarce winter rain. Thus, it is the most saline of the northern Egyptian lakes (Zahran and Willis 1992). Soil salinity, moisture content, calcium carbonates and the human disturbance may be considered the most important environmental factors that affect the distribution and the abundance of halophytic vegetation and other habitats of the lake. There is a significant variation between the distribution of the halophytic species and the elevation. The halophytic plant communities surrounding the salt marshes along the Mediterranean coast of Egypt often exhibit large zonal pattern of species composition (Zahran and Willis 2002). The plant species found in the salt marshes of Bardawil Lake are similar but not identical to those in the delta lakes and the western Mediterranean salt marsh habitat of Egypt (Khedr 1999). The salt marsh habitat at Bardawil Lake, comprises the lower elevation islands, the sand bar and interdunal Sabkhas. Twenty-nine halophytic species are recorded in these habitats. In Bardawil Lake the salt marshes are either close to the lake's water (on the lake shores or on the low elevated islands). Five groups of halophytic plants are identified (Khedr and Zahran 2002) in Bardawil Lake salt marshes which are dominated by: 1 – *Salsola tetrandra*, 2 – *Nitraria retusa*, 3 – *Zygophyllum album* and *Halocnemum strobilaceum*, 4 – *Arthrocnemum macrostachym* and *Suaeda aegyptiaca*, 5- *Sarcocornia fruiticosa*.

Salt-encrusted Sabkhas of Bardawil Lake have almost no vegetation due to the extremely high salinity only about 2–5% of the area in wet Sabkhas is vegetated (Shaheen 1998). An extensive salt production system has been constructed in the eastern part of the lake. The lake is a very important habitat for a variety of wildlife, especially of a prime bottleneck for migratory birds. Due to the extreme drought, heavy grazing and trampling by Bedouin herds and the severe sand shifting are some special features of the lake salt marshes. It is generally poor in halophytic species and communities.

2.3 Oases Salt Marshes

The Western Desert of Egypt covers about two-thirds of Egyptian area. It is characterized by large depressions and oases occupying about 36% of its area. Siwa, Moghra, Wadi EL-Natrun and Bahariya represent the northern oases, while Kharga, Dakhla and Dungul represent the southern oases. El-Khouly and Zahran (2002) reported that the soils of the halophytic communities in such oases are characterized by high percentage of $CaCO_3$ and low percentage of organic carbon. Four main types of vegetation are characterized in these oases and depressions (El-Khouly and Zahran 2002), namely: 1 – iced swamp vegetation, 2 – salt marsh vegetation, 3 – sand dune vegetation, and 4 – desert plain vegetation. Eleven families are recorded in the halophytic vegetation of these oases which comprise 25 genera including 32 species. Only two of these families, Gramineae and Chenopodiaceae contribute about half (46.9%) of the total number of the recorded halophytes. The most common genera was *Tamarix* (four species/genera) in the halophytic vegetation of the Egyptian Oases.

The halophytic vegetation is represented by communities dominated by the following species: *Juncus rigidus*, *J. acutus*, *Cyperus laevigatus*, *Pharagmities asulrails*, *Arthrocnemum macrostachyum*, *Alhagi graecorum*, *Sporobolus spicatus*, *Zygophyllum album*, *Tamarix nilotica*, *Suaeda monica*, *Salsola imnbricate*, *Nitraria retusa*, *Sarcocornia fruticosa*, *Desmostachya bipinnata*, *Cressa cretica*, *Imnperata cylindrical* and *Aeluropus lagopoides*.

Nine growth forms can be distinguished in the halophytic species of these oases (El-Khouly and Khedr 2000), namely: (1) rhizomatous growth form, is represented by 31.3% of species, e.g. *J. rigidus*, *T. domingensis*, *C. laevigatus and J. clayndrica*, (2) leaf less succulent shrub growth form, is represented by 25% of species, e.g. *S. aegyptiaca, S. tetrandra* and *A. macrostachyum*, (3) leafy non-succulent shrub growth form, is represented by 12.5% of species, e.g. *T. nilotica* and *A. graecoriem*, (4) *Stoloniferous* growth form, is represented by 9.4% of species, e.g. *P. australis* and *A. lagopoides*, (5) non succulent herb growth form, is represented by 6.3%, e.g. *T. amplixcaulis*, (6) leafy non-succulent subshrubs, (7) sthorny shrub, (8) leafless tree and (9) succulent herb growth forms each is represented by 3.1% of species, e.g. *N. retusa*, *T. aphylla* and *C. cretica*, respectively. The growth forms of the halophytes reflect the high stress in the salt lands; also, some of species modified its organs to store the water (El-Khouly and Khedr 2000; El-Khouly 2001). The changes in hydrology and increase of number of lakes and lakes area together resulting from the agricultural development in Siwa Oasis may be the cause of the dominance of these halophytic species.

3 Salt Marshes Habitats and Its Potentialities

3.1 Reeds and Rushes

Reeds and rushes form extremely important component of the wetlands in Egypt, namely: River Nile, estuaries (Damietta and Rosetta) and coastal and inland reed swamps. The reeds are represented by *Phragrmites* and *Typha* and the rushes are represented by two *Juncus species*: *J. acutus* and *J. rigidus*. It is noted that *Phragrmites* is the highest species tolerate soil salinity followed by *Juncus rigidus*, *J. acutus* and *Typha domingensis*; *Juncus rigidus* tolerate soil salinity more than *Juncus acutus* (Serag et al. 2002).

Reeds and rushes provide habitat (e.g. nesting sites, substratum, feeding materials) for a wide range of organisms, from invertebrates to birds. They are very important as producers in nutrient rich freshwater habitats such as River Nile, Manzala, and Burulius Lakes and the irrigation and drainage canals. Moderate growths of macrophytes can help oxygenate the water, assisting the survival of fish and invertebrates. They are also important to process such as nutrient cycling within freshwater ecosystems (Serag et al. 1999; Zahran and Willis 2002).

3.2 Phragrmites

The common reed *Phragrmmites* is widespread in Egypt, where it occurs in water and wetlands of all phytogeographical regions namely: the Nile region, Oases, Mediterranean coastal region, Desert region, Red Sea coastal region and Sinai (Tackholm 1974; Zahran and Willis 1992; Boulos 1995). *Phragrmites* is an economically valuable resource. Its stems have been used for centuries as fishing rods, weavers spools, and musical instrument mouth pieces; for roofing, fencing, and basket weaving, as animal fodder, and as fuel. An international attention is being directed towards the capacity of constructed reed wetlands to control water pollution and to treat municipal and industrial wastewater (Wittgren and Maehlum 1997).

3.3 Typha Species

There are two Typha species in Egypt: *Typha domingensis* and *Typha elephantina*. The first Typha is recorded wherever marshy conditions prevail (Tackholm 1974) especially along the margins of drains, main Nile branches and northern lakes of Egypt (Khedr 1999; Serag et al. 1999). Typha is a beneficial plant in temperate zone of world, but it can get out of hand and is considered a pest when it blocks canals and ditches. Because of the widespread and rapid growth rates of cattails, these plants can be environmentally and economically important. For example, they can provide a highly desirable habitat type for fishes. These plants can serve as nutrient "scrubbers in polluted aquatic systems, thereby playing a key role in nutrient cycling. However, because of their ability to propagate vegetative and extend their coverage, they may dominate the plant community and degrade the habitat of some native species.

3.4 Rushes

Nine species belonging to the macus (family *Juncaceae*) are recorded in the salt marshes; viz: *Juncus acutus*, *J. bufbnius*, *J. fontanesili*, *J. inflexus*, *J. littoralis*, *J. punctorius*, *J. hybridus*, *J. rigidus* and *J. subulatus* (Mohammed 1980). Most of them are salt tolerant and are growing in areas with hot and dry climate. The rush *Juncus acutus* shows a wide range of distribution in the marsh and saline habitats compared with *Juncus rigidus* which is mainly dominated in saline habitats. The most common associates with both species are obligate halophytes, e.g. *Halocnemum strobilaceum, Inula crithmoides*, *Arthrocemum macrostachyum*, *Suaeda vera* and *Atriplex portulacoides. Juncus rigidus* is more tolerant to aridity and soil salinity (Zahran 1993).

Juncus species are fiber producing and if successfully managed to produce good quality paper; they would provide a non-conventional crop to be cultivated on salt affected soils of Egypt (Zahran 1993). In ancient Egypt, *J. rigidus* leaves and/or culms were used as pens for writing on papyrus (Snogerup 1993). Macus rigidus has been used for making mats, baskets and sandals (Tackholm and Drar 1954). In some rural areas of Egypt, such mats are used for manufacturing local handmade cheese. Moreover, in the Bahariya Oasis and probably elsewhere such mats are used as prayer mats. The seeds of both Juncus species contain flavonoids, glycosides, tannins and unsaturated steroids (Zahran et al. 1972). Saponins were present only in *J. rigidus* seeds, but alkaloids were absent from both species as reported by Zahran et al. (1972). The same authors also indicated that the protein constituent of *J. rigidus* was slightly higher (11.12% than that of *J. acutus* (10.54%) seeds. *Juncus* seeds are found to be rich in oils, proteins, amino-acids, carbohydrates, etc. This finding might suggest the possibility of their use as a raw material in various chemical industries; e.g. drug, oils, etc. Care must be taken in the case of human consumption; they might have toxic effects. Accordingly, one may conclude that both *J. rigidus* and *J. acutus* have agro-industrial economic potentialities. However, as reported by Serag et al. (2002), J. *rigidus* is superior to *J. acutus* for the following reasons:

1. *J. rigidus* is resistant against fungal infection, an advantage that increased its vegetative yield.
2. The salt tolerance and desalination effects of *J. rigidus are* higher than those of *J. acutus*.
3. The seeds of *J. rigidus* are, relatively, richer in their chemical constituents than those of *J. acutus*.

3.5 Other Halophytes

Halophytes are the plants capable of growing and surviving in the saline environment. Halophytes as a group have one or more of several physiological adaptations that allow for the survival in the saline environment. They are belonging to 550 genera and 117 families and 1,560 species; later estimate reports that the numbers of species are almost 2,600 (Menzel and Leith 1998). However, differences in taxonomic classification and defining them salt limits has been a major debate and as result, the number of species are reported to vary between 2,000–3,000 species having diversified characters (Squires and Ayoub 1994). Halophytes grow in many arid and semi-arid regions around the world and are distributed at salt mashes from coastal areas to mountains and deserts.

The feasibility of growing halophytes on salt marshes can be maximized with plant species that in addition to its primary product can also provide 'indirect' and economical benefits. Factors affecting the input costs, like establishment costs of the plants, cost of water and value of product (feed quality in case of forage species) in addition to indirect benefits could certainly increase the value of halophytes to be used commercially.

3.6 Potential Use of Halophytes

Many potential role of halophyte species have been listed (Squires and Ayoub 1994) including land rehabilitation, as forage/fodder, oil-seed crops, medicinal plants, fuel wood and even sequestration of CO_2. Aronson et al. (1989) enumerated 1560 halophyte species which are already in use. Some halophytes are used as building materials (e.g. *Avicennia marina* and *Prosopis tamarugo*), others as wood for furniture, timber, charcoal, fire wood (e.g. *Tamarix spp.*). Many halophytes are used in medical purposes as drugs; these include: *Annona glabra*, *Juncus acutus* and *Salsola kalli*. Moreover, halophytes have been reported to be used as fertilizers (e.g. *Seshania speciosa* and *Zostera mnarilia*). Many of them are edible and utilized as vegetables such as *Aster tripolium*, *Sueda glauca* and *Salicornia fruticosa*, which are used for oil production. Other uses of halophytes include the utilization in laundry detergent, paper production, herbal tea, sea floor fixation, as a green cover, as ornamental plants and as hedges. However, economical prospects for halophytes can be evaluated when they are looked beyond their salt tolerance values. Factors related to establishment, productivity, water requirements, forage quality could qualify the species for agricultural purposes. There is also a need to evaluate species for multiple-use which will increase the economical values of halophyte crops, like *Salicornia* where both foliage and seed has their individual economic values. Shrubs (*Atriplex lentiforimis*, *A. halimus* and *A. nummularia*) and *trees* (*Acacia*, *Prosopis* and *Tamarix spp.*) that can have both forage and wood values should be also considered as cash crops.

3.7 Prospects of Halophytes as Animal Feeds

The shortage of animal feeds is the main constraint to increase indigenous animal production, particularly in arid and semi-arid regions. Animal production, as the main source of income for nomads, is based mostly on the natural vegetation for feeding sheep, goats and camels. The palatable plant species are always over grazed due to the grazing pressure of animals (El Shaer 1981). However, unpalatable and less-palatable halophytes are widely distributed in salt marshes throughout Egypt. Halophytic plants such as *Suaeda* species, *Atriplex spp.*, *Nitraria retusa* and *Salsola spp.*, are generally the most palatable chenopod shrubs in several salt marshes (El Shaer 1981) which always disappear fast due to overgrazing. Such plants are extremely valuable as a fodder reserve particularly during drought season. Factors influencing grazing and nutritive values of halophytes are the plant species, ecotypes, stage of growth (Abd El-Aziz 1982; El Shaer 1986), season of use (wet season versus dry season), environmental factors (El Bassosy 1983), and location (Gihad and El Shaer 1994). Halophytic plants differ in their nutritive value as from one species to another. Chemical composition of saltbush species differ widely. Table 23.1 shows some of the differences that exist between some halophytic species, grown naturally in most of salt marshes in Egypt, in terms of chemical composition, *in vitro* digestibility and palatability, on over all average bases, for sheep, goats and camels.

The chemical analysis of some halophytic plants reveals that halophytic species have the potential as an animal fodder (Gihad and El Shaer 1994). However, the highest forage values are found during the wet season of the year (El Shaer 1981; Le Houerou 1993). The process of aging and maturation of halophytes is found to be associated with a decline in digestibility, and crude protein(CP) content, consequently the nutritive value (El Shaer 1981; El Bassosy 1983). It was found

Table 23.1 Nutritive values (%, DM basis) and palatability rate (PR) of most dominant halophytes

Plant species[1]	PR[2]	DM	CP	EE	CF	Ash	NFE	DMD[3]
Allage maurorum	S,G	44.0	9.45	4.42	29.5	25.9	30.73	46.4
Arhth. glaucum	C	229.9	3.38	1.26	12.1	51.9	31.36	50.4
Atriplex halimus	A	34.2	12.6	2.28	2.4	22.7	37.02	66.7
Atriplex leucoclada	S,G	25.6	15.1	2.69	27.4	31.7	23.11	52.2
Atriplex nummularia	A	21.7	13.3	5.09	24.2	26.7	30.71	58.8
Haloc. strobilaceum	C	29.7	6.69	2.22	7.04	40.3	43.75	63.0
Haloxylom salicornicum	Nil	42.2	14.8	6.11	24.1	15.9	30.09	46.5
Jancus acutus	A	35.0	7.11	2.35	28.5	12.3	49.94	34.4
Nitraria retusa	A	37.6	10.2	2.46	32.6	33.0	21.74	61.8
Salcornia fruticosa	C	37.6	13.5	1.89	18.9	14.3	51.41	70.5
Salsola tetraundra	A	37.1	6.32	2.37	36.1	35.9	19.31	68.0
Suaeda fruticosa	A	25.0	10.0	5.00	33.2	16.1	35.7	70.4
Limonias. monopetahum	A	48.6	11.5	3.49	14.6	23.6	46.81	68.2
Tamarix aphylla	G,C	34.9	12.9	3.99	13.6	20.1	49.41	48.7
Tamarix mannifera	A	40.0	8.19	3.57	11.6	24.9	51.74	59.6
Zygophyllum album	Nil	24.7	7.76	2.46	11.2	34.2	44.38	65.3
Zygophyllum simplex	C	40.5	11.1	2.12	16.7	29.8	40.28	56.9
Zygophyllum decumbens	C,G	37.7	9.37	1.80	24.1	26.9	37.83	49.6
Overall average		34.3	10.1	3.08	21.7	27.0	38.12	58.2

[1]Data cited from El-Shaer (1981), Abd El-Aziz (1982), El-Bassosy (1983), El Shaer and Zahran (2002).
[2]PR: Palatability for animal species; S: sheep; G: goats; C: camels; A: all these animals species.
[3]DMD%: in vitro dry matter digestibility (%).

that dry matter intake (DM1) and dry matter digestibility (DMD%) of halophytic forages were higher in grazing season than in drought season with both sheep and goats (El Shaer 1981). Le Houerou (1993) reported that sheep became adapted to saltbush and increased their intake of forage over a 3–5 month period. Therefore, it seems that feeding halophytes need more time than other feeds for animals to be adapted on such feed materials, particularly the less palatable forages.

Organic constituents of halophytes are moderately digestible by goats and sheep and that mixed diets containing halophyte forages are acceptable to sheep and goats. Most of the halophyte shrubs are high in protein content and of moderate digestibility. In Southern Sinai, Hassan et al. (1982) indicated that sheep fed a mixture of indigenous halophytic ranges without any feed supplements lost weight during all months studied (12 months), but losses were minimal in the spring season (26 g day) as compared to sheep summer season (134 g day). Therefore, halophytes should not be offered as sole diets to animals for long time feeding period because of the adverse affects on animals (El Shaer and Attia Ismail 2002). However, feeding halophytic fodders supplemented with any energy supplements, or mixing halophyte forages with other feed materials or processing of halophytic feed materials in sort of silage or haylage can significantly improve their nutritional value and utilization (Warren et al. 1990; El Shaer 1997). Moreover, the performance of animals on the range particularly the halophytic ranges depends on several factors mainly: animal species, the season of the year, abundance of forages, and nutritional values of forage species. The poor intake of some halophytic species could be attributed to three main factors: (1) high Sodium (Na),Calcium(Ca) and silica contents, (2) higher levels of acid detergent lignin(ADL) and neutral detergent fiber(NDF) and (3) many shrubs contain higher levels of plant secondary metabolites (El Shaer and Attia Ismail 2002). In addition, there are several factors, which could considerably limit the feeding values of halophytes such as physical and chemical defenses of the plants (Gihad and El Shaer 1994).

In general speaking, halophytes that are used as forage species will have better cash value if they have better forage quality. High palatability, digestibility and good nutritional value (high protein and lesser fiber, ash and oxalate contents) would significantly improve the forage quality. Feeding small ruminants on good quality halophytic feed materials would reduce the feed costs not less than 30% (El Shaer 1997). On the other hands, it is estimated that with the same nutrient level of berseem, halophyte forage can increase the profitability of a farm by 17%.

4 Conclusion

It is concluded that salt marshes are an integral component of the Egyptian coastal and inland ecosystems, serving as important areas of primary production for coastal food chains. They are also an important habitat for the production of grazing animals, waterfowls and fish. However, the diversity of the halophytes and other natural recourses in the salt marshes in Egypt are, unfortunately, facing dangerous impacts due to the uncontrolled human interference. Many plant, animal, bird's species are either endangered or even exterminated. Such bad environmental situation and interference need urgent solutions through the conservation and sustainable use of the halophytic vegetation and its ecosystem in the salt marshes by applying several approaches such as: 1 – preserve the genetic resources of these species in the Egyptian National Gene Bank, 2 – restoring the endangered species in its habitats, 3 – cultivating the economic halophytes species or crops of salinity resistant in habitat by using saline and brackish water of the lakes and springs of the oases, and 4 – cultivating the multi-purposes halophytic species.

The feasibility of growing halophytes on salt marshes can be maximized with plant species that in addition to its primary product can also provide indirect and economical benefits.

References

Abd El-Aziz D M (1982) A study of the nutritive value of some range plants in the North-Western Coastal Desert. Ph.D. thesis, Fac. Agric., Ain Shams University, Egypt

Aronson J A, Pasternuk D, Danon A (1989) HALOPH: A database of salt tolerant plants of the world. In Whitehead E E (ed) 77 pp., Office of the Arid Land States, University of Arizona, Tucson, AZ

Boulos L (1995) Flora of Egypt Checklist. Al Hadara Publishing, Cairo,203 pp.

El Bassosy A A (1983) A study of the nutritive value of some range plants from Sallom to Mersa Matrouh. Ph.D. thesis, Fac. Agric., Ain Shams University, Egypt

El-Khouly A A (2001) Plant diversity in the dry land habitats of Siwa Oasis, in the Western desert of Egypt. J Environ Sci Mansoura Univ 22: 125–143

El-Khouly A A, Khedr A A (2000) Species diversity and phenology in the wetland vegetation of Sewa Oasis, in the Western desert of Egypt. Desert Inst Bull Egypt 50: 239–258

El-Khouly A A, Zahran M A (2002) On the Ecology of the Halophytic Vegetation of the Oases in Egypt. Proceedings of International Symposium on "The Optimum Resources Utilization in Salt-Affected Ecosystems in Arid and Semi-Arid Regions", pp. 277–286, 8–10 April, 2002, Cairo, Egypt

El Shaer H M (1981) A comparative nutrition study on sheep and goats grazing Southern Sinai desert range with supplements. Ph.D. thesis, Fac. Agric., Ain Shams University, Egypt

El Shaer H M (1986) Seasonal variation of the mineral composition of the natural pasture of Southern Sinai grazed by sheep. Proceedings of the 2nd Egyptian - British Conference on Animal and Poultry Production, 26–28 August 1986, Bangor, Gwynedd, UK

El Shaer H M (1997) Practical approaches for improving utilization of feed resources under extensive production system in Sinai. Proceedings of International Symposium on Systems of Sheep and Goat Production, 25–27 October 1997, Bella, Italy

El Shaer H M, Attia Ismail S A (2002) Halophytes as animal feeds: potentiality, constraints, and prospects. Proceedings of International Symposium on "The Optimum Resources Utilization in Salt-Affected Ecosystems in Arid and Semi-Arid Regions", pp. 411–418, 8–10 April, 2002, Cairo, Egypt

El Shaer H M, Zahran M A (2002) Utilization of halophytes in Egypt: an overview. Proceedings of the International Conference on "Halophyte Utilization and Regional Sustainable Development of Agriculture", 14–20 September 2001, Huanghua, Shijiazhnag, China

Gihad E A, El Shaer H M (1994) Nutritive value of halophytes. In: Squires V R, Ayoub A T (ed) Halophytes as a Resource for Livestock and for Rehabilitation of Degraded Lands, pp. 281–284, Kluwer, Dordrecht

Hassan N I, El Shaer H M, Kandil H M (1982) Performance of lambs and kids on salty pastures supplemented with different levels concentrates. World Rev Anim Prod 18: 37–42

ICARDA (1997) Annual Report: International Center for Agricultural Research in the Dry Areas (ICARDA), Aleppo, Syria

Kassas M, Zahran M A (1967) On the ecology of the Red Sea littoral salt marsh, Egypt. Ecol Monogr 37: 792–316

Khedr A A (1999) Floristic composition and phytogeography in a Mediterranean deltaic lake (Lake Burollos), Egypt. Ecologia Mediterr 25: 1–11

Khedr A A, Zahran M A (2002) The salt marsh visitation of lake Bardawil, North Sinai, an overview. Proceedings of International Symposium on "The Optimum Resources Utilization in Salt-Affected Ecosystems in Arid and Semi-Arid Regions", pp. 339–345, 8–10 April, 2002, Cairo, Egypt

Krumglaz B S, Hornung, H, Oren OH (1980) The study of a natural hyper saline lagoon in a desert area (the Bradawil Lagoon in northern Sinai) Estuar. Coast Mar Sci 10: 403–415

Le Houerou H N (1993) Salt tolerant plants for the arid regions of the Mediterranean isoclimatic zone. In: Leith H, El-Masoom A (ed) Towards the Rational Use of High Salinity

Tolerant Plants, Vol. 1, pp. 403, Kluwer, Dordrecht, The Netherlands
Menzel U, Leith H (1998) Tabulation of halophytes reported as utilized in different publications and handbook. In: Hamdy A, Leith, Fl, Todorovic M, Moschenko M (eds) Halophytes Uses in Different Climates. Biometerology Ii, 127–133. Bbackuys Publications, Leiden
Mohammed N E (1980) Studies in the genus *Juncus* in Egypt. M.Sc. thesis, Fac. Sci., Cairo University, Egypt
Serag M S, Khedr A A, Zahran M A, Willis A J (1999) Ecology of some aquatic plants in polluted water courses, Nile Delta, Egypt. J Union Arab Biol 9: 85–89
Serag M S, Zahran M A, Khedr A A (2002) Ecology and economic potentialities of the dominant salt-tolerant reeds and rushes in the Nile Delta. Proceedings of International Symposium on "The Optimum Resources Utilization in Salt-Affected Ecosystems in Arid and Semi-Arid Regions", pp. 245–256, 8–10 April, 2002, Cairo, Egypt
Shaheen S E (1998) Geoenvironmental studies on Bardawil lagoon and its surroundings, North Sinai, Egypt. Ph.D. thesis, Fac. Sci., Mansoura University, Mansonra, Egypt
Squires V R, Ayoub A T (1994) Halophytes as resource for livestock and for rehabilitation of degraded lands. In: Squires V R, Ayoub A T (eds) Halophytes as a Resource for Livestock and for Rehabilitation of Degraded Lands, pp. 315, Kluwer, London
Snogerup S (1993) A revision of *Juncus* subgen, Juncus (juncaceae). Willdenowia 23: 23–73
Szabolcs I (1994) Salt affected soils as ecosystem for halophytes. In: Squires V R, Ayoub A T (eds) Halophytes as a Resource for Livestock and for Rehabilitation of Degraded Lands, 19 pp., Kluwer, London
Tackholm V (1974) Students' flora of Egypt. Cairo University Press, Egypt
Tackholm V, Drar H (1954) Flora of Egypt. Bulletin Fac. Sci. III. University of Cairo, 66 pp.
Warren B E, Bunny C I, Bryant E R (1990) A preliminary examination of the nutritive value of four saltbush (*Atriplex* species). The Camel. Longmans, London
Wittgren H B, Maehlum T (1997) Wastewater treatment wetlands in cold climates. Water Sci Technol 35: 45–53
Younes H A, Zahran M A, El Qurashy M E (1983) Vegetation-soil relationships of a sea – inlandward transect, Red Sea coast, Saudi Arabia. J Arid Environ 6: 349–356
Zahran M A (1993) *Juncus* and *Kochi* fiber-and fodder-producing halophytes under salinity and aridity stress. In: Pessarakli M (ed) Plant and Crop Stress, pp. 505–528, New York-Basal-Hong Kong>
Zahran M A, Willis A J (1992) The Vegetation of Egypt. Chapman & Hall, London
Zahran M A, Willis A J (2002) The Plant Life of the River Nile in Egypt. Marc Publishers, Reyadh, Suadi Arabia
Zahran M A, Boulos S T, Kamal El-Din H (1972) Potentialities of fiber plants of Egyptian Flora in National Economy. *Juncus rigidus* and paper industry. Bull Inst Desert Res Cent Egypt 22: 193–203

Abbreviations

ADF	Acid detergent fiber
ADL	Acid detergent lignin
Ca	Calcium
CF	Crude fiber
CP	Crude protein
DM	Dry matter
DMD	Dry matter digestibility
DMI	Dry matter intake
EE	Ether extract
ICARDA	International Center for Agricultural Research in the Dry Areas (ICARDA), Aleppo, Syria
Na	Sodium
NDF	Neutral detergent fiber
NFE	Nitrogen free extract

Author Index

T

U

V

Subject Index

SPRINGER NATURE

GPSR Compliance

The European Union's (EU) General Product Safety Regulation (GPSR) is a set of rules that requires consumer products to be safe and our obligations to ensure this.

If you have any concerns about our products, you can contact us on ProductSafety@springernature.com

In case Publisher is established outside the EU, the EU authorized representative is:

Springer Nature Customer Service Center GmbH
Europaplatz 3
69115 Heidelberg, Germany

FSC
www.fsc.org
MIX
Papier | Fördert
gute Waldnutzung
FSC® C083411

Kolibri 360 GmbH
Ferdinand-Jühlke-Straße 7
99095 Erfurt, Deutschland
produktsicherheit@kolibri360.de